CHEMISTRY OF ENGINE COMBUSTION DEPOSITS

CHEMISTRY OF ENGINE COMBUSTION DEPOSITS

Edited by
Lawrence B. Ebert

Exxon Corporate Research Laboratories
Clinton, New Jersey

PLENUM PRESS • NEW YORK AND LONDON

Library of Congress Cataloging in Publication Data

Main entry under title:

Chemistry of engine combustion deposits.

"Expanded version of the proceedings of a symposium on chemistry of
engine combustion deposits, held at the 181st American Chemical Socie-
ty meeting, March 30, 1981, in Atlanta, Georgia"–T.p. verso.
  Bibliography: p.
  Includes index.
  1. Combustion deposits in engines–Congresses. I. Ebert, Lawrence, B. II.
American Chemical Society. Meeting. (181st: Atlanta, Georgia)
TJ756.C48 1985   621.43   85-3688

ISBN-13:978-1-4612-9498-6   e-ISBN-13:978-1-4613-2469-0
DOI:10.1007/978-1-4613-2469-0

Expanded version of the proceedings of a symposium on Chemistry of Engine
Combustion Deposits, held at the 181st American Chemical Society meeting,
March 30, 1981, in Atlanta, Georgia

PREFACE

On March 30, 1981, a symposium entitled "Chemistry of Engine Combustion Deposits" was held at the 181st American Chemical Society National Meeting in Atlanta, Georgia, under the sponsorship of the Petroleum Division. This book is an outgrowth of that symposium, including papers from all of the Atlanta presentors, as well as from others who were invited to contribute.

Research on engine deposits has not been as "glamorous" as in the related fossil fuel areas of petroleum, coal, or oil shale, and publications in the field have been largely confined to combustion and automotive engineering journals. One objective of this book is to bring a large body of work on the chemistry of deposits into more general accessibility. We hope to make people more familiar with what deposits are, with what problems they cause, and with what present workers are doing to solve these problems.

The creation of the book has involved many people. Patricia M. Vann of Plenum Publishing Corporation gave guidance in planning. We thank Claire Bromley, Ellen Gabriel, and Halina Markowski for the preparation of many of the Exxon contributions. Finally, we thank Joseph C. Scanlon for his useful advice and encouragement.

Lawrence B. Ebert

Exxon Research and Engineering Company
Route 22 East
Annandale, New Jersey 08801

CONTENTS

INTRODUCTION

John P. Longwell

Department of Chemical Engineering
Massachusetts Institute of Technology
Cambridge, MA 02139

The symposium reported in this book was held in early 1981 and is based on work performed during the preceding few years when soaring fuel prices and impending shortages had stimulated studies of all factors affecting the efficiency of fuel use and manufacture. Deposits formed in the engine combustion chamber are of major interest because they affect engine efficiency by changing fuel octane number requirement and by changing heat loss to the cooling system. Physical interference with piston ring and gas flow through the valves can also occur. Also, the porous nature of the deposits is believed to be responsible for cyclic accumulation and release of hydrocarbons, thus increasing hydrocarbon emissions.

The papers, presented here, highlight the complex nature of sources of deposits and their effects on engine performance.

Recent changes in lubricant and fuel composition as well as engine design have forced re-evaluation of the nature of the problem, and the classic engine and fleet testing remain an important component of an overall program. Confirming older work, deposits in the end gas zone are held responsible for octane requirement increase. Both fuel and lubricant can contribute. Oil consumption is important since oil can contribute inorganic components and reactive molecules. The relative contribution of fuel and lubricant depends strongly on their composition and on engine design and condition. Inorganic fuel components are becoming less important with the drastic reduction in lead tetraethyl content, but the accompanying increase in aromatics tends to increase deposit formation. A new observation is the reported large increase in fuel economy resulting from reduced heat loss to cooling water when the combustion chamber is insulated on the inside by combustion deposits or by a teflon coating. The implications of this observation obviously require further study since deposits in the end zone are probably deleterious because of ORI, while the insulating effect of deposits elsewhere may be desirable.

A highlight of the symposium is the application of modern analytical and computer modeling methods to advance toward a more detailed understanding of the chemistry and physics of deposits and their effect of ORI on performance. A computer model, taking into account the cyclic heat transfer through an insulating layer of deposit, shows the effect of increasing end gas temperature (and, therefore, knocking tendency), the effect of lowered heat loss on efficiency and the small effect of deposit volume on compression ratio. While in their early stages, such models are an important tool for quantitative testing of various hypotheses for explanation of observed effects.

Relatively little work has been done on the chemistry of the polymerization process by which deposits are formed. Studies of composition have, however, advanced by adding ^{13}C NMR to the array of tools being applied to this problem. The bulk of carbon in deposits from both engines and a laboratory flame was found to be aromatic in nature. This correlates with the observation that aromatic compounds in fuel or lubricant greatly increase deposit formation. Studies using a laboratory laminar flame produced deposits from benzene fuel, but none from an aliphatic fuel. Formaldehyde and acetylene diffused to the cold surface for both flames; however, phenol was a major diffusing species for the deposit forming benzene and is implicated as a contributor to deposit formation.

Research on the spark ignition engine deposit problem has diminished with the current decrease in concern over fuel price and supply; however, the economic problem of continuing high cost for fuel to the consumer, and international competition for improved engine performance calls for continued effort to understand and solve deposit related problems.

2

CHEMISTRY OF ENGINE COMBUSTION DEPOSITS: LITERATURE REVIEW

Lawrence B. Ebert

Corporate Research-Science Laboratories
Exxon Research and Engineering Company
P. O.Box 45
Linden, NJ 07036

I.   INTRODUCTION

As an autombile engine runs, its fuel quality requirements, as measured by the octane number of the fuel needed to inhibit knocking, may change in time. Historically, this phenomenon is referred to as the "ORI" problem, standing for octane requirement increase.

Octane requirement increase can be quantified. Knock can be perceived, by either people or instruments, by an audible "pinging" sound emanating from the engine, caused by an approximately 5000 Hz vibration of the engine structure induced by gas pressure waves within the combustion chamber. These waves are created by "auto ignition" of the gas ahead of the normal flame front.   By using a variety of test fuels, of known octane rating, one determines the octane requirement of the engine as that octane for which knock is "borderline".   The octane requirement increase is simply the difference between the final octane requirement and the initial octane requirement of an engine.   Figure 1 illustrates actual examples of the "ORI" problem.

The single most important operational variable affecting "ORI" is the presence of combustion chamber deposits. As the flame front progresses across the chamber, it never actually touches the relatively cool wall, and carbonaceous materials coat the inside surfaces of the combustion volume, illustrated in Figure 2. Modification or elimination of these deposits can have a beneficial impact on the octane requirement increase problem.

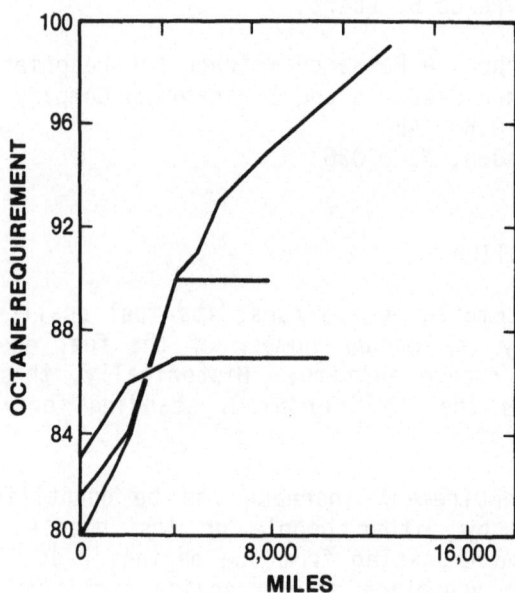

Figure 1. Examples of "octane requirement increase" for automobile engines.

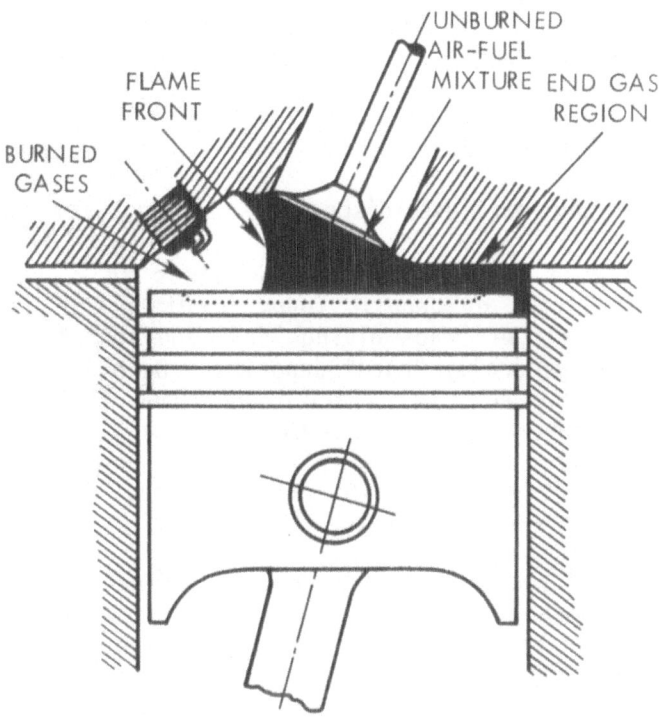

Figure 2. A diagram of the combustion chamber. The reader should note the "end gas region", an area which has been suggested to yield "bad" combustion chamber deposits.

Nevertheless, before a systematic improvement of the situation can be achieved, one must understand what the deposits are, in particular what they are made of and how they respond to various perturbations. In this book, we shall seek to determine the chemical nature of deposits formed within internal combustion engines. Before proceeding to the areas of current interest, we review some of the past work in the field. Because the nature of the deposits is dependent on the presence, or absence, of lead additives in the fuel, the review is divided into three parts, covering the period to 1930 (no lead additives), to the early 1970's (lead additives), and the most recent work on deposits formed from gasoline without lead additives.

II. EARLY WORK IN DEPOSITS

One of the earliest discussions of deposits is that by Parish in an article on lube oils for World War I aircraft.[1] In the final figure of the article, one finds an illustration of the different forms of deposit found in different parts of the engine, with cylinder heads giving "soft and sooty" deposits and pistons giving generally "hard" deposits. Developing specifications for Liberty Aero Oil, Parish concluded that there was a correlation between Conradson carbon of the oil (residue following distillation) and carbon left in the engine. The final specification was that suitable oil would not show a Conradson carbon residue greater than 1.5%.

Orelup and Lee studied the factors influencing carbon formation in automobile engines.[2] As seen in the following, they correctly recognized the relationship between carbon deposits and knocking:

> Carbon in an engine is like scale in a boiler. It is one of the best heat insulators. The temperature of fuel burning in a heat insulated cylinder soon rises above its critical temperature and thereupon breaks down in an extraordinary manner. This is manifested by knocking or detonating. Sudden large pressures are produced. To avoid the knocking, the operator must run with the spark retarded, thus losing power. There is an increased fuel consumption and a tendency for the engine to overheat. The car is no longer able to climb steep hills in high gear and in general lacks that desirable activity described as "pep".

Orelup and Lee recognized that engine deposits were not like graphite in character, more closely resembling "highly condensed hydrocarbons of a bituminous nature". They further delineated the nature of the deposits:

> The larger portion of the deposit consists of heavy hydrocarbons, some of them of a coke-like nature, possibly having a structure of interlinked carbon rings, and about 15% of a lighter hydrocarbon similar to lubricating oil. The mineral matter varies from 6 to 25%, according to the conditions of formation.... The organic portion contains small amounts of nitrogen and sulfur. Very few organic or inorganic solvents have any effect on cylinder carbon. A few of the most effective dissolve only a small amount of the lighter hydrocarbon. The best solvents found are monochloro-benzene, xylene, and cymidine, and to a lesser degree pyridine. Carbon deposits have the property in common with ordinary asphalts of forming a dark, soluble product with aqueous caustic soda, indicating material of an acidic nature resembling salts of weak bases.

They concluded that the key factor in creating deposits in automobile engines was the amount of lubricating oil projected into the combustion chamber. Although cars today consume far less oil than in 1925, it is still true that lubricating oil does contribute to deposit formation. However, one notes that the typical engines in 1925 consumed one quart of oil for every 150 miles, corresponding to the consumption of one quart of oil for every ten gallons of gasoline (at the 1925 mileage figure of 15 miles per gallon).

Certainly, in this time period, lubricating oil was considered responsible for deposit formation. W. A. Gruse performed more extensive studies implicating it in deposit formation.[3] Noting that oil-derived deposits formed in the crankcase, on the cylinder walls, and in the combustion chamber, Gruse proposed different mechanisms for the formation of each.

- crankcase - slow oxidation to form resins and asphaltenes.

- cylinder walls - low temperature coking in the piston ring grooves.

- combustion chamber volume - thermal decomposition of the lube, with little chance of oxidation, because the fuel consumes the oxygen present in the combustion volume.

However, with the introduction of lead into gasoline,[4,5] the relative contributions of fuel and lube to the deposit began to change, with the fuel becoming more important.

III. ENGINE DEPOSITS CONTAINING LEAD

L. F. Dumont reported that the introduction of 1.5 ml tetraethyl lead per gallon fuel caused a fourfold increase in deposit weight over that for an unleaded fuel.[6] Doubling the tetraethyl lead content to 3.0 ml/gal. had only negligible effect on the deposit weight, suggesting that deposit attrition mechanisms might be operative. This possibility was confirmed by analysis of deposit accumulation through microscopy and x-ray diffraction which showed deposit growth to occur in three phases:

1. Accumulation of essentially carbonaceous material.

2. Accumulation of lead salts.

3. Attainment of equilibrium deposit quantity.

The distribution of lead salts was heterogeneous, with volatile, low melting point materials near the engine metal wall and with nonvolatile, high melting point materials at the deposit surface. Such a heterogeneity can create a thermal gradient with correlative thermal stresses, ultimately leading to deposit flaking.

Dumont postulated three possible mechanisms by which combustion-chamber deposits could increase octane requirement:

1. By occupying volume, the deposits increase the compression ratio of an engine and thus increase octane requirement. Dumont estimated the volume-effect to account for 10 to 47% of the total deposit knocking harm, with leaded fuel deposit volume contributing twice as much to ORI as the volume of unleaded fuel deposits.

2. By possessing an active surface, the deposits might "catalyze" various knocking reactions. No positive evidence for this effect was found.

3. By being a thermal insulator, the engine deposit could cause substantial increase in end gas temperatures, and therefore octane requirement by a) heating the charge drawn into the cylinder and b) by reducing the rate of heat dissipation from the hot compressed working fluid during compression of the end gas by the piston and by the advancing flame front. Dumont considered the thermal insulation effect to be the most important.

To actually experimentally separate the volume effect from

the thermal insulation effect, Dumont coated the inside surfaces of a single cylinder engine with a dispersion of Teflon in water, followed by a 750°F fusion. The octane requirement increased in direct proportion to the thickness of the Teflon film, with 25% of the ORI due to the volume effect, 75% of the ORI due to the insulating effect.

Consistent with previous work, Dumont believed that the carbonaceous portion of the deposit depended on the type of crankcase oil used and on the amount of oil reaching the combustion-chamber surfaces. This lube dependence was further confirmed in a comparison of ORI results between synthetic lubes (as a polyether and a polybutene, which gave low values of ORI) and conventional lubes (as neavy white oil and Coastal B which gave high values of ORI).[7]

However, this lube dominance of the carbonaceous portion of the deposit came into question as radiotracer experiments were used in engine research. The first radiotracer experiment performed on an engine deposit involved ^{210}Pb, and was used to demonstrate that new deposits will form at a more rapid rate on thick deposit surfaces than on thin surfaces or bare metal walls. Later, P. C. White reported ^{14}C results using either labeled benzene or toluene as a fuel and a synthetic polyglycol as a lubricant.[9] In the case of the benzene, deposits from the cylinder head showed 94% of the activity of the fuel and deposits from the piston in crowns showed 70% of the activity of the fuel, demonstrating that the fuel, not the lube, created the majority of the carbonaceous deposit. Similar results were obtained for toluene.

L.B. Shore and K. F. Ockert performed ^{14}C studies on labeled fuel with an L-head CFR engine employing a high quality 10W-30 motor oil.[10,11] They found that nearly all of the combustion chamber deposit came from the fuel; additionally, they noted:

1. The deposit-forming tendency of hydrocarbons goes up strongly with boiling point.

2. At a given boiling point, aromatics are more harmful than paraffins, while the olefins appear to be intermediate.

3. Boiling point is not merely an important factor in determining the deposit-forming tendency of aromatics; it seems to be the only factor. That is, as long as a compound contains at least one aromatic ring, its deposit-forming tendency can be predicted from its boiling point alone.

4. In an aromatic, all carbons in the molecule are equivalent with regard to their tendency to make deposits. That is to say, the carbon in the methyl group of toluene is as likely to end up in the deposits as the aromatic carbons.

5. On any given surface, the ratio of ^{14}C labeled material to unlabeled material does not change systematically as the deposits build up.

6. There is, however, an effect of deposit location on concentration ratio, arising from variations of surface temperature within the engine.

Observation 4 of the above led Shore and Ockert to propose that engine deposits are formed via a pyrolysis of liquid-phase material. This hypothesis was considered to explain the difference in deposit-forming tendencies of aromatics and paraffins, for aromatics are more likely to "carbonize" than are paraffins. Nevertheless, Shore and Ockert were not able to reconcile observation 6 with this liquid-phase mechanism. The highest radioactivity concentration ratio occurred on the piston top (which showed the highest average temperature but the lowest total carbon). Although one might be tempted to propose that contributions from the lube "diluted" the intake value deposits, Shore and Ockert considered this unlikely.

The single experiment arguing against significant lube contribution was performed in the CFR engine using benzene and labeled benzene as the fuel. The average concentration ratio of the deposit was 0.92, suggesting that the lube could contribute 0.08g/g, deposit. Shore and Ockert stressed that this result might not carry over to automotive engines under typical operating conditions. In fact, Sechrist and Hammen[12] reported a ^{14}C radiotracer study involving labeled benzene in benzene with an uncompounded SAE 30 petroleum oil lubricant (standard CFR overhead-valve cylinder and piston) which showed the lube to contribute 0.5g/g, deposit. The message here is most likely that different lubes can contribute differently to deposit weight, with the key contributors to the deposit being the higher molecular weight, less volatile, components of both fuel and lube.

Other mechanisms interrelating engine deposits and octane requirement increase were proposed in the 1950's. Moxey[16] discussed the possibility of autoignition, the setting off of any flame front other than that initiated by the normal ignition, which would result in more rapid compression of the end gas, and thus more knock. Mikita and Bottoney[14] divided this concept into preignition (ignition before the spark is passed, equivalent to advancing the spark) and postignition (ignition of some of the charge after the spark is passed). By injecting engine deposits into the combustion chamber of a six-cylinder engine, they were able to induce preignition. Such an effect could have a dramatic effect on ORI, for inflammation of a charge 10° before spark passage can increase octane requirement 5 to 10 units. Mikita and Bottoney noted that in some cases preignition, or postignition, occurred to such an extent that the engine continued to operate with the ignition turned off, a phenomenon they called run-on.[14] In addition to the previously proposed thermal insulation effect, Mikita and Sturgis[15] suggested that the finite heat capacity of the deposits would also adversely affect octane requirement by heating the incoming charge.

Very little of the early work on deposits dealt with chemistry. The statements of Orelup and Lee[2] were prescient, and White[9] presented carbon, hydrogen, and oxygen analyses for unleaded fuels. At the end of the 1950's, Lauer and Friel[17] presented work which not only contained detailed chemical analyses but also advocated the theory of deposit formation from the gas phase, rather than from the previously assumed liquid phase.

Lauer and Friel conducted their experiments on a single cylinder, L-head Lauson engine, initially using a leaded commercial fuel and a multigrade lubricating oil. Consistent with the work of Dumont,[6] they noted carbonaceous deposits to accumulate first, followed by increasing amounts of lead compounds after 15 hours. The deposits had H/C ratios ranging from 0.72 (valve area) to 0.84 (piston top). To focus on the carbonaceous portion of the deposit, they used two fuel/lube combinations with no metallic species: iso-octane fuel (=2,2,4 trimethylpentane) with hydrocarbon lubes [to study lube contributions to the deposit] and toluene fuel with synthetic (polyether) lubes [to study fuel contribution to the deposit]. To facilitate examination of the deposits, removable plugs were inserted into the combustion chamber head, thereby allowing both deposit thickness measurements and electron microscopic analysis.

The combination of toluene fuel with a paraffinic lube gave the deposit of most weight (1.2g) and lowest density (0.23 g/mL). The iso-octane/naphthenic lube combination gave less weight (0.4g) but a high density (0.65 g/mL), thus suggesting that different hydrocarbon combinations did yield different deposits. However, electron microscope studies suggested all deposits to be similar, consisting of agglomerates of small spheres of varying density and packing. Noting that carbon formed in the gas phase from diffusion flames was also spherical, Lauer and Friel suggested the deposits to be gas phase in origin.[17]

These deposits were analyzed in terms of chemical composition, Soxhlet extractability, and infrared spectroscopy. They contained between 20 and 36% by weight oxygen, with H/C ratios ranging from 0.55 to 0.91. The highest extractability was for the iso-octane/naphthene combination in pyridine (40%). Two features in the infrared spectrum were associated with oxygen functionality: a peak at 1724 cm^{-1} (saturated monobasic carboxylic acids, and possibly aldehydes, ketones, esters, or anhydrides) and at 1587 cm^{-1} (ionized salts of carboxylic acids). Consistent with the interpretation of the infrared in terms of acids, Lauer and Friel noted that decarboxylation seemed to account for most of the weight loss on heating to ca. 610°C. Furthermore, Lauer and Friel generated "simulated engine deposits" of comparable oxygen content by torching lube oil which had been placed on a rotating metallic disk.

In terms of previous theories interrelating ORI and deposit characteristics, Lauer and Friel pointed out that observed ORI was inversely dependent on the bulk density of the deposit, a result consistent with both the volume and thermal insulation effects proposed by Dumont.

In addition to affecting the octane requirement of an engine, combustion deposits can influence the chemistry of exhaust gases leaving the engine. With increased concern about pollutants as NO, NO_2, CO, and unburned hydrocarbons, a number of studies were made on the relationship between deposits and gas phase emissions.[18-22] Both wall temperature and surface roughness affect hydrocarbon emissions, with lower temperatures and rougher surfaces enhancing hydrocarbons in the exhaust.[23]

This concern about gas phase pollutants has caused a return to unleaded fuels. The catalytic converters of most new cars, which are designed to lower amounts of CO and unburned hydrocarbons, are poisoned by lead fuel. Thus, the combustion chamber deposits formed within these new cars are composed primarily of carbon, just as they were in 1925, when Orelup and Lee wrote their classic paper!

IV. RECENT WORK ON DEPOSITS FROM UNLEADED FUELS

With the return to unleaded fuel in the 1970's, the deleterious effect of deposits on octane requirement again became an issue. It is difficult, from both an economic and energy conservation standpoint, to produce high octane unleaded fuel, with the size and composition of the clear octane pool allowing a 91 RON unleaded gasoline.[24] With this lower octane rating, and correlative diminished compression ration of new automobiles, the octane requirement increase problem is again with us. Should the final octane requirement of a new car exceed the octane of available unleaded fuel, engine knock is inevitable.

Benson has reported octane requirement increases for General Motors cars equipped with emission control hardware and low compression engines designed to run on unleaded fuel.[25] He found that greater octane requirement increases occurred with:

1. An engine oil containing bright stock compared to an engine oil without bright stock. (bright stock is a high molecular weight, aromatic-containing, material derived from deasphalted vacuum residuum).

2. An unleaded fuel which contains a polymeric detergent-dispersant additive relative to conventional additive packages.

3. Customer-type driving relative to rapid mileage accumulation on a dynamometer.

With respect to the relationship of ORI to engine deposits, Benson stated that all of the ORI with unleaded fuel could be eliminated by removing combustion chamber deposits, with about two-thirds of the ORI specifically caused by deposit accumulated in the end-gas region of the chamber.[25] With reference to the mechanisms of Dumont,[6] Benson stated that ten percent of the ORI is due to the volume effect, the other 90 percent probably due to the thermal effect.[25] Benson noted that octane requirement increase was not affected by the lead content of the fuel.

In a study of the effect of lubricant viscosity index improvers on ORI, Bachman and Prestridge discussed physical and chemical characteristics of deposits formed within a 1975 model car (350 cid OHV V-8) using commercial unleaded fuel.[26] Electron microscopy indicated the carbon to be amorphous (non-graphitic), extremely porous, and characterized by a heterogeneous granular structure. They noted that the observed morphology was consistent with a pyrolytic mechanism, and thus distinct from the spherical deposits reported by Lauer and Friel.[17] Heat capacity values were found to be 0.40 and 0.44 cal/(g°C), and the density was 1.53 ± 0.02 g/cc. Microanalysis of deposits showed about 63 wt% carbon, 5 wt% hydrogen and 2 wt%

nitrogen. Chemical composition did not vary significantly between different engines or between different locations within the combustion chamber. In terms of deposit weight, Bachman and Prestridge reported values ranging from 2.64 to 3.00 grams per cylinder. The piston tops generally had the most deposit, with values from 1.28 to 1.60 grams. If the pistons run at higher temperatures than the cylinder head, any further increase in surface temperature (as by the deposits) could be particularly harmful.[26]

Lee and Thomas considered two lubricant effects on octane requirement increase: the instantaneous effect of the presence of slight amounts of lubricant in the fuel/air charge and the effects of lube-created combustion-chamber deposits.[27] They made the following conclusions:

1. The decrease in fuel octane number of unleaded gasoline due to instantaneous consumption of lube was one octane number or less.

2. Less than or equal to 19% of the combustion chamber deposit derived ORI was due to the fuel, with the remainder due to the lube.

3. The amount of lubricant consumed by the knocking cylinder determined the ORI of a multi-cylinder engine. Assessing the combustion chamber deposit contribution ot ORI of multi-cylinder engines by using the check-back ORI (=final octane requirement after removal of combustion chamber deposits) was superior to using the final minus initial octane requirement as the ORI.

4. The variation of ORI between six motor oils was due to their different ash levels. The lowest sulfated ash oil had the lowest ORI; the highest sulfated ash oil had the highest ORI.

5. Of the components in an SE quality, high sulfated ash motor oil, the overbased petroleum sulfonate accounted for most of this oil's ORI.

Nakamura, Yonekawa, and Okamoto discussed the relationship between combustion chamber deposits and octane requirement increase for Japanese cars of 4 and 6 cylinders.[28] Octane requirement increases ranged from 3.5 to 11 octane units, corresponding to final octane requirements from 86.5 to 95.5 RON. Of eleven cars studied, three cars did not approach an equilibrium value of octane requirement even after 15,000 km! The authors noted that there was a linear relation between octane requirement increase (ORI) and initial octane requirement (IOR) of the form:

$$(ORI) = 43.44 - 0.44 (IOR)$$

This behavior was first reported by Kunc in 1951, who suggested that cars with higher initial octane requirements would have higher temperatures and pressures of operation, thereby inhibiting the formation of deposits.[29] If deposits are more difficult to form, the magnitude of octane requirement increase will be lower, and one will find that cars with higher initial octane requirements will have lower octane requirement increases. This hypothesis is consistent with the observation that the octane requirement of an engine which has developed much deposit weight in low-speed stop-and-go driving can be reduced by subjecting the engine to hard, high-speed, operation for a short period of time.[29] Wimmer, Holman, and Alquist also found a linear relation between ORI and IOR.[30] Their value of slope was -0.43, close to the -0.44 of Nakamura, Yonekawa, and Okamoto although their intercept was two RON units higher, presumably the result of the use of American cars of larger engine displacements.

Contrary to other work on lube variables, Nakamura, Yonekawa, and Okamoto found that neither bright stock nor sulfated ash raised octane number requirements.[28] They suggested that the volatility and viscosity of the lube might be significant variables and that lube consumption must be considered in assessing lubricant-induces ORI. Of this latter point, their average number for lube consumption was 240 $g/10^3$ km, which equals 4.62×10^{-4} quart/mile. Corresponding to 0.07 quart consumed per 150 miles, this number is a factor of 14 lower than the lube consumption reported by Orelup and Lee in 1925, definitely suggesting a change in the balance of fuel and lube contributions to combustion deposits between the years 1925 and 1979!

With respect to the weight of deposits, Nakamura, Yonekawa, and Okamoto reported between 1.15 g and 2.07 g per cylinder for four cylinder cars. No direct correlation between total deposit weight and ORI was found nor was there a relation between weight of a specific deposit (e.g., piston top deposit or end gas deposit) and observed ORI. Previously, Meguerian, Smith, Tracy, and Keller and suggested a relation between the weight of cylinder head deposit and ORI.[31,32]

With respect to thermal conductivity, unleaded fuels gave values of 0.16 to 0.17 W/(m-°C), somewhat lower than the values of 0.24 previously reported by Mikita and Sturgis.[15] Heat capacity values ranged from 0.1 to 0.4 cal/(g°K), and depended on the carbon content of the sample, the lower the carbon content, the lower the heat capacity.

V. WHAT ARE ENGINE DEPOSITS?

Research performed over the last sixty years has established that the presence of combustion chamber deposits has an unfavorable effect on the octane requirement of internal combustion engines. Yet, one must recognize that many variables are involved in the problem of "knock", including not only deposits but also compression ratio, spark timing, combustion chamber design, air-fuel ratio, coolant temperature, engine speed, throttle opening, exhaust diluent, inlet air temperature, pressure, and humidity.[24] In this book, we shall try to determine the chemistry of engine deposits, which is an important component of the "knock" problem. In analyzing the total problem, one must of course take into account all the options at one's disposal, including not only operational parameters as fuel and lube but also design options as prechamber stratified charge or direct injection stratified charge engines.[33]

REFERENCES

1. W. F. Parish, Liberty Aero Oil, J. Am. Soc. Naval Eng., 32:45 (1920).
2. J. W. Orelup and O. I. Lee, Factors Influencing Carbon Formation in Automobile Engines, Ind. and Eng. Chem., 17:731 (1925).
3. W. A. Gruse, Summary of Factors Causing Formation of Carbon Deposits in Internal Combustion Engines, Oil and Gas Journal, pp. 13-14 (November 23, 1933).
4. H. F. Williamson, R. L. Andreano, A. R. Daum, and G. C. Klose, The American Petroleum Industry: The Age of Energy 1899-1959, Northwestern University Press, Evanston, 1963.
5. D. Kolb and K. E. Kolb, Petroleum Chemistry, J. Chem. Ed., 56:465 (1979).
6. L. F. Dumont, Possible Mechanisms by Which Combustion-Chamber Deposits Accumulate and Influence Knock, SAE Quarterly Transactions, 5:565 (1951).
7. J. D. Bartleson and E. C. Hughes, Combustion Chamber Deposits As Related to Carbon-Forming Properties of Motor Oils, Ind. and Eng. Chem., 45:1501 (1953).

8. H. P. Landerl and B. M. Sturgis, Tetraethyl Radiolead Studies of Combustion Chamber Deposit Formation, Ind. and Eng. Chem., 45:1744 (1953).
9. P. C. White, Proc. Fourth World Petr. Congr., Section VI/F, pp. 376-377, Rome, 1956.
10. L.B. Shore and K.F. Ockert, Radiotracer Study Points Way to Make Clean-Burning Gasoline, SAE Journal 65:18 (1957).
11. L. B. Shore and K. F. Ockert, Combustion-Chamber Deposits-- a Radiotracer Study, SAE Trans., 66:285 (1958).
12. C. N. Sechrist and H. H. Hammen, Radiotracer Studies of Engine Deposit Formation, Ind. Eng. Chem. 50:341 (1958).
13. H. J. Gibson, C. A. Hall, and D. A. Hirschler, Combustion Chamber Deposition and Knock, SAE Trans., 61:361 (1953).
14. J. J. Mikita and W. E. Bottoney, Oil and Gas Jurnal, pp. 34-35(July 6, 1953).
15. J. J. Mikita and B. M. Sturgis, The Chemistry of Combustion Chamber Deposits, Proc. Fourth World Petr. Congr. Section VI/F, pp. 357-376, Rome, 1956.
16. J. G. Moxey, Autoignition and Knock Seen as Combined Problem, SAE Trans., 61:376 (1953).
17. J. L. Lauer and P.J. Friel, Some Properties of Carbonaceous Deposits Accumulated in Internal Combustion Engines, Comb. and Flame, 4:107 (1960).
18. M. W. Jackson, W. M. Wiese, and J. T. Wentworth, The Influence of Air-Fuel Ratio,Spark Timing, and Combustion Chamber Deposits on Exhaust Hydrocarbon Emissions, Automotive Engrs. Tech. Prog. Ser., 6:175 (1964).
19. J. C. Gagliardi, The Effect of Fuel Antiknock Compounds and (Combustion-Chamber) Deposits on Exhaust Hydrocarbon Emissions, SAE Paper No. 670128 (1967).
20. J. F. Conte and A. J. Pahnke, Effect of Combustion Chamber Deposits and Driving Conditions on Vehicle Exhaust Emissions, SAE Paper No. 690017 (196?).
21. H. E. Leikkman and E. W. Beckman, The Effect of Leaded and Unleaded Gasolines on Exhaust Emissions as Influenced by Combustion Chamber Deposits, SAE Paper No. 71-843 (1971).
22. F. H. Robinson, P. M. Kerschner, R. P. Doelling, A. F. Gerber, and M. S. Rakow, Additives Can Control Combustion Chamber Deposit Induced Hydrocarbon Emissions, SAE Paper No. 720500 (1972).
23. J. B. Heywood, Pollutant Formation and Control in Spark-ignition Engines, Prog. Energy Combustion Sci., 1:135 (1976).
24. J. Benson, What good are octanes?, Chemtech, 6:16 (1976).
25. J. D. Benson, Some Factors Which Affect Octane Requirement Increase, SAE Paper No. 750933 (1975).
26. H. E. Bachman and E. B. Prestridge, The Use of Combustion Deposit Analysis for Studying Lubricant-Induced ORI, SAE Paper No. 750938 (1975).

27. R. C. Lee and S. P. Thomas, Some Lubricant Effects on Octane Requirement Increase, Preprints, Div. Petr. Chem., Amer. Chem. Soc., 23:974 (1978).

28. Y. Nakamura, Y. Yonekawa, and N. Okamoto, The Relationship Between Combustion Chamber Deposits and Octane Number Requirement Increase, J. Japan Petrol. Inst. 22:105 (1979).

29. J. F. Kunc, Effect of Lubricating Oil on Octane Requirement Increase, SAE Quant. Trans. 5:582 (1951).

30. D. B. Wimmer, G. E. Holman, and H. E. Alquist, Some Observations of Factors Affecting ORI, SAE Paper No. 750932 (1975).

31. G. H. Meguerian, J. B. Smith, C. B. Tracy, and B. D. Keller, ORI of Today's Vehicles, SAE Paper No. 760195 (1976).

32. B. D. Keller, G. H. Meguerian, J. B. Smith, and C. B. Tracy, ORI of Today's Vehicles--2, SAE Paper No. 770195 (1977).

33. R.E. Baker, Satisfying the Appetites of 1980 cars, Chemtech 10:375 (1980).

EFFECTS OF COMBUSTION CHAMBER DEPOSIT

LOCATION AND COMPOSITION

Karen M. Adams and Richard E. Baker

Engineering and Research Staff
Ford Motor Company
Dearborn, Mi 48121

ABSTRACT

Combustion chamber deposits were analyzed for chemical composition in an attempt to explain the octane requirement increase (ORI) and exhaust hydrocarbon (HC) increase observed during mileage accumulation in two vehicle test fleets. Comparisons are made of deposit composition differences within different areas of a combustion chamber, between cylinders, and between engines. Deposits accumulating in the end gas areas, on exhaust valves, and in cylinders with high oil consumption can be distinguished by comparison of H/C atom ratios, carbon content and inorganic compound content. Differences in deposit composition can be observed between engine families and between driving cycles with the same engine family. Composition differences between deposits from two engine families suggest a possible explanation for the ORI trends observed in the fleet test. Chemical composition alone is not sufficient to identify a unique relation between deposits and ORI. The flame end gas region was located with ionization gaps and was compared with deposit mass distribution to examine the influence of chamber geometry on ORI. Similarly, the HC increase mechanism requires a more complete definition than available from chemical characterization alone. A vehicle was tested with fuel containing MMT followed by operation and tests with clear fuel. The results suggest that HC are controlled by the deposits on the exposed surface only and that deposit equilibrium takes place primarily on the exposed surface rather than by massive deposit removal and replacement.

INTRODUCTION

Combustion chamber deposits must be considered in the design and operation of an internal combustion engine. The effects of chamber deposits have been studied extensively in engine and vehicle tests and are described in the literature. Among the most significant of these effects are an increase in octane requirement (1, 2, 3), an increase in exhaust hydrocarbon emissions (2, 4, 5, 6) and a decrease in durability of some components (7, 8). The octane requirement increase (ORI) is significant since it prevents the use of optimum compression ratio and ignition timing, and consequently limits the engine efficiency well below the maximum possible with fuel of a given octane used in a clean engine. Similarly, combustion chamber deposits may cause exhaust hydrocarbon (HC) emissions to increase so much with mileage that other controls, such as decreased compression ratio or spark retard, are necessary. As with ORI, these changes cause a loss in fuel economy.

Although the consequences of ORI and HC increase are similar, the characteristics of deposits which cause these effects are different. Deposits cause the largest ORI when concentrated in the end gas region, i.e., the last fuel mixture to burn, and when the composition provides high thermal insulation and possibly catalysis of fuel self-ignition (3). Decrease in chamber volume is not substantial with unleaded fuels unless oil consumption is excessive. Exhaust HC emissions are increased most by deposits located near the exhaust valve that permit collection of fuel and oil during compression and then release the fuel during the exhaust process (4). Therefore, the significance of chamber deposits in causing ORI, HC increase or both depends upon their location, composition and structure.

Excessive chamber deposits can prevent valves from seating properly or rings from flexing freely (7). Under certain operating conditions, deposits will also foul spark plugs (8). Most of these durability problems are minor with the use of unleaded fuel. The principal exception is increased wear of exhaust valve seats, which has been corrected by induction hardening the valve seats (9).

Previous studies have presented elemental composition of deposits and have described general correlations of ORI with metals content (10, 11). The results are scattered and are not always conclusive. A more complete characterization of deposits is needed in order to fully describe the conditions causing ORI and HC increase. The intent of this study was to determine whether any relation could be established between the observed engine effects and an initial description of deposit composition and structure, as distributed through the combustion chamber, between cylinders and between engines.

PROGRAM DESCRIPTION

Deposit Samples

Combustion chamber deposit samples were taken from two vehicles from an ORI fleet (1) and six vehicles from a three-way catalyst (TWC) durability fleet. The vehicles are described in Table 1. This selection allowed a comparison of engine type, driving cycle, and the antiknock fuel additive methylcyclopentadienyl manganese tricarbonyl (MMT).

The combustion chamber deposits were thoroughly scraped from selected areas, as shown by the shaded areas in Figure 1. The samples have been labeled for identification according to fleet, engine, region of the the combustion chamber, cylinder number, and fuel. For example, ORI-2.3-HEG-1 is the sample from the cylinder head end gas region of cylinder no. 1 from the 2.3L engine of the ORI fleet; if an "M" follows the cylinder number, this indicates MMT fuel was used.

Chemical Characterization

Selected samples were characterized by quantitative elemental analyses and by x-ray diffraction. The quantitative analyses for C, H, N, and S were done using standard procedures by Spang Microanalytical Laboratory, Eagle Harbor, Michigan. The analyses of other elements, which include P, B and all metallic elements, was done using inductively coupled plasma torch emission spectroscopy. The samples analyzed by the emission spectroscopy were dissolved with nitric and perchloric acids and then volumetrically diluted with nitric acid. The analyses should be accurate to within 5% of the reported values. The x-ray diffraction analyses utilized the Debye-Scherer powder method. Patterns for both ashed and unashed samples were taken. Ashing at $600^{\circ}C$ eliminated the carbonaceous material to facilitate identification of additional inorganic compounds. The samples from the exhaust valves did not need to be ashed.

DEPOSIT COMPOSITION

Clear Fuel

Deposit weight for each section of the combustion chamber was presented in reference 1 for two engines from the ORI fleet. Total deposit weight varied considerably among the cylinders for both the 2.3L and the 5.0L engines. Within each cylinder, the deposits were more uniformly distributed, with the percent of deposits in the end gas region generally between 20-30% which is about equal to the

Table 1. Vehicle Description

Fleet	'76 ORI	'76 ORI	'79 TWC	'79 TWC	'79 TWC	'79 TWC	'79 TWC	'79 TWC
Engine Intertia Wt. Class (lb.)	2.3L 3000	5.0L 4000	2.3L 2750	5.0L 4000	6.6L 5000	2.3L 2750	5.0L[a] 4000	6.6L 5000
Engine Modification	RCT[b]	RCT[b]	Prod.	Prod.	Prod.	Prod.	Prod.	Prod.
Drive Cycle	Ford ORI Drive Cycle		Certification Emissions Durability Cycle					
Duration (Miles)	12K	18K	50K	50K	50K	50K	50K	50K
Fuel	91 RON Clear Unleaded						91 RON Unleaded + 0.125 gMn/gal	
Oil	SE 10W30 Cert. Durability Mg base 1.0 wt. % ash		SE 10W30 Cert. Durability Mg base		SE 10W30 Experimental Mg base 0.8 wt. % ash			
Fuel Economy (mi/gal)	20.5	13.5	23.5	13.7	10.2	23.4	13.8	10.8
Oil Consumption (mi/qt)	9,300	4,000	12,000	6,100	6,000	12,800	3,700	8.900

a/ This vehicle was operated with clear fuel for the final 10K miles.

b/ RCT, reduced coolant temperature, see reference 1.

22

ORI FLEET

Head end gas area (HEG)

Head combustion chamber area (HCC)

Piston end gas area (PEG)

Piston combustion chamber area (PCC)
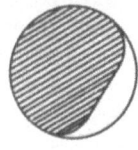

TWO FLEET

Head end gas area (HEG)

Head combustion chamber area (HCC)
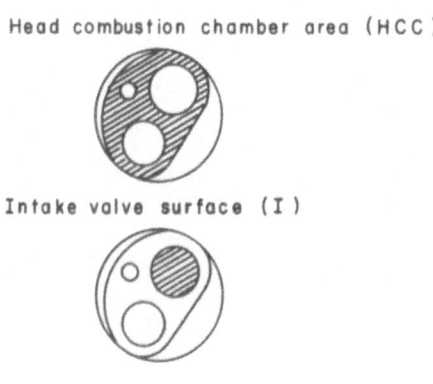

Exhaust valve surface (E)

Intake valve surface (I)

Piston surface (P)

Figure 1.   Deposit Sample Definition

surface area fraction in this region. The quantitative analyses for the deposit samples taken from these two engines are listed in Tables 2 and 3. The weight percent of carbon (C) is on the order of 40 to 55%, and for hydrogen (H), 3 to 5%. The inorganic portion of the deposits consists mainly of compounds containing Mg, Zn, P, Pb, S, Fe and Al. The amounts of these inorganic elements has been determined to be on the order of 1 to 8 wt. %. Trace elements, i.e., those found with less than 0.5 wt. %, are B, Ba, Ca, Cr, Cu, K, Mn, Na, Ni, Si, Sn, and Ti. In general, the data in Tables 2 and 3 show that the deposits are primarily highly oxidized organic material.

These deposits come from both the lubricant and the fuel. To what degree these organic deposits are from the lubricant or the fuel cannot be determined unless carbon tracers are introduced. The sources of the inorganic elements are more obvious. The Fe and Al are from wear and corrosion of the heads and pistons plus an undetermined amount due to sampling. The Mg, Zn, and P are from the oil additives. Small amounts of Pb are typical of residual levels in unleaded fuels. The source of S is from both oil additives and natural occurrence in the fuel. The N is part of the organic material in the fuel additive. While the inorganic compounds in the combustion chamber deposits must come from elements in the fuel and oil, the quantity and specific compounds do not necessarily reflect the source, but are the result of thermodynamics of the compound formation and their stability within the deposits. Thus, it is not possible to determine relative amounts of the total deposits due to the fuel or to the oil based upon the relative amounts of certain inorganic elements.

Inorganic compounds were identified by x-ray diffraction, Table 4. A white or tan deposit commonly seen on the exhaust valves, as for sample C7-5.0-EX-4, give a diffraction pattern resembling $(Zn, Mg)_3(PO_4)_2$. The same diffraction lines are also found for the deposits scraped from the 2.3L combustion chamber cylinder head (HCC) area, which includes the valves. Lead sulfate is the lead compound identified in all the sample areas of the combustion chamber. It is most likely that the α-Fe metal found in the head deposits and the Al metal found in the piston deposits are a result of abrasion during sample collection. All the other compounds found are metal oxides. It may be noted that the metal ions in these oxides are all in their highest oxidation state, e.g., Fe_2O_3 and SnO_2. Several inorganic oxides SiO_2 and SnO_2 have been identified after ashing the deposits, even though these elements were each found in less than 0.5 wt. % of the unashed sample. It is assumed that ashing the deposits did not alter the inorganics formed in combustion, so the diffraction data would identify the crystalline inorganic compounds as they existed in the deposit.

Tables 2 and 3 include several collections of the elemental analyses useful to identify and compare deposit characteristics.

Table 2

Deposit Composition - ORI Fleet 2.3L

wt. %

Sample No.	C	H	N	Mg	Zn	P	Pb	S	Fe	Al	Mg+ Zn+P	H/C	Atom Ratios	
													Zn+Mg/P	P/Zn
ORI-2.3														
HEG - 1	50.9	5.2	1.5	0.8	1.0	0.7	2.5	0.9	2.6	0.1	2.5	0.9	1.8	1.6
HEG - 2	54.6	4.2	1.7	1.2	1.7	1.3	3.4	1.0	4.9	0.2	4.2	1.1	2.2 $\bar{x}$=2.0	1.5 $\bar{x}$ 1.5
HEG - 3	54.4	4.8	1.6	0.8	1.0	0.6	2.2	0.8	6.2	0.1	2.4	1.1	1.7 s= .2	1.3 $\bar{x}$= .1
HEG - 4	57.9	5.4	1.5	2.0	2.6	1.8	1.5	0.9	5.1	0.1	6.4	1.1	2.2	1.4
HCC - 1	45.9	3.0	1.5	3.7	4.9	5.0	7.5	1.5	2.1	0.8	13.5	0.8	1.4	2.2
HCC - 2		2.8		2.8	3.9	3.7	3.4		2.6	0.6	10.4		1.5 $\bar{x}$=1.4	2.0 $\bar{x}$-2.2
HCC - 3			1.5	2.7	3.5	3.1	2.1	1.7	2.8	0.3	9.3		1.6 s= .3	1.9 s= .4
HCC - 4	37.0	3.3	1.0	3.0	5.3	6.8	0.3		1.4	0.3	15.0	1.1	0.9	2.7
PEG - 1	42.6	3.9	1.3	2.0	2.7	1.6	2.1	2.0	0.8	0.4	6.3	1.1	2.4	1.3
PEG - 2				1.6	2.1	1.1	3.0		0.8	1.2	4.8		2.9 $\bar{x}$=2.5	1.1 $\bar{x}$=1.4
PEG - 3				1.8	2.1	1.2	3.4		0.7	1.6	5.0		2.3 s= .3	1.5 s= .2
PEG - 4				4.8	5.2	3.8	0.8		0.3	0.1	13.7		2.3	1.5
PCC - 1	44.0	3.2	1.4	2.2	2.7	1.7	3.0	2.0	0.9	1.3	6.6	0.9	3.8	0.5
PCC - 2				2.2	2.6	1.9	3.6		0.8	0.4	6.7		2.1 $\bar{x}$=2.4	1.5 $\bar{x}$=1.4
PCC - 3				2.2	2.5	1.9	2.9		0.7	0.4	6.7		2.1 s=1.0	1.6 s= .6
PCC - 4				3.4	4.7	4.2	1.1		0.4	0.2	12.3		1.6	1.9

25

Table 3

Deposit Composition - ORI Fleet 5.0L

Sample No.	Wt. %											Atom Ratios		
	C	H	N	Mg	Zn	P	Pb	S	Fe	Al	Mg+Zn+P	H/C	$\frac{Zn+Mg}{P}$	$\frac{P}{Zn}$
ORI-5.0														
HEG - 1	54.7	3.8	1.4	1.5	1.8	1.4	1.8	1.2	3.6	0.3	4.7	0.8	2.0	1.7
- 2	53.1	3.6	1.5	1.3	1.8	1.5	3.3	1.0	3.8	0.2	4.7	0.8	2.0	1.8
- 3	52.7	3.5	1.4	1.7	2.1	1.8	3.0	1.2	3.0	0.3	5.7	0.8	1.8	1.8
- 4	47.2	4.4	1.0	3.1	5.5	5.1	1.1	1.2	3.7	0.1	13.6	1.1	1.3	2.0
- 5	54.1	3.7	1.4	1.6	2.0	1.6	2.4	1.2	3.7	0.5	5.3	0.8	1.9	1.7
- 6	55.1	3.7	1.6	1.7	2.0	1.6	2.3	1.2	2.5	0.3	5.3	0.8	1.9	1.7
- 7	55.2	3.1	1.5	1.3	1.6	1.5	2.9	0.9	1.3	0.1	4.4	0.7	1.6	2.0
- 8	54.4	3.8	1.6	1.8	2.2	1.8	3.1	1.0	2.5	0.1	5.8	0.9	1.8	1.8
													$\bar{x}$=1.7 s=.2	$\bar{x}$ 1.8 s= .1
HCC - 1	49.4	3.3	1.5	2.0	2.8	2.6	5.4	0.9	3.8	1.0	7.3	0.8	1.5	1.9
- 2				1.6	2.5	1.5	5.3		0.8	0.7	6.2		1.5	1.8
- 3				1.9	2.9	2.6	5.3		4.1	0.9	7.3		1.5	1.9
- 4				2.1	4.2	6.7	0.2		0.9	0.1	13.0		0.7	3.3
- 5				1.5	2.8	2.6	3.8		10.1	2.7	5.3		1.6	1.8
- 6				2.0	2.5	2.3	5.0		7.6	2.2	7.3		1.5	2.0
- 7	45.8	2.9	1.6				4.2	1.2	4.7	1.8	6.6	0.8	1.7	2.0
- 8	45.0	3.1	1.4	2.4	3.2	2.7	5.0	1.1	3.7	2.1	8.3	0.8		1.8
													$\bar{x}$=1.4 s= .3	$\bar{x}$=2.1 s= .5
PEG - 1				1.8	2.1	1.5	1.2		1.0	0.4	5.5		2.2	1.5
- 2				1.9	2.3	1.5	1.5		0.8	0.5	5.7		2.3	1.4
- 3				2.0	2.4	1.8	1.7		0.9	0.4	6.2		2.1	1.5
- 4				4.1	5.6	5.9	0.5		0.3	0.1	15.6		1.3	2.2
- 5				1.8	2.2	1.7	1.4		0.4	0.7	5.7		2.1	1.6
- 6				1.8	2.2	1.3	1.2		0.7	0.2	5.3		2.0	1.3
- 7				2.0	2.5	1.9	2.0		0.7	0.2	6.4		2.0	1.6
- 8	55.0	4.1	1.5	2.0	2.4	1.9	1.7	1.3	0.9	0.2	6.3	0.9	2.0	1.6
													$\bar{x}$=2.1 s= .4	$\bar{x}$=1.6 s= .3
PCC - 1				1.6	2.1	1.4	1.8		1.5	0.8	5.1		2.3	1.4
- 2				1.9	2.7	1.8	2.3		1.1	0.5	6.5		2.1	1.4
- 3				1.7	1.9	1.9	2.3		1.0	0.6	6.2		1.8	1.6
- 4				2.3	4.4	6.5	0.2		0.3	0.4	13.2		0.8	3.1
- 5				1.9	2.4	1.7	1.5		2.0	0.9	6.0		2.1	1.5
- 6				1.7	2.1	1.7	1.7		0.8	0.2	5.1		2.3	1.4
- 7				1.6	2.1	1.8	2.3		0.8	0.2	5.5		1.9	1.8
- 8	53.1	3.5	1.7	1.9	2.5	1.9	2.6	1.3	0.9	0.2	6.2	0.8	1.9	1.6
													$\bar{x}$=1.9 s= .5	$\bar{x}$=1.7 s= .5
C7-5.0-EX-4	53.1	3.5	1.7	11.4	20.3	19.0	5.7	1.3	0.4	0.2	50.6	0.8	1.3	2.0

Table 4

X-Ray Diffraction Results - ORI Fleet

Sample No.	Compounds Characterized	Weight Loss after ashing @ 600^{0}C
ORI-2.3-HEG-3	$PbSO_4$ α-Fe γ-Fe_2O_3	83%
ORI-2.3-HCC-3	$PbSO_4$ $(Zn, Mg)_3(PO_4)_2$*	68%
ORI-2.3-PEG-3	$PbSO_4$ ZnO SnO_2 Al	82%
ORI-2.3-PCC-3	$PbSO_4$ ZnO MgO γ-Fe_2O_3 α-SiO_2 SnO_2	78%
C7-5.0-EX-4**	$(Zn, Mg)_3(PO_4)_2$*	

*Zn^{2+}, Mg^{2+} ions similar, randomly distributed

**C7≡Deposits removed after durability. This deposit accumulated during octane rating.

Inorganics from the oil have been summed and selected atom ratios were calculated to compare with the compounds observed with x-ray diffraction. H/C ratio is a measure of the degree of oxidation. Note that water absorption will increase the H weight %, but this is estimated to be less than 0.3%.

Fuel with MMT

The deposits accumulated with fuel containing MMT are very distinctive, both visually and chemically. The exhaust valves always have a characteristic light-brown deposit. This deposit also appears on top of the black carbonaceous deposit in random areas throughout the rest of the combustion chamber. The x-ray pattern identified this deposit as Mn_3O_4 only. The quantitative analysis is consistent with the x-ray data. The exhaust valve samples from engines run with MMT fuel show very high Mn concentrations, Tables 5 and 6. The Pb and lubricant inorganics are such minor components of these exhaust valve deposits that in most cases they are undetectable by x-ray diffraction. Similarly, the quantitative data for the MMT deposits show the inorganic components are almost exclusively Mn.

DEPOSIT LOCATION AND KNOCK

Flame Propagation

Octane requirements are increased most substantially by deposits in the end gas region (3). However, the end gas region is defined by flame front position near the end of combustion, not by combustion chamber geometry. To locate the flame front and identify the end gas, ionization gaps were installed in a 2.3L cylinder head. Figure 2 shows the flame arrival time at each gap in terms of crankshaft degrees of rotation before top center piston position (BTC). Negative numbers indicate flame arrival after piston top center. This condition corresponds to borderline knock in the cylinder observed to knock first. There is some offset from a spherical flame starting at the spark plug. The flame reaches the exhaust valve side of the main chamber earlier than the intake valve side due to clockwise swirl generated by the intake port. However, the last portion to burn is in the squish area. Defining the total squish area opposite the plug as the end gas region appears to be a reasonable approximation, Figure 1. This was found to be true at a different speed and spark timing, and in another cylinder with a different intake port approach angle.

Chamber Location

Several trends were observed with the composition data given in Tables 2, 3, and 4.

Table 5

Deposit Composition - TWC 2.3L

Sample No.	wt%C	H	Mg	Zn	P	Pb	Mn	Fe	Al	Cu	wt% Mg+Zn+P	H/C Atom ratio
TWC-2.3												
HEG - 1	59.8	3.4	0.6	0.9	0.6	0.4	0.3	0.3	0.0	0.1	2.1	0.7
- 2	58.1	3.5	0.8	1.1	0.8	0.4	0.3	0.4	0.0	0.1	2.6	0.7
- 3	58.9	3.2	0.5	0.7	0.7	0.6	0.4	0.4	0.0	0.1	2.0	0.6
HCC - 1	32.1	1.5	3.6	5.6	4.7	1.3	0.7	0.9	0.1	0.9	13.9	0.6
- 2	30.6	2.5	2.9	4.5	3.8	0.9	0.5	0.6	0.1	0.3	11.2	1.0
- 3	37.7	1.6	3.0	4.4	3.2	1.9	0.9	1.4	0.1	0.6	10.6	0.5
P - 1	45.4	2.1	2.2	3.1	2.6	1.1	0.5	0.5	0.1	0.2	8.0	0.6
- 2	32.2	2.4	3.5	5.1	4.4	1.9	0.7	0.6	0.1	0.2	13.0	0.9
- 3	44.1	1.8	2.4	3.1	2.6	2.3	1.1	0.8	0.2	0.3	8.1	0.5
E - 1	0.4	<.1	7.7	9.5	12.9	11.0	3.8	2.8	3.1	0.6	30.1	
HEG -1-M	50.2	3.3	0.5	0.7	0.5	0.5	11.4	0.5	0.1	0.1	1.7	0.8
-2-M	46.2	3.4	0.5	0.6	0.5	0.7	11.4	0.3	0.1	0.1	1.6	0.9
-3-M	52.3	2.8	1.1	1.6	1.5	1.2	0.8	0.8	0.0	0.2	4.2	0.6
HCC -1-M	30.3	1.8	1.3	1.8	1.3	1.5	7.5	0.6	0.1	0.6	4.5	0.7
-2-M	27.3	2.2	1.1	1.6	1.1	1.9	7.5	0.4	0.1	0.3	3.8	1.0
-3-M	31.9	1.8	0.7	1.2	0.8	0.9	4.2	0.5	0.1	0.2	2.7	0.7
P -1-M	36.0	2.0	0.9	1.3	1.0	0.8	6.3	0.3	0.1	0.1	3.2	0.7
-2-M	35.4	2.4	1.0	1.3	0.9	1.3	6.4	0.3	0.1	0.1	3.2	0.8
-3-M	39.7	2.1	0.6	0.9	0.6	1.3	7.3	0.3	0.1	0.1	2.0	0.6
E -1-M	0.7	<.1	0.5	0.7	0.6	4.0	33.9	1.3	2.7	0.2	1.9	
I -1-M			0.2	0.5	0.3	0.8	18.4	0.7	0.1	0.5	1.0	

Table 6

Deposit Composition - TWC Fleet, 5.0L and 6.6L

Sample No.	wt%C	H	Mg	Zn	P	Pb	Mn	Fe	Al	Cu	Mg+Zn+P	H/C Atom Ratio
TWC-5.0												
H-1	36.5	2.6	1.9	2.1	3.2	1.2	0.1	0.8	0.1	0.7	7.2	0.9
E-1	0.6	0.1	4.4	6.6	11.0	29.4	2.9	2.4	1.3	1.3	21.9	
I-1												
H-1-M	34.1	2.6	1.9	2.5	1.9	1.5	8.8	0.5	0.1	0.7	6.3	0.9
E-1-M	0.8	<.1	2.0	3.0	4.5	14.5	53.9	0.9	1.7	1.2	9.5	
I-1-M	9.8	1.1	1.1	1.7	2.0	6.2	39.5	0.5	0.1	0.7	4.7	1.3
TWC-6-6			4.8	7.9	12.1	21.1	3.5	2.8	2.9	1.3	24.8	
E-M			1.5	2.0	1.7	4.8	49.1	1.2	2.7	1.4	5.2	

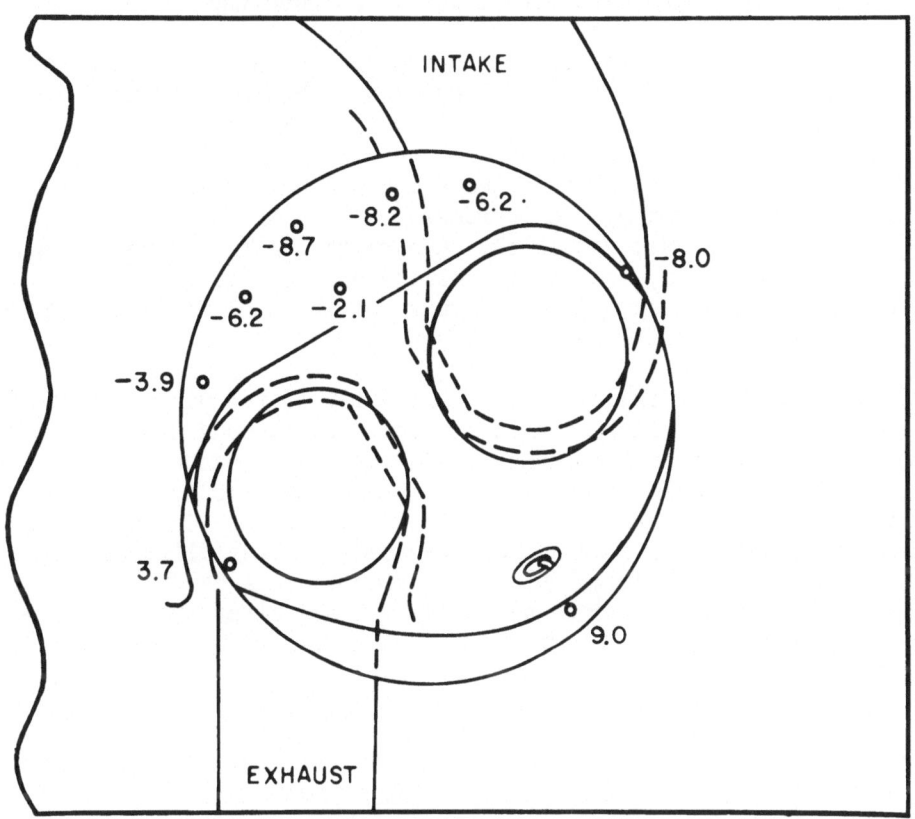

CYLINDER   4
ENGINE SPEED   2450 RPM @ WOT
SPARK ADV   34.5° BTC - BL KNOCK
AIR / FUEL   12.3

INTAKE

-6.2
-8.2
-8.7
-8.0
-6.2
-2.1
-3.9
3.7
9.0

EXHAUST

Figure 2.   2.3L Engine Flame Mapping

. End gas deposits are distinguished from deposits in other
  locations by having
  - about 10% wt. more C content
  - higher H/C atom ratio
  - lower inorganic content, Mg + Zn + P

. The inorganic (Zn, Mg, P) compounds are different on the
  exhaust valve than on all other chamber surfaces sampled,
  as determined by x-ray diffraction and inferred from atom
  ratios.

. Pb is uniformly distributed except for generally higher
  concentrations in samples from the open chamber portion
  of the cylinder head (HCC), which includes the valves.

The end gas deposits are more carbonaceous, and less oxidized
as indicated by the H/C ratio. Since less carbon was consumed
during combustion, it follows that there is a slightly lower concentration of lubricant inorganics in this area. This slightly
different chemical character of the end gas deposits is probably a
result of flame quenching and increased cooling in this region. The
flame may not propagate completely through the close clearance area
of the end gas region, leaving incomplete products to be deposited on
the walls. In addition, close proximity of the coolant in this area
keeps wall temperature lower, which retards combustion.

The face of the exhaust valve is usually coated with a white or
tan colored deposit, instead of the black carbonaceous deposits found
throughout the rest of the chamber. The accepted reason for this
is that exhaust valves reach very high temperatures such that any
carbonaceous material is "burned off". The inorganic deposits that
remain on the valve have been characterized by x-ray diffraction
as predominantly $(Zn, Mg)_3(PO_4)_2$ when clear fuel is used,
Table 4. This phosphate does not appear in the x-ray characterization for any of the other sample areas. Similarly, Zn and Mg appear
as oxides in the other sample areas.

The atom ratios given in Tables 2 and 3 support the x-ray diffraction observations. For the formulation $(Zn, Mg)_3(PO_4)_2$, the atom
ratio of (Zn + Mg)/P is about 1.5. Samples from the cylinder head
containing deposits from the exhaust valve have an average value of
1.3 to 1.4 for this ratio, which is well within one standard deviation of 1.5. Samples from other locations have values for this ratio
significantly higher than 1.5. This would occur if much of the Zn
and Mg were oxides rather than the phosphate.

The P/Zn atom ratio also shows this distinction. The source of
P and Zn is the oil additive zinc dialkyldithiophosphate, which has
a P/Zn atom ratio of 2.0. Samples containing exhaust valve deposits
have P/Zn ratios between 2.0 and 2.2, while this ratio is consistently

lower in samples from other areas. Therefore, the inorganic compounds on the exhaust valve are different from those on other surfaces. This is probably due to the much higher temperature of the exhaust valve surface.

Cylinder Differences

An objective of this study was to determine if the chemical composition of the deposits can account for ORI. One approach is to determine if the chemical composition of the knocking cylinder deposits differs from the other cylinders. The knocking cylinder, which establishes the octane requirement was determined after the deposits had stabilized with the use of spark plug pressure transducers. For the 2.3L, knock was detected in cylinder no. 4; for the 5.0L, cylinder no. 8. For the 2.3L engine, the deposits of cylinder no. 4 have a significantly higher H/C ratio, higher wt. % Mg + Zn + P, lower wt. % Pb and higher total deposit weight than any other cylinder. But this difference does not follow for cylinder no. 8 of the 5.0L CID. In fact, the deposits for cylinder no. 4 of the 5.0L engine, rather than no. 8, are very similar chemically to those of cylinder no. 4 of the 2.3L. It was observed that there was high oil flow from the intake valve guide of cylinder no. 4 of the 5.0L. This chemical trend in the deposit data is probably characteristic of a cylinder with high oil consumption and is not necessarily associated with knock. In Tables 2 and 3, there does not appear to be any data that indicate a unique chemical composition for the knocking cylinder.

Engine Family Differences

An additional objective was to determine if the deposit chemical composition can account for differences in trends of ORI with mileage. The plots of ORI (and OR) vs. mileage for the two ORI fleet engine families have been reported elsewhere (2). There is a difference in the trends of ORI with miles for the 2.3L and the 5.0L engines. The octane requirement (OR) for the 2.3L levels off by 2K miles and then remains constant. For the 5.0L, there is a gradual increase in OR to about 8K miles and then it oscillates about a mean value.

Two chemical differences in the deposits for the 2.3L and 5.0L have been noted in Tables 2 and 3 respectively: the overall average H/C atom ratio for the 2.3L ($\bar{x} = 1.0$, $s = 0.1$) is significantly higher than for the 5.0L ($\bar{x} = 0.8$, $s = 0.1$); and 2) the amount of P relative to Zn is higher for the 5.0L (1.60 - 1.81), in all the areas except where the valves are located (HCC), than for the 2.3L (1.35 - 1.47).

The observed differences in H/C ratio suggest mechanisms for deposit stabilization in the two engines, as measured by the ORI trends. The deposits from the 2.3L, having higher H/C ratio, are less highly oxidized hydrocarbon material and are expected to become

33

fluid at a lower temperature than deposits from the 5.0L engines. The possibility of a fluid character of the deposits has been demonstrated when deposit samples were heated in a glass capillary; subtle phase changes were observed to start in the temperature range from 190°C to 280°C. Temperatures in this range can occur on the chamber surfaces. Although the fluid character relative to the degree of oxidation has not been demonstrated, an analogy can be made with the parafinic hydrocarbons which have higher melting and boiling points when they are oxidized to aldehydes, ketones, acids, etc. The less highly oxidized deposits of the 2.3L engine are expected to flow somewhat uniformly on the chamber surface during engine operation and be less subject to brittle cracking. Deposit stabilization by fragmented breaking may account for the fluctuating ORI trend of the 5.0L engine.

Comparison of Fleets

There are some noticeable differences in the quantitative composition of the deposits from the TWC Durability Fleet engines compared to the deposits from the ORI Fleet, Tables 2, 3, 5 and 6. Sampling was very similar for both fleets except that for the TWC fleet engines, deposit samples from the exhaust and intake valves have been collected separately. The TWC fleet valve deposits are almost entirely inorganic; less than 1% C on the exhaust and 10% C on the intake. Excluding the valves, the wt. % C ranges from 25 to 60%. This is a wider range but generally lower than observed for the ORI fleet deposits. The H content also has a lower range, 1.5 to 3.5%. The same inorganic elements were identified as in the ORI fleet, but in lower concentration. With the exception of Mn and in exclusion of the valves, the concentration of inorganic elements ranges up to 6 wt. %.

Exhaust HC with MMT

An experiment was conducted with one vehicle in the TWC fleet to investigate the source of increased exhaust HC due to the use of MMT. A vehicle that had used MMT fuel for 40K miles was changed to clear fuel for the final 10K miles. The HC dropped substantially, nearly equaling the HC from a vehicle which had used only clear fuel, Table 7. The composition data in Table 6 indicate that these deposits have a very high Mn concentration characteristic of deposits accumulated with MMT fuel. Visual inspection of the intact deposits revealed only a few traces of the brown deposit; i.e., the exposed surface of the deposits had an appearance very similar to deposits accumulated with clear fuel. In this case, new deposits had collected over the old deposits without removing them or mixing with them. This is relevant to the mechanism of deposit formation and equilibrium. It also suggests that by physically covering the Mn_3O_4 deposits, the source of increased HC is eliminated. The active mechanism could be either chemical or physical, but it is apparent that the increased

Table 7

Effect of MMT on Feedgas HC

	Feedgas HC (g/mi)		
Fleet:	TWC	TWC	
Veh. No.:	-70P	-92P	
Fuel:	Clear	MMT	Mileage
	2.4	2.8	"0" K
	2.1	4.1	5 K
	2.0	4.5	10 K
	3.0	4.6	15 K
	2.5	5.3	20 K
	2.6	5.1	25 K
	2.6	5.3	30 K
	2.1	5.4	35 K
	2.3	5.3	40 K
		Clear	
	3.3	5.2	45 K
	3.0	3.8	50 K

HC are due to properties of the exposed surface and not the total
deposit composition.

CONCLUSIONS

- Combustion chamber deposits consist mainly of carbonaceous
  material with inorganic compounds from both the fuel and
  the oil. The source of the carbonaceous material cannot be
  determined uniquely from inorganic analysis.

- In most cases the H/C ratio is higher for deposits removed
  from the end gas region. This suggests these deposits are
  less highly oxidized due to differences in surface temperature or exposure to the flame.

- Deposits removed from the exhaust valve are almost entirely
  inorganic compounds, metal phosphates from the oil and
  manganese oxide if the fuel contains MMT. The composition
  of these deposits and the lack of carbon reflect the high
  surface temperature of the exhaust valve.

- The cylinder with the highest octane requirement, and hence
  the first cylinder to knock, cannot be distinguished from
  the other cylinders by chemical composition of the deposits.

- Cylinders with high oil consumption could be identified by
  the high concentration of oil inorganics and low concentration of fuel inorganics compared to the other cylinders.

- Differences in deposit composition were observed between
  engine families and between driving cycles with the same
  engine family. The source of these differences is believed
  to be the engine operating conditions but a specific relation
  has not been established. Composition differences between
  deposits from the 2.3L and the 5.0L suggested a possible
  explanation for the ORI trends observed in the fleet test.

- Mn_3O_4 was present in the deposits when MMT was used in the
  fuel. No specific explanation linking Mn_3O_4 and increased
  HC emissions can be offered; however, properties of the
  exposed surface were shown to be relevant.

- Deposit accumulation and stabilization involves more than a
  single mechanism. Change of fuel in a 5.0L TWC fleet vehicle
  proved that layered deposits are possible, while heated
  samples from a 2.3L ORI fleet vehicle indicated that the
  deposits could be partially fluid at operating temperature.

REFERENCES

1. D. J. Daly, R. D. Anderson and R. E. Baker, SAE 780595, June 5, 1978.
2. R. E. Baker, CHEMTECH, June 1980, page 375.
3. J. D. Benson, SAE 750933, October 13, 1975.
4. J. T. Wentworth, SAE 720939, October 31, 1972.
5. J. C. Gagliardi and F. E. Ghannam, SAE 690015, January 13, 1969.
6. J. D. Benson, R. J. Campion, L. J. Painter, SAE 790706, June 11, 1979.
7. G. H. Amberg and W. S. Craig, SAE 62-554D, August 13, 1962.
8. Y. Kimbara, Y. Noguchi, T. Ishiguro, SAE 800832, June 9, 1980.
9. W. Giles, SAE 710368, October 22, 1970.
10. B. D. Keller, G. H. Meguerian, C. B. Tracy, J. B. Smith, SAE 760195, February 23, 1976.
11. B. D. Keller, G. H. Meguerian, J. B. Smith, C. B. Tracy, SAE 770195, February 28, 1977.

FUEL-RELATED FACTORS AFFECTING ENGINE OCTANE REQUIREMENTS

Leonard B. Graiff

Fuels Department
Shell Development Company
Houston, Texas

INTRODUCTION

An investigation was undertaken to determine the effect of various engine deposits on octane requirements (OR). Our major finding, namely, that intake valve and intake port deposits can have as great an effect on OR as combustion chamber deposits, has been previously reported.[1] It has long been known[2] that maldistribution of fuel and air among cylinders of an engine can have a substantial effect on the octane requirements of the cylinders. Hence, it was of interest to determine if engine deposits have an effect on mixture distribution among cylinders (which in turn could affect the octane requirements of the individual cylinders). Accordingly, this paper first discusses the effect of air and fuel distribution among cylinders on octane requirements and then summarizes our overall observations of the effect of fuel factors on OR, using a recent engine test as an example.

EXPERIMENTAL

Test Engine, Fuel and Lubricant

A 1978 model 301 cubic-inch displacement (4.9ℓ) V-8 engine with a two-barrel carburetor and automatic transmission was used for this study. This engine, which had exhaust gas recirculation (EGR) and a breakerless electronic ignition system as standard equipment, was installed on a dynamometer test stand equipped with flywheels to simulate the inertia of a car. This engine was selected for the program because it was a new engine, designed with exhaust emission and fuel economy standards (and driveability problems) in mind. The engine operating conditions and cycle were the same as those used in earlier work.[1] Deposits were accumulated in the engine while operating on an unleaded gasoline without a detergent additive. Typical inspection properties of the gasoline are given in the Appendix. The engine lubricant was a multigrade of API SE quality.

Octane Requirement Determinations

Octane requirements were determined while operating the engine at 2500 rpm wide-open-throttle with the automatic transmission in second-gear. Full-boiling-range unleaded (FBRU) reference fuels of one Research octane number increments were employed for the ratings. The octane requirements of the individual cylinders were determined using our Selective Ignition Retard Device.[3]

Air-Fuel Ratio Measurements

Air-fuel ratios (AFR) of the individual cylinders and of the engine were measured using a Lamdascan Air-Fuel Ratio Analyzer which is a rapid-response unit made by Sensors Inc. Exhaust gas for the Lamdascan was sampled through stainless steel tubes inserted in the exhaust ports of the individual cylinders. The overall AFR of the engine was obtained from exhaust from a sample line located in the exhaust pipe.

Deposit Removal Procedure

The following three deposit areas were investigated as to their effects on OR and AFR: (1) combustion chamber; (2) intake port area, that is, intake valve tulips and stems and intake ports (excluding the intake manifold); and (3) all valve seats and faces. Deposits were removed by scraping, followed by buffing the metal surfaces with a power-driven rotary wire brush and then solvent cleaning. During cleaning of the cylinder heads, the spark plugs were removed, but not cleaned before reinstalling. It was observed that noticeable but low levels of deposits accumulated in the short time (about two to four hours) between tests. Accordingly, in the deposit removal tests, the areas which were previously cleaned of deposits were cleaned again when the engine was disassembled to remove deposits from another location.

RESULTS AND DISCUSSION

The octane requirements of the engine and of the individual cylinders were determined periodically during deposit buildup from clean engine conditions, as shown in Figure 1. Note that there was a spread in octane requirements between cylinders of up to 12 numbers. The effect of deposit location on the octane requirements and air-fuel ratios of the individual cylinders of the engine was determined after the end of the test shown in Figure 1, that is, after the engine

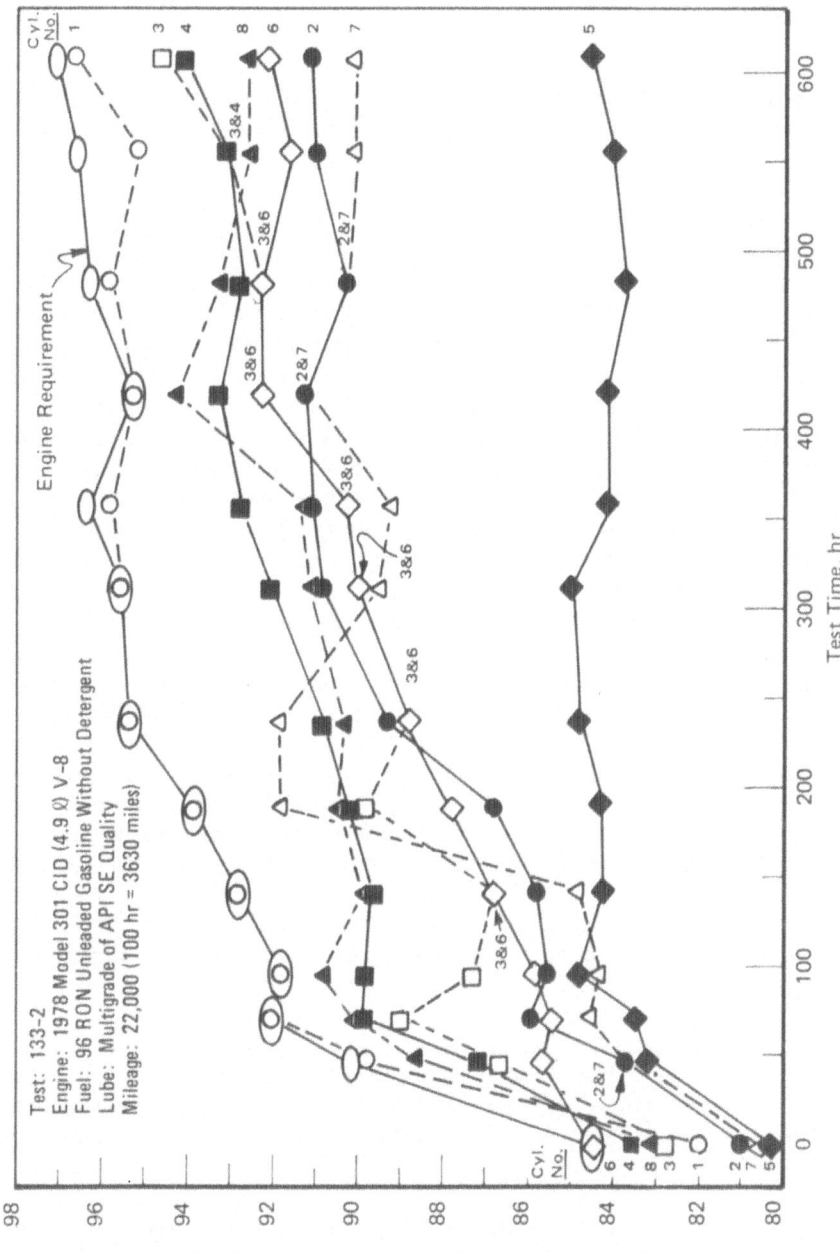

Fig. 1. Effect of test time on the octane requirements of the individual cylinders of an engine.

41

accumulated 609 hours (equivalent to about 22,000 miles) on the test fuel. The mixture distribution (or air-fuel ratio) study is discussed first.

Engine Octane Requirements as Affected by Mixture Distribution

This study was conducted in two parts: the first part was to determine the effect of engine speed and deposit location on air-fuel ratios of individual cylinders of the engine; the second part was to determine the effect of AFR on the knocking characteristics of a cylinder using a CFR Research Method knock test engine.

Effect of Engine Deposits on Mixture Distribution. After the V-8 engine operated for 609 hours on the test fuel, the air-fuel ratios of the individual cylinders and engine were determined at five level-road-load speeds (30, 35, 45, 55, and 65 mph) and at wide-open-throttle (WOT), 2500 rpm in second gear - the condition under which the engine was octane rated. The heads of the engine were then unbolted, and all combustion chamber deposits were physically removed. After reassembling the engine, the AFR's were determined again. The intake port and intake valve deposits were removed next. The last part of the test involved grinding all valve faces and seats, installing new spark plugs, and cleaning the carburetor parts to determine their effects, if any, on AFR distribution. The detailed results of the investigation are given in Table 1, where the AFR's from each bank of cylinders from the V-8 engine are grouped together.

It is evident from Table 1 that, under constant speed, road-load conditions: (a) all cylinders were operating very lean relative to the stoichiometric AFR (14.35); (b) all cylinders in a given bank at a given speed had similar AFR's, that is, mixture maldistribution was minimal in each bank of cylinders, considering that the accuracy of the Lamdascan is given as ±0.1 AFR; (c) removal of deposits had no significant effect on the AFR's of the individual cylinders; and (d) the cylinders in the right bank of the engine were operating leaner (up to one AFR) than those in the left bank. This latter observation was unexpected, and it appeared to be a characteristic of this particular two-barrel carburetor, since AFR measurements with other same model engines showed no differences in AFR between cylinder banks. This recent-design V-8 engine is unusual in that the cylinders have siamese intake ports; that is, the intake manifold has only four runners, instead of eight, and the right barrel of the carburetor supplies the four cylinders in the right head and the left barrel those in the left head.

Considering the AFR's in Table 1 under the knock-rating conditions (WOT, 2500 rpm, second gear), it is seen that removal of deposits from the engine had no significant effect on the AFR of each cylinder. This observation is important since it indicates that our earlier finding[1] of the large effect of intake-port-area deposits on octane requirements

Table 1. Effect of Engine Deposits and Car Speed on Air-Fuel Mixture Distribution Among Cylinders of a 1978 Model 301 CID V-8 Engine

Speed, mph	Engine Condition[a]	AIR-FUEL RATIOS Left Cylinder Head					AIR-FUEL RATIOS Right Cylinder Head					Ave 1-8	Overall Engine
		1	3	5	7	Ave	2	4	6	8	Ave		
30	A	17.5	17.1	17.2	17.5	17.3	18.6	18.5	18.6	18.5	18.6	17.9	17.5
	B	17.2	17.2	17.2	17.6	17.3	18.4	18.0	18.3	18.3	18.2	17.8	17.8
	C	17.2	17.0	17.2	17.5	17.2	18.5	18.0	18.1	18.2	18.2	17.7	17.6
	D	17.5	17.2	17.0	17.2	17.2	17.7	18.0	18.1	18.0	18.0	17.6	17.9
35	A	17.0	16.8	16.6	17.0	16.8	18.0	17.7	17.8	17.7	17.8	17.3	17.0
	B	17.0	16.9	16.8	17.0	16.9	17.8	17.5	17.6	17.7	17.6	17.3	17.3
	C	16.8	16.4	16.4	16.8	16.6	17.4	17.2	17.3	17.3	17.3	17.0	16.8
	D	16.8	16.6	16.6	16.7	16.7	17.7	17.4	17.5	17.5	17.5	17.1	16.8
45	A	16.4	16.2	16.0	16.2	16.2	16.9	16.7	16.8	16.6	16.8	16.5	16.3
	B	16.7	16.3	16.2	16.4	16.4	17.0	16.6	16.7	16.6	16.7	16.6	16.4
	C	16.2	16.1	15.9	16.2	16.1	16.7	16.5	16.6	16.4	16.6	16.3	16.4
	D	16.2	16.0	15.8	16.1	16.0	16.7	16.6	16.6	16.4	16.6	16.3	16.2
55	A	16.3	15.9	15.7	16.0	16.0	16.7	16.5	16.3	16.1	16.4	16.2	16.1
	B	16.4	15.9	15.6	15.8	15.9	16.6	16.3	16.2	16.1	16.3	16.1	16.0
	C	15.9	15.7	15.5	16.0	15.8	16.4	16.2	16.2	16.0	16.2	16.0	15.9
	D	16.0	15.7	15.5	15.8	15.8	16.6	16.3	16.2	16.0	16.3	16.0	16.0
65	A	16.2	15.7	15.4	15.8	15.8	16.3	16.2	16.1	16.0	16.2	16.0	16.0
	B	16.7	16.0	15.6	16.3	16.2	16.5	16.3	16.2	16.3	16.3	16.2	16.0
	C	16.1	15.7	15.4	16.0	15.8	16.3	16.3	16.2	16.1	16.2	16.0	15.9
	D	16.0	15.7	15.3	15.9	15.7	16.4	16.2	16.1	16.0	16.2	16.0	15.9
WOT, 2500 rpm, 2nd Gear[b]	A	12.2	11.3	11.0	11.0	11.4	12.2	11.6	11.6	11.4	11.7	11.5	11.5
	B	12.6	11.7	11.1	11.0	11.6	12.0	11.7	11.5	11.4	11.6	11.6	11.5
	C	12.1	11.3	11.0	11.0	11.4	12.1	11.6	11.5	11.2	11.6	11.5	11.3
	D	12.1	11.2	11.0	11.0	11.3	12.3	11.7	11.6	11.3	11.7	11.5	11.3

[a] A - After 609 hours of deposit accumulation (about 22,000 miles).
B - After all combustion chamber deposits removed.
C - After all intake port and intake valve deposits removed.
D - After all valve faces and seat ground, carburetor cleaned, and new spark plugs.
[b] Wide-Open-Throttle, 2500 rpm in second gear: engine octane rating conditions.

is not the result of the deposits changing the mixture strength of the cylinders. It is also seen from Table 1 that the cylinders operated much richer than stoichiometric AFR (14.35), with the front cylinder in each bank (cylinders one and two) running noticeably leaner than the other cylinders. Hence, mixture maldistribution does exist in this engine under the knock-rating conditions, and since it is known[2,4] that AFR does affect the knocking tendency of a fuel, it was of interest to determine the magnitude of this effect.

Effect of Air-Fuel Ratio on Apparent Octane Quality of a Fuel. For this study a single-cylinder CFR Research Method Engine was used. The engine was operated under standard Research Method knock test conditions and the exhaust gas was monitored with the Lamdascan. The results of the evaluation, using the same test fuel employed in the engine tests, are shown in Figure 2. The curve was obtained by adjusting the compression ratio of the engine to give standard knock intensity (as determined by the knock meter) at each AFR. The full-boiling-range unleaded (FBRU) reference fuels, which were used to determine the octane requirements of the engine, were then knock-rated under standard Research Method conditions to obtain the octane scale. It is seen from Figure 2 that the apparent octane quality of the test fuel increased rapidly as the AFR was richened or leaned from the maximum knock AFR (13.5, about 6% richer than stoichiometric).

Figure 2 can be used to estimate the effect of fuel maldistribution on the octane requirements of an engine cylinder. For example, cylinder number one operated at an AFR of 12.2 under the WOT octane-rating conditions (see Table 1); however, if there were no maldistribution and the AFR for cylinder one was the average value (11.5) for the engine, the octane requirement of the cylinder would have been about one octane number lower. In this case, since cylinder one determined the OR of the engine (that is, it was the cylinder with the highest OR), the octane requirement of the engine also would have been lower by about one octane number.

Effect of Deposit Location and Air-Fuel Ratio on Octane Requirements

Octane requirements of the engine and of the individual cylinders were determined after each deposit removal step (as were the air-fuel ratio measurements discussed in the preceding section). The results of the OR tests are shown in Figure 3.

The tops of the bars in Figure 3 represent the octane requirements of the engine (first bar) and individual cylinders at the end of the 609-hour test. Note that the OR of the engine at that point was 97, and that the engine requirement was controlled by cylinder one, the cylinder with the highest OR. The arrows in Figure 3 indicate the estimated OR of the engine and cylinders if there were no maldistribution of the air-fuel mixture among cylinders under the engine knock-rating conditions and the AFR for each cylinder was the average value

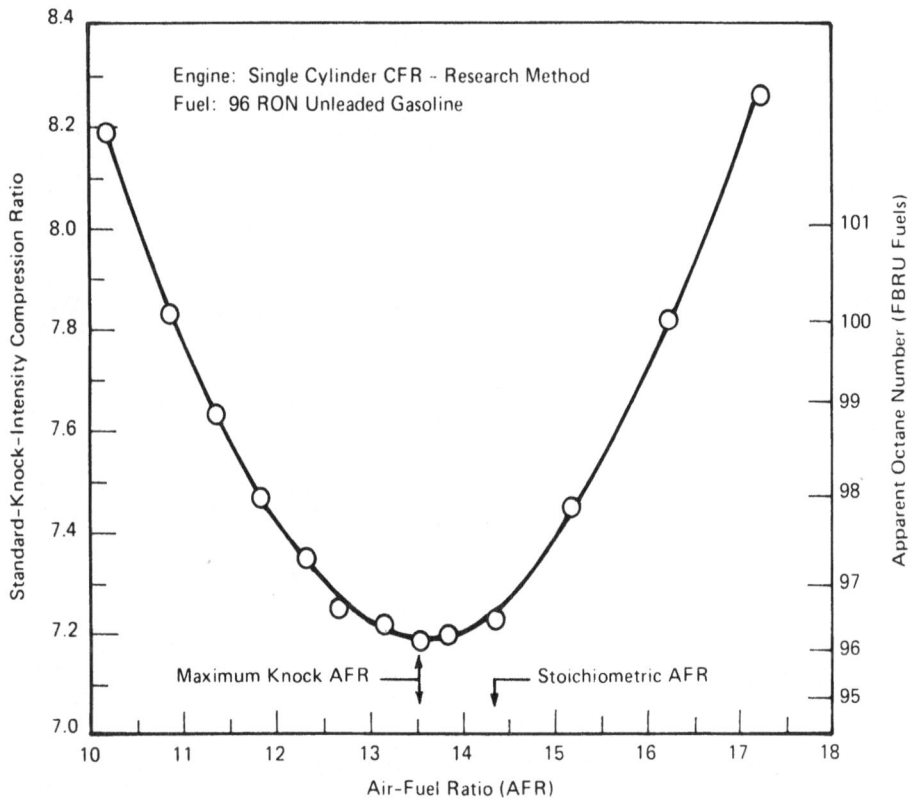

Fig. 2.   Effect of air-fuel ratio on apparent octane quality of a fuel.

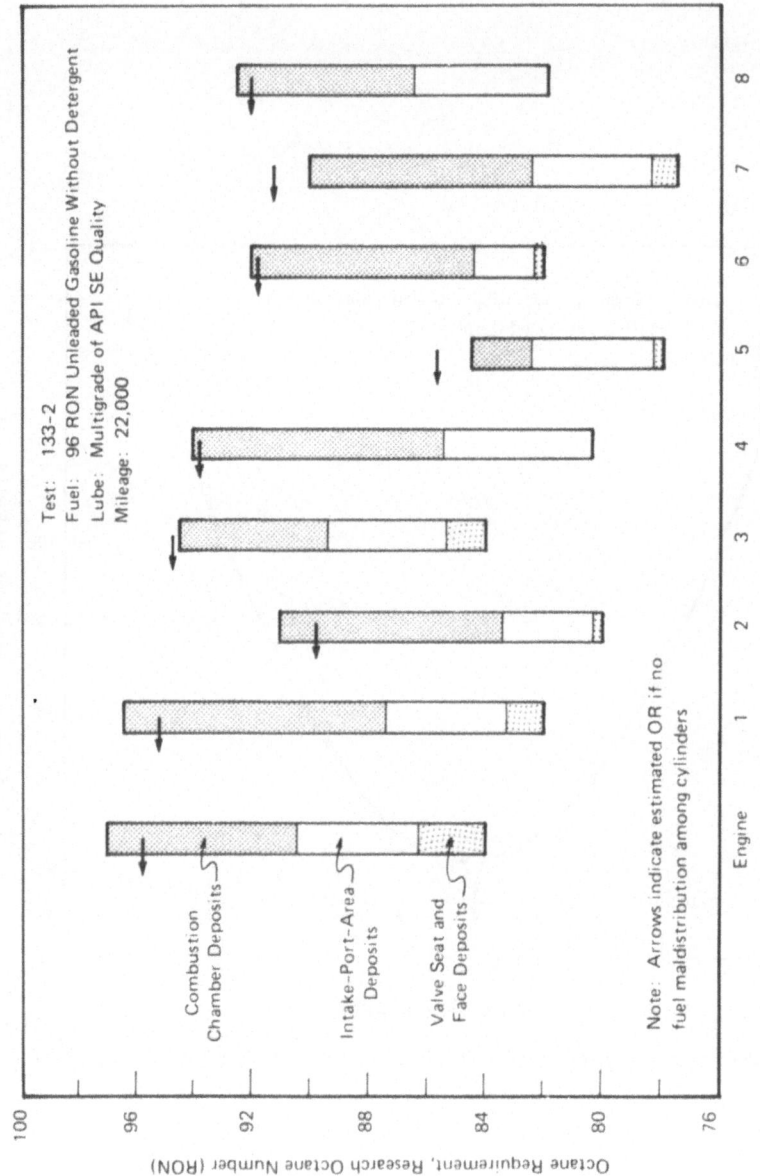

Fig. 3. Effect of engine deposits and other fuel-related factors on the octane requirements of the individual cylinders of a V-8 engine.

46

for the engine. These OR values were based on test results discussed in the preceding section and are only estimates* (because of the non-linearity of the octane scale). Note that with better mixture distribution, the OR of the engine would have been about one octane number lower.

The top shaded parts of the bars in Figure 3 represent the contribution of combustion chamber deposits to the octane requirements of the engine and individual cylinders. From measurements of compression ratios (CR) of the individual cylinders, both clean and with deposits, we estimate that up to 20% of the ORI due to combustion chamber deposits is the result of higher CR with deposits. Thermal insulating and heat capacity effects of the combustion chamber deposits appear to be the biggest contributors to the ORI phenomenon; catalytic effects of deposits on ORI appear to be negligible.[5,6]

The clear parts of the bars in Figure 3 represent the effects of intake-port-area deposits on the octane requirements. Earlier detailed studies[1] of such deposits showed that ridge-type deposits in the intake ports are the biggest contributors of the port deposits to ORI; also, intake valve deposits contribute to octane requirements. Photographs of the intake valve and port deposits are shown in Figure 4.

The bottom shaded sections of the bars in Figure 3 represent the contribution of exhaust valve seat and face deposits to ORI. For this particular phase of the test, all exhaust and intake valve seats and faces were ground, the carburetor was cleaned and new spark plugs were installed. Subsequent tests have shown that of these variables only grinding the exhaust valve faces and seats affects octane requirements. We speculate that deposits on the exhaust valve seats and faces reduce heat transfer from the very hot valves to the cooler engine surfaces, thereby resulting in higher valve surface temperatures in the combustion chamber and consequently higher octane requirements. From Figure 3, it is seen that the OR of only some cylinders respond to the valve grinding operation (OR changes of less than one octane number are considered to be not significant). This observation suggests that exhaust valve rotators may be effective in minimizing engine ORI by keeping the seats polished.

Note in Figure 3 that there is a large spread (up to 6.5 octane numbers) in the "clean" octane requirements of the cylinders. This indicates that there are other factors, probably mechanical, contributing to the differences in OR among cylinders. Compression ratios

---

* Better estimates could be obtained by operating the carburetor of the V-8 at both a somewhat richer and a somewhat leaner mixture than its normal setting under knocking conditions. This would give the mixture response (AFR versus octane number) at the octane requirement level of each cylinder.

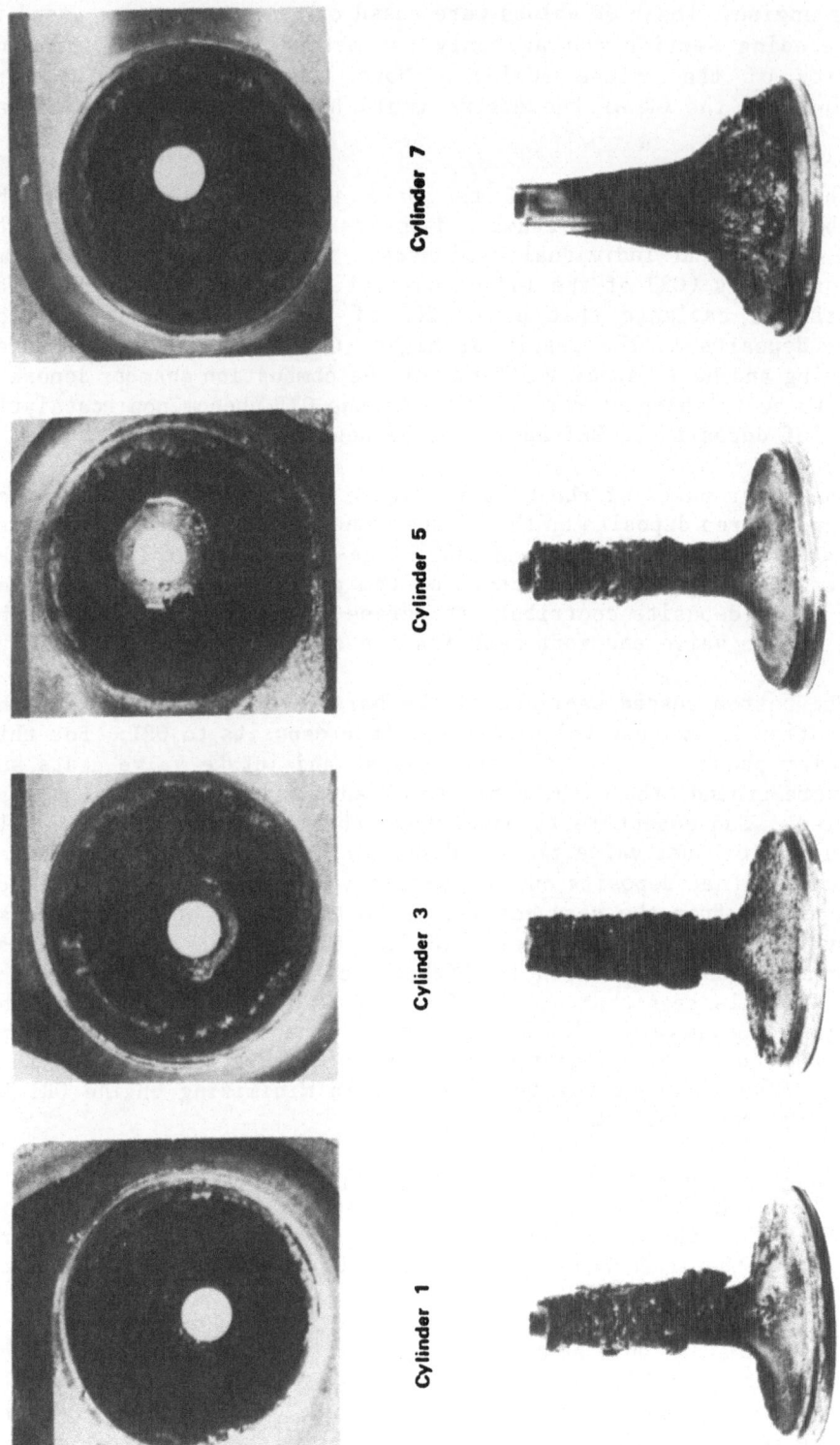

Fig. 4. Intake valve and port deposits after 22,000 miles.

Cylinder 1

Cylinder 3

Cylinder 5

Cylinder 7

48

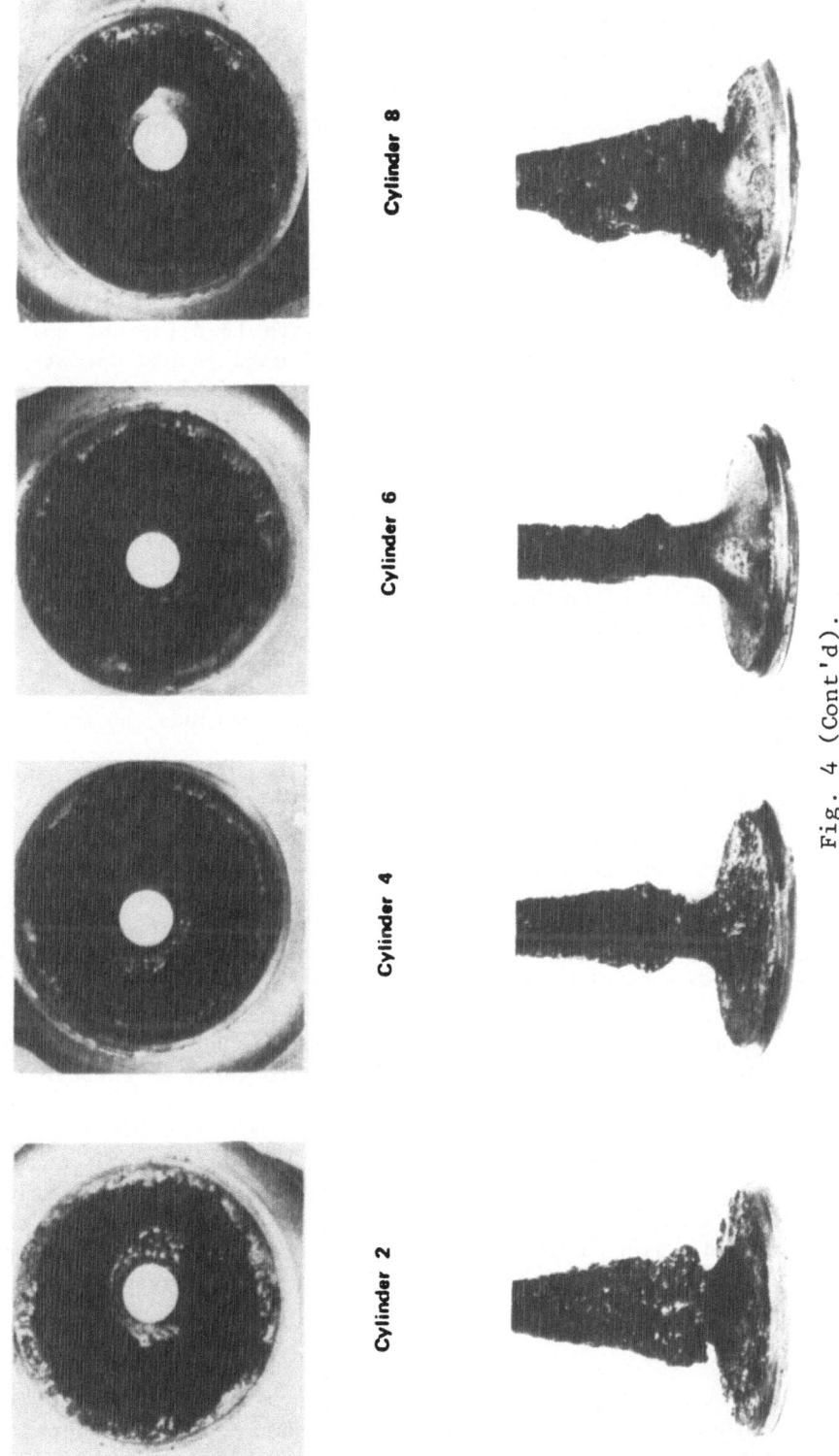

Cylinder 8

Cylinder 6

Cylinder 4

Cylinder 2

Fig. 4 (Cont'd).

49

(CR) of the clean combustion chambers of this engine were measured and found to vary from 7.88 to 8.04 - a spread of only 0.16 CR (or only 0.8 octane number spread, considering that a one CR unit increase raises the OR by about five numbers[7]). Other factors which contribute to the difference in clean octane requirements among cylinders include possible variations among cylinders in (a) spark timing, (b) mixture charge density, (c) jacket water temperature and (d) mixture turbulence and swirl.

It should be mentioned that although the test results given in this paper are from only one engine, similar results were observed in 42 other deposit location/OR tests conducted with 13 different engines. Most of these tests were conducted after the engines had operated for the equivalent of 14,000 to 22,000 miles on test fuels.

CONCLUSIONS

Studies conducted in a 1978 model V-8 engine illustrated the contribution of various engine deposits and mixture strength to the octane requirements of an engine and of its individual cylinders. The results comfirmed our earlier observations[1] of the substantial effect of intake port and intake valve deposits on OR.

Some new observations from the current work include the following:

a. Exhaust valve seat and face deposits can have an adverse effect on OR.
b. Intake manifold and carburetor deposits have little or no effect on engine octane requirements.
c. Engine deposits (that is, intake port, intake valve and combustion chamber deposits) do not affect the distribution of air and fuel among the cylinders of an engine. This finding indicates that the previously observed effect of intake port and intake valve deposits on the octane requirement of an engine does not result from these deposits changing the mixture strength of the cylinders. (Our earlier speculation[1] that these deposits interfere with mixture swirl or turbulence in the cylinder still appears plausible.)
d. Mixture maldistribution was negligible (at least among cylinders fed by the same carburetor venturi) under normal cruising conditions. However, under wide-open-throttle conditions, maldistribution was observed and a quantitative measure of its effect on OR was determined using mixture strength versus knock-limited compression ratio data obtained from a CFR knock test engine.

ACKNOWLEDGMENT

The author is indebted to Mr. E. J. Haury, who conducted the multicylinder engine tests and contributed valuable suggestions to the conduct of the program.

REFERENCES

1. L. B. Graiff, "Some New Aspects of Deposit Effects on Engine Octane Requirement Increase and Fuel Economy," SAE Trans., 88:3162 (1979); SAE paper 790938.
2. E. S. Taylor, W. A. Leary, and J. R. Diver, "Effect of Fuel-Air Ratio, Inlet Temperature and Exhaust Pressure on Detonation," NACA Report No. 699 (1941).
3. L. B. Graiff, W. M. Ehrhardt, and E. J. Haury, "A Device and Technique for Determining the Octane Requirements of Individual Cylinders of an Engine," SAE paper 801353.
4. D. Frazier and H. F. Hostetler, "Air-Fuel Ratio Control - A Minimal Fix for Octane Ratings Over 100," SAE paper 285B (1961).
5. L. F. Dumont, "Possible Mechanisms by Which Combustion Chamber Deposits Accumulated and Influence Knock," SAE Quarterly Trans., 5:565 (1951).
6. J. Warren, "Combustion Chamber Deposits and Octane Number Requirement," SAE Trans., 62:582 (1954).
7. T. O. Wagner and L. W. Russum, "Optimum Octane Number for Unleaded Gasoline," SAE paper 730552.

APPENDIX

TYPICAL TEST FUEL PROPERTIES

Gravity, °API	53.8
RVP, lb	6.9
Research octane number	96.0
Motor octane number	87.0
Sensitivity	9.0
(R+M)/2	91.5

ASTM D-86 Distillation, °F

IBP	92
10%v evaporated	121
20%v evaporated	158
30%v evaporated	190
40%v evaporated	213
50%v evaporated	229
60%v evaporated	240
70%v evaporated	251
80%v evaporated	272
90%v evaporated	318
EP	393
Recovered, %v	94.5
Residue, %v	1.0
Loss, %v	4.5

Hydrocarbon Type by FIA, %v

Saturates	60.0
Olefins	3.0
Aromatics	37.0
ASTM Gum, mg/100 ml, - Unwashed	1
- Washed	1
Gum Time, minutes	200+
Induction Period, hours	8+
Corrosion @ 122°F	1B
Sulfur, %w	0.0054

ON THE CHEMICAL COMPOSITION AND ORIGIN OF ENGINE DEPOSITS

Walter O. Siegl and Mikio Zinbo

Engineering and Research Staff
Ford Motor Company
Dearborn, Michigan 48121

INTRODUCTION

Knowledge of the chemical composition of engine deposits is essential to establishing the mechanisms of deposit formation and to understanding the influence of in-cylinder deposits on such problems as ORI and increased hydrocarbon emissions. Of particular concern is the identification of those deposit components which may be promoting deposit accumulation by functioning as binders.

A long standing hypothesis states that incomplete combustion of fuel produces highly oxygenated carbonaceous substances which contribute to deposit accumulation, in combination with resinous material from the lubricant, by functioning as binders to hold the more inert deposit components.[1] The higher boiling components of both fuels and lubricants have been shown to have a greater influence on deposit accumulation than the lower boiling components. Although the high boiling components which were once present in both fuels and lubricants have largely been removed, in some cases they have been replaced with synthetic materials of even higher molecular weights. The capacity of high molecular weight fuel and lubricant additives to influence deposit accumulation has been a subject of some concerns recently.[2-4] The polymeric additives are present in lubricant formulations as detergents, dispersants, friction modifiers, and VI improvers, and in some fuel formulations as intake system detergents. It is certainly possible that highly oxygenated polymer fragments, formed from partial combustion of the polymers, could function as binders to enhance deposit accumulation.

Efforts to attain an improved understanding of the mechanism of deposit accumulation have been hindered by a lack of convenient

53

techniques to obtain useful information on the chemical composition of deposits. Methods of deposit analysis currently employed usually provide quantitative information on the elemental composition and the qualitative identification of certain crystalline metal salts and oxides. However, with the gradual elimination of organometallic fuel additives, deposits from vehicles run on unleaded fuel are primarily carbonaceous; elemental analysis and x-ray fluorescence provide limited information on the chemical composition of these deposits.

Our current study has been concerned primarily with determining the composition and possible causes of excessive deposit accumulation in the intake port areas of certain test and fleet engines. In addition to the concern over the contribution of intake port area deposits to ORI,[5] the accumulation of significant amounts of deposit on the intake ports and on the valve tulips could eventually lead to valve malfunction. As part of our study on these deposits, a dual-gas thermogravimetric analysis method was developed which provides considerable information on the chemical composition of deposit materials and in particular on the presence of polymeric additives in the deposit. Thermogravimetric analysis (TGA) had been employed previously for the compositional analysis of synthetic resin and rubber formulations,[6] but to our knowledge had not been reported for the analysis of engine deposits. It seemed possible that TGA could provide information on the relative amounts of components such as volatile fuel or oil, synthetic polymer, soot, and ash in the deposit.

In TGA the weight of the sample is continuously recorded as the temperature is raised at a controlled rate. The analysis may be carried out in air or under an inert atmosphere such as argon. Initial weight losses usually correspond to the volatilization of lower molecular weight compounds whereas higher temperature weight losses may involve the thermal degradation (breaking of chemical bonds) of large organic and/or thermally unstable inorganic compounds. The transition from one thermal phenomena to another is usually indicated by a change in slope of the weight loss versus temperature curve. The time required for sample preparation and analysis is on the order of two hours per sample.

This paper describes the application of TGA to the chemical characterization of certain engine deposits. Several chemically distinct components were evident, and based upon the similarity of the thermal behavior of these components with those of known materials, tentative identifications were made. Ancillary analyses were carried out to substantiate the identifications.

EXPERIMENTAL

Samples were removed from engine locations with care not to remove any of the metal surface; samples were weighed and the deposit location was noted. The deposit samples were homogenized by grinding in an agate mortar before analyses were carried out.

Elemental analysis was performed by Spang Microanalytical Laboratory (Eagle Harbor, MI) using the techniques indicated: carbon and hydrogen (Pragel), nitrogen (Dumas), oxygen (LECO), and sulfur (Wagner). In some cases a semi-quantitative analysis for phosphorous and metals was carried out by x-ray fluorescence. Quantitative determination of phosphorous and metals was carried out by inductively coupled plasma torch emission spectroscopy.

Infrared spectra were recorded with a Perkin-Elmer 283 infrared spectrophotometer. Samples were analyzed either as thin-coated films on NaCl plates or as KBr pellets.

TGA measurements were performed on sample sizes of 10 to 20 mg, utilizing a DuPont 951 Thermogravimetric Analyzer at a heating rate of $20^{\circ}C$/min. with a gas (air or argon) flow rate of 40 ml/min. In general the sample was heated under argon to $550^{\circ}C$ or until the weight loss versus temperature curve had leveled off. The sample was then allowed to cool to $300-350^{\circ}C$ under argon; the gas was switched to air; and the sample was heated to ca. $700^{\circ}C$ or until no further weight loss was observed.

Under identical conditions, self-carbonization was found to be negligible during heating under argon in TGA studies on new and also on used engine oils containing soot; it seems unlikely that any significant amount of self-carbonization occurs with the deposit samples under the TGA conditions.

Solvent extractions of deposit samples were carried out at room temperature. In a typical extraction, 4 - 5 ml of solvent was added to ca. 200 mg of deposit sample. The mixture was agitated and then allowed to settle for several hours. The supernatant solution was removed and filtered through a 0.2 μm Millipore filter (FG). This extraction procedure was repeated 3 - 6 times as necessary.

A Waters Associates Model 150-C ALC/GPC supported by a Model 730 Data Module was used for gel permeation chromatographic (GPC) analysis of toluene extracted deposit samples. The column set used consisted of two Altex μ-spherogel columns, for which permeability limits were designated as 500 and 1000 A°. The column set was operated at $30^{\circ}C$ with "Baker Analyzed" reagent grade THF as the eluting solvent, flowing at 1.0 ml/min. Each sample solution was prepared by exactly weighing the extract sample (20 to 40 mg) into a 4 ml screw top vial and adding 3 ml of THF. A 30 μl sample solution was injected into the column by an automatic syringe pump.

RESULTS AND DISCUSSION

The analytical methods described here were developed while analyzing a number of intake port area deposits; however, they should be applicable to the analysis of deposits from a variety of engine locations. The techniques will be illustrated by their application to several intake valve deposits and an EGR valve deposit.

Our approach was based upon the use of TGA to classify the deposit materials as chemically distinct components which differ in their volatilities and/or thermal stabilities. A typical weight-loss versus temperature curve for a deposit is shown in Fig. 1. From the slope changes, it was possible to identify tentatively several general classes of chemical compounds.

As the sample is heated under argon, the initial weight loss (ambient → 200°C) can be attributed to the volatilization of moisture and largely unoxidized low and medium boiling hydrocarbons. Continued heating brings a change in slope at approximately 200°C. In the 200-350°C range the volatilization of the high boiling oil fraction as well as partially oxidized fuel and oil components is the most likely process occurring. A change in slope above 350°C is frequently observed which presumably corresponds to the thermal degradation of polymeric components as well as to the vaporization of highly oxidized hydrocarbons. Occasionally, a further gradual weight loss is observed under argon in the 500-650° range which has not yet been identified but perhaps is due to the thermolysis and volatilization of semicarbonized hydrocarbons and/or unstable inorganics. Usually the weight loss under argon is complete at 500-550°C, leaving a residue consisting primarily of amorphous carbon and inorganic oxides and salts.

If the residue is allowed to cool below 400°C under argon and is subsequently reheated under air, the carbon black is rapidly oxidized to CO_2 as the temperature rises above 450°C. The resulting residue consists only of inorganic salts and oxides (ash).

It is not always possible to distinguish between the various hydrocarbon fractions from the TGA curve obtained under argon alone and in such cases it is beneficial to compare the TGA curves obtained under argon and air. For example in Fig. 2, the TGA carried out under air exhibits a distinct slope change in the 300-400°C region, which is not observed under argon. This change in slope probably reflects differences in thermo-oxidative stability of two dissimilar hydrocarbon materials. It might be expected that poly-

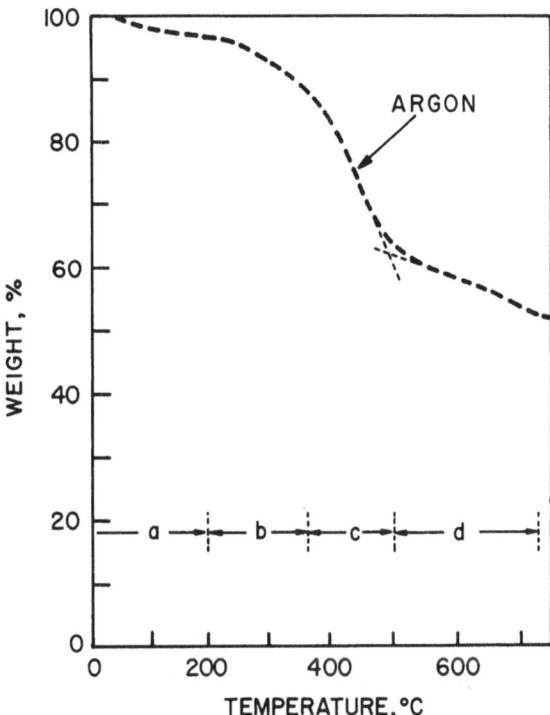

Fig. 1. TGA Weight-Loss Temperature Ranges of Engine Deposit Components under an Argon Atmosphere. (a) Moisture and low and medium boiling hydrocarbons; (b) High boiling hydrocarbons (engine oil); (c) Oxidized hydrocarbons and polymeric additives; (d) Semicarbonized hydrocarbons and volatile or thermally unstable inorganics.

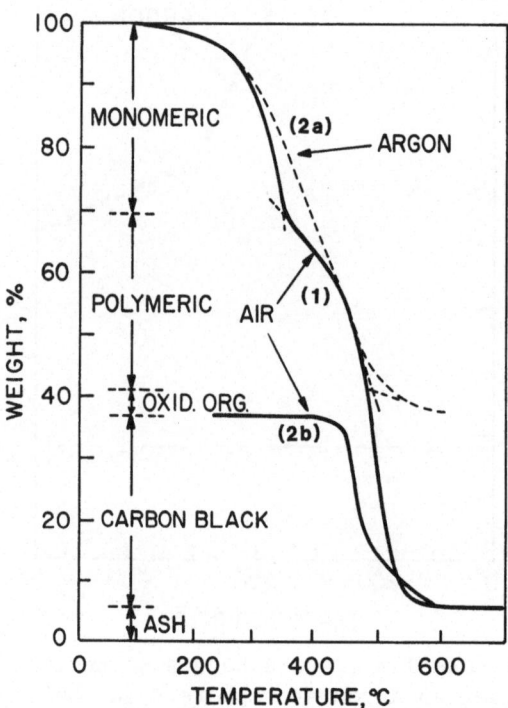

Fig. 2. A Typical Dual—Gas TG Analysis of an Engine Deposit.
Curve: (1) under air; (2a) under argon; (2b) reheated the
residue from (2a) under air.

meric hydrocarbons would exhibit thermo-oxidative weight loss at a slower rate than monomeric lubricant molecules, even if chemical structures and thermal stabilities were similar, because of the lower volatility of fragments formed.

Most of the tentative assignments made above on the basis of volatilities and thermal stabilities were substantiated by the application of various combinations of techniques such as infrared, gel permeation chromatography (GPC), elemental analysis, and various x-ray techniques.

Beginning with the least ambiguous assignment, that of the inorganic ash obtained from TGA carried out under air, infrared analysis indicated the absence of hydrocarbons and frequently showed the presence of metal phosphates, sulfates, or oxides. X-ray diffraction was used to identify these compounds if they were crystalline, which they frequently were, and x-ray fluorescence could be used to identify the relative amounts of various elements present.

The residue obtained from a deposit sample heated to 600°C under argon was tentatively labeled a mixture of carbon black and inorganic ash. Both the infrared spectrum and the elemental analysis are consistent with this label. The infrared exhibits no C-H or C-O bands; carbon was found to be the most abundant element with oxygen being the next most abundant. The level of oxygen in the residue was on the order of that expected for the inorganic ash (metal salts and oxides) alone, therefore, it appeared unlikely that there was much oxygen associated with the carbon and more likely that the carbon was of elemental form.

The presence of hydrocarbons in the bulk deposit was evident from the infrared spectrum and the elemental analysis. Information about specific hydrocarbon components was obtained by extracting a deposit with progressively more polar solvents, analyzing the extracts by IR and GPC, and analyzing the intractable residue by TGA. Infrared spectra of the extracts provided information on the types of hydrocarbons and the extent of oxidation and GPC provided information on the relative molecular size distribution. TGA on the intractable residue yielded information on which components had been extracted.

Most of the unoxidized and mildly oxidized fuel and lubricant hydrocarbons as well as some polymeric additives (e.g. some methacrylates) were extractable into toluene; whereas more highly oxidized hydrocarbons and some other polymers were only extracted by more polar solvents or not at all.

Intake Valve Deposits - Extended Oil Drain Fleet

An examination for cleanliness of 5.0L engines obtained from an extended oil drain fleet indicated that intake port area deposit accumulation had a marked dependence on the fuels and crankcase oils employed. Intake valve tulip deposits were removed from engines run on three different fuel and oil combinations, as shown in Table 1, and were analyzed to determine if the chemical composition of the deposits differed and if so, in what manner. Engine III, run on fuel X (without a polymeric detergent) with oil B (a 10W40 mineral oil based oil), had significantly greater intake valve deposits than engine I, run on Fuel X with oil A (a synthetic oil based oil), and engine II, run on Fuel Y (with a polymeric detergent) with oil A. The cleanliness ratings for the intake valves of the three engines are given in Table I. The relative oil consumption of the three engines was III $\cong$ I > II, where oil consumption for engine III was approximately 18% greater than for II.

Elemental analyses were obtained on all three deposits for C, H,O, and N and on deposits I and III for those other elements thought most likely to be present; the results are presented in Table 2. A summation of the weight percent for the individual elements indicated that essentially the entire mass of the deposit can be accounted for. All three deposits were primarily carbonaceous with greater than 90% of the deposit mass accounted for by the elements C, H, O, and N. As a rough estimate, perhaps 1/3 of the total oxygen in the bulk deposit was present as part of the inorganic components in the form of metal phosphates, sulfates, and oxides. From Table 2 it can be seen that deposit III differed from I and II in having a higher C and H content and a lower content of O and other inorganic elements.

TGA was carried out for all three deposits under argon and air; the curves for deposit II and III are shown in Figs. 3 and 4 and the results are summarized in Table 1. In accord with the elemental analysis data, the TGA results indicated that deposit III differed by having higher concentrations of the fractions we have labeled "monomeric" hydrocarbons and "polymeric" materials. The carbon black/ash ratio was similar for all three deposits. The greater volume of deposit III was due to the greater accumulation of hydrocarbon material and in particular of the component associated with polymeric additives.

The classification of deposit components, which had been made from TGA on the basis of similarities of thermal behavior with those of known materials, was substantiated by the application of complementary analytical techniques. The component labeled carbon black had a very high thermal stability under argon. A sample of deposit I was heated in the TG furnace under argon to 700°C and the residue, which was labeled as carbon black and inorganic ash in Fig. 2, was analyzed for C,H,N and O. The analysis indicated 77.3% C, 2.1% H, and 5.6% O (12.7% other elements). The oxygen content was in line with that expected for the inorganic ash component alone,

TABLE 1. Intake Valve Deposits from an Extended Oil Drain Fleet

	Engine No.[a]			
	I	II	III	III[e]
Fuel[b]	X	Y	X	–
Crankcase Oil[c]	A	A	B	–
Intake Valve Cleanliness Rating[d]	4.68	6.54	2.80	–
TG Analysis:				
Monomeric Hydrocarbons	25%	18%	31%	3%
Polymeric Materials	11	17	28	44
Highly Oxidized Organics	5	5	4	8
Carbon Black	50	51	31	38
Inorganic Ash	9	9	6	8

[a] 5.0L engines with 50,000 miles accumulated.

[b] Fuel Y contains a polymeric detergent whereas fuel X does not.

[c] Oil A is a synthetic ester type lubricant and oil B is a mineral oil based 10W40 SE oil.

[d] Based on a scale of 1-10 where 10 is most clean.

[e] Residue from deposit III after sequential extraction with toluene, CH_2Cl_2, MEK, and THF.

TABLE 2. Elemental Composition of Intake Valve
Deposits from Extended Oil Drain Fleet

	Deposit from Engine (wt. %)		
Element	I	II	III
C	67.8	69.3	72.2
H	5.8	5.8	8.0
N	2.1	2.0	2.2
O	16.9	15.8	13.3
S	1.3	–	0.9
P	0.9	–	0.4
Ba	0.2	–	0.1
Ca	0.9	–	0.4
Cu	0.2	–	–
Fe	0.7	–	0.3
Mg	< 0.1	–	0.3
Pb	0.1	–	1.2
Zn	1.2	–	0.6
Total	98.1	92.9	99.9

61

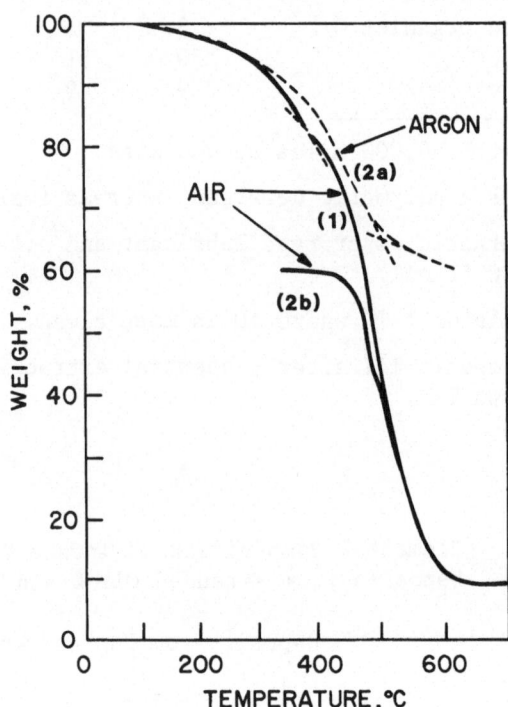

Fig. 3. TG Analysis of an Intake Valve Deposit from Engine II.
Curve: (1) under air; (2a) under argon; (2b) reheated
the residue from (2a) under air.

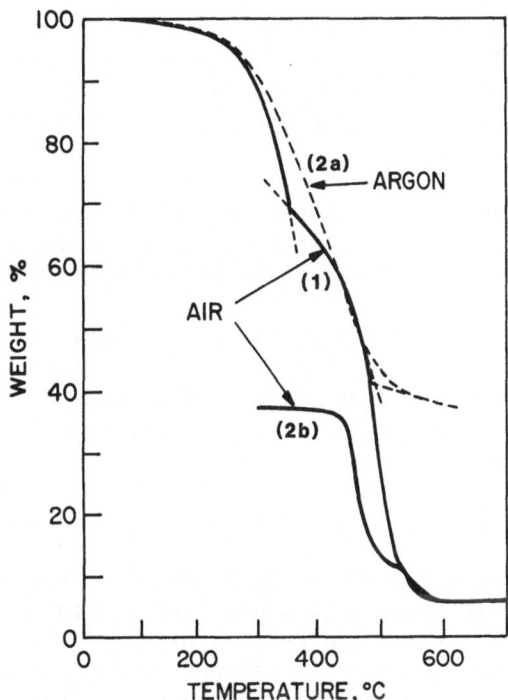

Fig. 4. TG Analysis of an Intake Valve Deposit from Engine III.
Curve: (1) under air; (2a) under argon; (2b) reheated
the residue from (2a) under air.

suggesting that little oxygen is bound to the carbon and that this thermally stable carbonaceous component had a composition in accord with that expected for carbon black. Although the H/C ratio dropped from 1:1 in the bulk deposit to 1:3 in the residue, it is still high for a carbon black; how much of the hydrogen was bound up with the inorganic ash remains uncertain.

To obtain more information on the hydrocarbon component of deposit III, a portion of the sample was extracted first with toluene and then with CH_2Cl_2, MEK, and THF in that order. The weight percent of the deposit extracted with each solvent in the sequence was: 28.5% (toluene), 3.6% (CH_2Cl_2), 1.9% (MEK), and 2.7% (THF). TG analysis of the intractable residue of deposit III indicated the following components: 3% monomeric hydrocarbons, 44% polymeric materials, 8% highly oxidized organics, 38% carbon black and 8% inorganics. Thus, it was mostly the "monomeric" hydrocarbon component which was extracted. IR analysis of the toluene extract indicated that it consisted primarily of slightly oxidized oil and some 2-ethylhexyl methacrylate polymer. GPC analysis of the same extract indicated the presence of relatively unoxidized as well as mildly oxidized oil in addition to polymeric material with a very broad molecular weight distribution.

Infrared analysis of the extracts with more polar solvents indicated the presence of highly oxidized hydrocarbons (carboxylic acids, lactones, etc.). However, most of the deposit component which has been labeled "polymeric", was not extractable under the conditions employed.

Intake Valve Deposits - PROCO Engine

Excessive deposit accumulation occurred in the intake port area of a PROCO test engine which had been run on 1980 certification fuel and with a 10W30 mineral oil based oil (Oil C). To obtain information on chemical factors influencing deposit accumulation, deposit samples were removed and analyzed for chemical composition. Although several samples were examined, only the results for a valve tulip deposit are described here.

TG analysis was carried out under both argon and air and the curves are shown in Fig. 5. Four major components were identified and are listed in Table 3 along with other analytical data. Of the four major components, the one most likely to function as a binder and hence to promote deposit accumulation was the one labeled "polymeric"; this component therefore received the most attention. The deposit was extracted with 4:1 toluene-dioxane at ambient temperature. IR analysis of the extract showed the presence of an ester-type material (with a carbonyl band at 1730 cm^{-1}) in addition to mineral oil. Small scale preparative GPC of the extract indicated the presence of a polymer of average molecular weight 2400 and

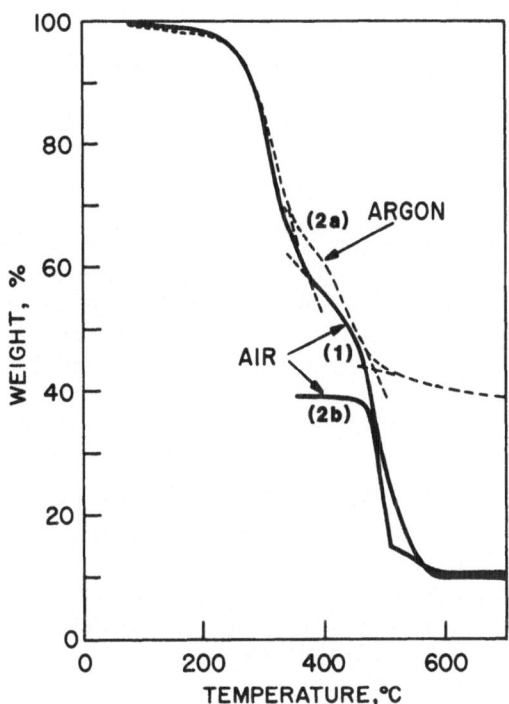

Fig. 5. TG Analysis of an Intake Valve Deposit from a PROCO engine.
Curve: (1) under air; (2a) under argon; (2b) reheated
residue from (2a) under air.

TABLE 3. Analytical Data for Intake Valve Deposit-PROCO Engine

Appearance: Black amorphous solid

Fuel: 1980 certification fuel (unleaded)

Lubricant: Oil C, 10W30 SE mineral oil based oil

Elemental Analysis:

Element	Weight %
C	70.3
H	7.5
N	0.7
O	12.4
Total C,H,N, & O	90.9

TG Analysis:

Temperature Ranges, $^{\circ}$C	Wt. Loss %	Type of Material Lost
25-360	24	Monomeric Hydrocarbons
360-490	24	Polymeric Materials
490-600 (argon)	5	Semicarbonized Organics or Unstable Inorganics
410-520 (air)	23	Carbon Black
Residue	17	Inorganic Ash

monomeric materials of molecular weight 400 and lower. The two
fractions were collected and analyzed by infrared from which they
were identified as a polyester-type polymer and a mineral oil respectively. In separate experiments, infrared analysis of Oil C
showed the presence of an ester-type additive and TGA indicated the
presence of 5% polymeric additives by weight. The polymer fraction
of oil C was concentrated by dialysis[7] and preparative GPC was carried out on the concentrate. The polymeric fraction collected from
GPC of the concentrate was essentially identical in molecular weight
distribution and IR spectrum to that isolated from the valve tulip
deposit. TG analysis of the extracted deposit residue indicated
that almost all of the component labeled "polymeric" had been removed by the extraction.

Although the extraction and subsequent GPC analysis of the
deposit were not carried out in a quantitative manner, it was clear
that the polymer additive represented a significant fraction of the
deposit and that the ratio of polymer/oil was much higher in the
deposit than in the lubricant itself. From this data it could be
concluded that a polymeric ester, which originated as an additive
in the lubricant, was being concentrated in the deposit and was

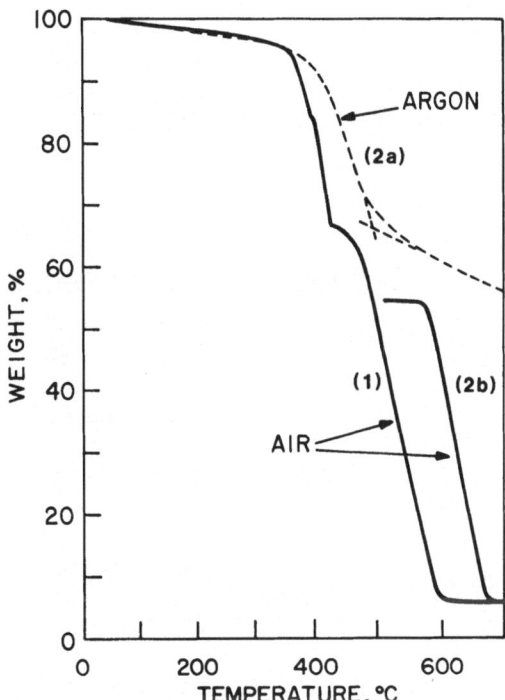

Fig. 6.   TG Analysis of an EGR Valve Deposit.   Curve: (1) under
          air; (2a) under argon; (2b) reheated the residue from (2a)
          under air.

TABLE 4. Analytical Data for an EGR Valve Deposit

Appearance: Black amorphous solid

Fuel: Certification fuel (containing polyisobutene-derived detergent)

Elemental Analysis:

Element	Weight %
C	72.2
H	5.8
N	2.4
O	14.3
Total C,H,N & O	94.7

Metals: Major: Mg, P
         Minor: Fe, Pb, Si, S, Zn

TG Analysis:

Temperature Range, °C	Wt. Loss %	Type of Material Lost
25-350	4	Monomeric Hydrocarbons
350-500	29	Polymeric Materials
500-750	12.5	Semicarbonized Hydrocarbon and Unstable Inorganics
570-680 (air)	48.5	Carbon Black
Residue	6	Inorganic Ash

possibly functioning as a binder to enhance deposit accumulation.

EGR Valve Deposit

A sample of black amorphous solid was removed from the pintel of a sonic EGR valve whose operation was impaired by deposit accumulation. To obtain information on the origin and mechanism of the deposit formation, the deposit sample was analyzed for chemical composition.

Infrared analysis of the bulk deposit indicated the presence of a significant amount of hydrocarbon component, most of which appeared to be polyisobutene. TGA was carried out under argon and air; the curves are shown in Fig. 6. From the TG analysis, five components, including a large amount of what appears to be polymeric material, were evident; the relative amounts are shown in Table 4. Oil and partially oxidized oil, as were observed in the intake valve deposits, were not major components in this deposit.

An authentic sample of the polyisobutene-based polymeric detergent, known to be present in the fuel, was obtained and the TGA curve recorded and found to correspond to the polymeric component of the deposit. Based upon the observed accumulation of fuel detergent in the deposit and knowledge of the fuel intake system hardware, it was concluded that raw fuel, which could reach the valve from the downstream side was evaporating on the hot valve and leaving behind the non-volatile polymeric additive.

Carbon black and inorganic ash, from the EGR, probably accumulated on the sticky surface created by deposition of the polymeric additive on the valve. It would appear unlikely that a significant accumulation of carbon black and inorganic ash would occur on the valve surface without the aid of a binder or adhesive.

SUMMARY

Engine deposits, mainly from the intake port area, were analyzed for chemical composition using thermogravimetric analysis as the primary analytical technique. Several chemically distinct components were evident from the TGA curves and based upon similarities of thermal behavior with those of known materials, the components were tentatively identified. The identification of most of these components was supported by data from elemental analyses, infrared spectroscopy, and gel permeation chromatography. The four largest deposit components included monomeric hydrocarbons, carbon black, metal salts and oxides, and a "polymeric" fraction. In some cases polymeric fuel or crankcase oil additives were identified in this last fraction. The polymeric component may be the most important component of the deposit because it is the most likely to function as a binder or adhesive to promote deposit accumulation.

ACKNOWLEDGEMENT

The authors gratefully acknowledge B. Joshi, J. L. Parsons, C. R. Peters, and D. Smolenski for analytical assistance rendered and R. E. Baker, S. Korcek, and L. R. Mahoney for helpful discussions.

REFERENCES

1. H. J. Gibson, C. A. Hall, and D. A. Hirschler, "Combustion Chamber Deposits and Knock," SAE Transactions, 61:361 (1953).

2. J. D. Benson, "Some Factors which Affect Octane Requirement Increase," SAE Paper 750933, 1975.

3. H. E. Bachman and E. B. Prestridge, "The Use of Combustion Chamber Deposit Analysis for Studying Lubricant-Induced ORI," SAE Paper 750938.

4. T. R. Erdman, "The Fate of Ashless Dispersants in Gasoline Engines as Followed by Carbon-14 Radiolabeling," SAE Paper 810015.

5. L. B. Graiff, "Some New Aspects of Deposit Effects on Engine Octane Requirement Increase and Fuel Economy," SAE Paper 790938.

6. J. J. Swarin and A. M. Wims, "Applications of Integral and Derivative Thermogravimetry to the Analysis of Rubber Formulations," Rubber Chem. Technol., 47:1193 (1974).

7. G. I. Jenkins and C. M. A. Humphreys, "The Analysis of Lubricants and Additive Concentrates Using Spectroscopy and Physical Methods of Separation," J. Inst. Petrol., 51:1 (1965).

THE CHEMISTRY OF INTERNAL COMBUSTION ENGINE DEPOSITS —

I. MICROANALYSIS, THERMOGRAVIMETRIC ANALYSIS, AND INFRARED
   SPECTROSCOPY

Lawrence B. Ebert, William H. Davis, Jr.,
Daniel R. Mills, and John D. Dennerlein

Corporate Research-Science Laboratory
Exxon Research and Engineering Company
P. O. Box 45
Linden, New Jersey 07036

I. INTRODUCTION

As an automobile engine runs, its fuel quality requirements, as measured by the octane number of the fuel needed to
inhibit knocking, may change in time. Historically, this phenomenon is referred to as the "ORI", standing for octane
requirement increase.

The single most important operational variable affecting
ORI is the presence of deposits within the combustion
chamber.[1] Once an engine is run, carbonaceous materials coat
the inside of the combustion volume, illustrated in Figure 1.

To minimize octane demands on the fuel, an obvious goal is
to modify or eliminate deposits, so that ORI is kept low. To
achieve this end, one must know something of the chemistry of
the deposits. In this and the following paper, we use classical
analytical tools to probe the nature of the deposit.

II. EXPERIMENTAL

The deposits examined in this study were generated by six-
cylinder engines employing various unleaded fuel and lube combinations. The entire deposit mass (piston top and head area of
all cylinders) was combined; while obvious artifacts such as
paint and bristles from brushes were removed, ferromagnetic
impurities were not.

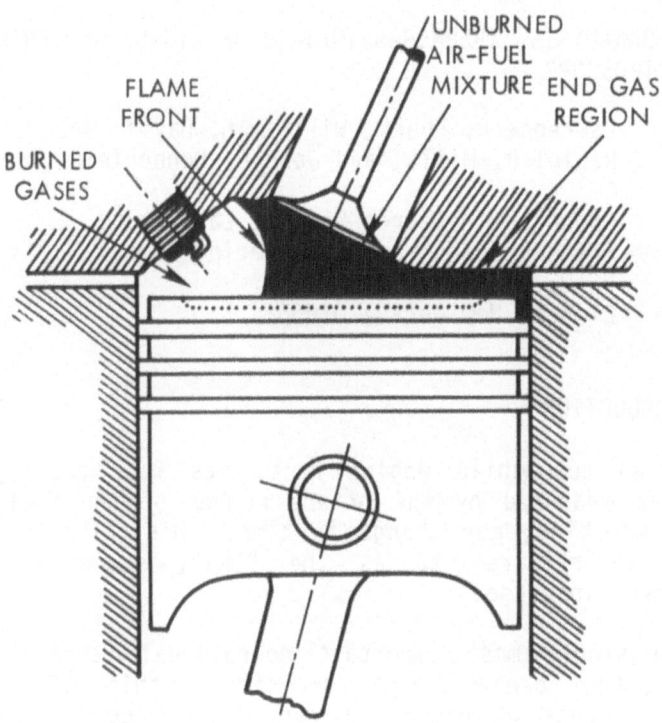

Figure 1. An illustration of the combustion volume. The deposits investigated came from all portions of the volume, including piston top and end gas region.

Elemental analysis was performed by Galbraith Laboratories (Knoxville, TN) using the techniques indicated: carbon and hydrogen (Pragl), nitrogen (Dumas), oxygen (LECO), and in-organics (wet chemistry/atomic absorption).

The weight of the sample as a function of time was deter-mined using a Dupont 951 thermogravimetric analyzer interfaced to a Dupont 990 thermal analyzer. Additionally, in this con-figuration, the derivative of the weight/temperature profile (the "rate" of weight loss) was instantaneously computed and displayed. The heating rate was 10°C/min, with argon (passed through Drierite) flowing across the sample. Typical sample sizes were 20 mg, and temperatures between ambient and 1000°C were recorded with a chromel/alumel thermocouple.

To obtain additional information on deposit pyrolysis, we set up our own apparatus. Heat was supplied by a Lindberg type 167 tube furnace (1" in diameter and 12" long), with the sample placed in a porcelain boat within a quartz tube in the fur-nace. The system was flushed with a 5-35 ml/min flow of inert gas (either He or N_2), which led first to U-tube immersed in an ice bath (for collecting evolved water) and then to a bubbler tower containing 0.1 N NaOH (Acculite) (to trap evolved CO_2). In some applications, the evolved gases were led to the sampling valves of an HP 5700 A gas chromatograph, equipped with a Spherocarb packed column (80/100 mesh, 9 ft. by 1/8"). Typical GC conditions were isothermal, with the detector, injector, and oven held at 150°C. For a helium flow rate of 30 ml/min, the retention time for CO is about 82 seconds, for CO_2 about 245 seconds.

Thermally evolved CO_2 was measured coulometrically in a set-up designed for "total organic carbon" (Huffman Laboratories, Wheat Ridge, CO).

Infrared spectra of deposits were run in pellet form (about 1 mg in KBr) on a Digilab FTS-14.

A variety of titration methods were used to characterize the as-received deposit. With the n-butyllithium method, a weighed amount of deposit (ca. 1 g) was placed in n-hexane (ca. 50 ml) and a known volume of n-butyllithium (approximately 1.6 N) was added. The mixture was allowed to react, with stirring, for about 1 week at room temperature in a He-filled VAC atmo-spheres glovebox, after which the mixture was filtered through a fine porosity frit. The liquid filtrate was quenched with water, extracted, and the aqueous phase titrated against 0.1 N HCl (Acculite) utilizing a Fisher automatic titrator. From this data one can calculate the amount of n-butyllithium consumed per gram of deposit, with typical numbers in the range 4 to 8 mmol. Titration with calcium acetate gives an estimate of the number of carboxyl groups. Herein, 0.1 to 0.3 g deposit (previously acidified with HCl) is mixed with 10 ml of 1 N calcium acetate and 50 ml H_2O. After reacting for 24 hours, the mixture is filtered, and the liquid filtrate titrated against 0.1 N NaOH. A related procedure for measuring acidity is the amount of hydrogen gas liberated on reaction with $LiAlH_4$, which should measure all acids. This was done for us by the Cabot Corporation, Billerica, MA. Finally, an additional method for the analysis of protic acids involves reaction with the silylating agent, BSTFA (Bis(trimethylsilyl)-trifluoroacetamide) with quantitation obtained by microanalysis for Si.

III. RESULTS

A. Elemental Analyses

The results of analysis for carbon, hydrogen, and oxygen are given for three deposits, of different fuel/lube origin, in Table 1. The fuels are either iso-octane (2,2,4-trimethylpentane) or whole fuels containing complete naphtha and aromatic fractions. The lubes are either the complete Lube A package or an ashless fraction thereof.

Table 1. Results of analysis for carbon, hydrogen, and oxygen

---

ELEMENTAL ANALYSIS

FUEL	LUBE	%C	%H	%O
WHOLE[a]	ASHLESS	69	4.0	25
WHOLE[b]	Lube A	59	4.5	26
I-OCT	Lube A	54	5.3	27

[a]AVERAGE OF TWO RUNS

[b]AVERAGE OF FIVE RUNS

In the first case of Table 1, in which no inorganics are present in the lube, one notes that the carbon, hydrogen, and oxygen account for 98% of the total deposit mass. In the second and third cases, in which inorganics are present in the lube, the carbon, hydrogen, and oxygen account for 86 to 90% of the total deposit mass. Table 2 demonstrates that a satisfactory mass balance can be obtained when all elements of the deposit are analyzed.

In terms of analogies to known systems, the elemental analysis is more similar to bituminous coals (Illinois #6, Monterey mine, dry basis: %C 69.39, %H 4.80, %O 11; Wyodak subbituminous, dry basis: %C 65.69, %H 4.92, %O 22) than to graphite (Union Carbide SP-1: %C > 99.5)[2]

B. Thermogravimetric Analysis

Thermogravimetric analysis confirms this analogy to volatile carbonaceous materials. Figure 2 shows the weight loss (top) and derivative weight loss (bottom) for a deposit arising from a base fuel/ashless oil combination.

To emphasize the degree of volatility, we have constructed a series of difference TGA profiles which interrelate this deposit (68.64 %C, 3.80 %H) with the Illinois #6 coal. In Figure 3 we plot the quantity (% mass loss of starting Illinois #6) - (% mass loss of a 500°C pyrolysis char of Illinois #6). As one would naively expect, the starting coal loses much more weight than the pyrolysis char at all temperatures between 300°C and 1000°C. If, however, we plot (% mass loss of starting coal) - (% mass loss of 300°C char), we see in Figure 4 that the difference is small.

With this background, the reader is directed to the difference plot of the ORI deposit and the initial Illinois #6 coal (Figure 5), which shows the deposit to lose more mass than the high volatile bituminous coal at all temperatures between 300°C and 1000°C.

Given that engine deposits do show considerable mass loss above 300°C, one would like to determine what types of molecules are evolved thermally. To focus on this issue, we switched from a TGA apparatus to a tube furnace, so that larger amounts of material could be used. To check reproducibility, we combined three sets of deposits from base fuel/Lube A conditions so that we would have sufficient material to perform several pyrolyses. The microanalytical data for this mixture is given in Table 3.

In the first experiment, 1.012 g deposit was heated for 30 minutes at 500°C under a nitrogen flow of 6.6 ml/minute. The outgas was bubbled through 0.1 N NaOH to trap CO_2 for

Table 2. Mass balance can be obtained when all elements of the deposit are analyzed

ELEMENT	249 T13 BASE/Lube A	243 T10 BASE/Lube A	249 T2 BASE/Lube B	249 T7 BASE/Lube B
%C	65.94	66.89	59.08	64.17
%H	3.56	4.14	3.74	3.95
%N	1.43	0.75	0.83	0.85
%S	0.63	0.51	1.24	0.68
%O	25.57	23.52	26.59	25.89
%P	0.44	1.36	1.28	0.73
%Mg	0.31	0.44		
%Ba	0.64	0.62		
%Ca			4.23	2.00
%Zn	0.42	0.65	2.08	1.09
%Fe	0.73	0.30	0.22	0.25
TOTAL	99.67	99.18	99.29	99.61

ELEMENT	261-13 BASE/ASHLESS	261-13 (CCD) BASE/ASHLESS	CHEV #3; PHASE 1 BASE/ASHLESS
%C	69.23	68.77	68.60
%H	3.96	3.66	4.16
%N	1.70	1.94	1.50
%S	0.61	0.57	0.24
%O	21.07	20.00	24.27
%P	0.27	0.27	0.30
%Fe		0.25	0.74
TOTAL	96.84	95.46	99.81

ELEMENT	CHEV 241-19/20 Blend #15/Lube A	256-2 BASE/Lube A
%C	60.99	59.95
%H	4.26	3.88
%N	3.09	1.49
%S	1.44	1.44
%O	26.16	26.18
%P	0.96	1.86
%Mg	0.45	0.94
%Ba	1.17	0.63
%Zn	0.77	1.56
%Fe	0.38	0.33
TOTAL	99.67	98.26

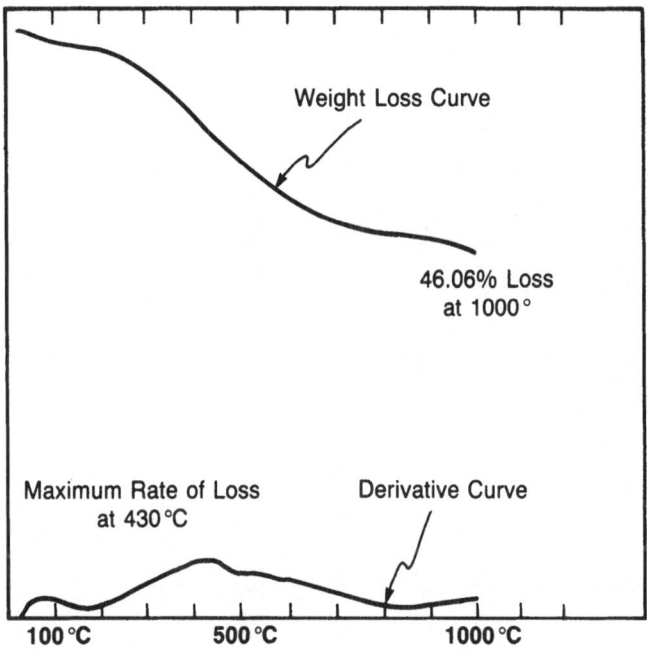

Figure 2.   Thermogravimetric analysis of a deposit from a
base fuel/ashless oil combination.

Table 3. Microanalytical results for tube furnace pyrolyses

Description	%C	%H	%N	%O	%S	%P	%Mg	%Ba	%Zn	%Fe
Initial deposit mixture	59.63	4.00	1.72	27.31	1.17	1.29	2.72	1.36	0.13	0.47
Initial deposit, duplicate analysis	61.91	4.17	1.73	25.65						
Residue, 500°C, 30 min	68.95	3.38	1.97	15.90	1.06	1.94	3.26	1.01	0.17	0.60
Residue, 500°C, 30 min Second Run	67.53	2.66	1.80	9.25	1.15	2.01	2.91	2.51	1.96	0.67
Residue, 500°C, 30 min Third Run	66.95	2.66	2.31	13.97	1.10	2.07	1.41	3.30	2.00	0.62
Residue, 500°c, 30 min Third Run, duplicate	69.62	2.62		10.55						
Oily liquid Third Run	76.96	9.44		[13.6]						
Residue, 500°C, 30 min Fourth Run	67.85	2.71	2.01	20.05	1.21	1.43	0.84	0.52	1.85	0.80
Residue, 500°C, 30 min Fourth Run, duplicate	69.11	2.93		10.92						
Oily liquid Fourth Run	77.89	9.63								

Description	%C	%H	%N	%O	%S	%P	%Mg	%Ba	%Zn	%Fe
Residue, 500°C, 30 min Fifth Run	68.71	3.07	1.97	12.58	0.99	1.69	0.78	--	1.89	0.60
Oily liquid Fifth Run	79.26	8.49								
Residue, 400°C, 30 min, 500°C, 30 min Sixth Run	69.06	2.89	2.24	11.62	0.86	1.78	0.81	--	1.91	0.61
Oily liquid Sixth Run	75.18	9.09								
Residue, 400°C, 30 min 500°C, 30 min Seventh Run	69.58	3.02	1.90	10.67	1.31	1.40	0.45	0.41	1.73	0.89
Residue, Variable temperature Eighth Run	70.51	2.74	1.94	8.67	1.26	1.37	0.44	0.44	1.65	0.75
Oily liquid Eighth Run	78.68	8.05								

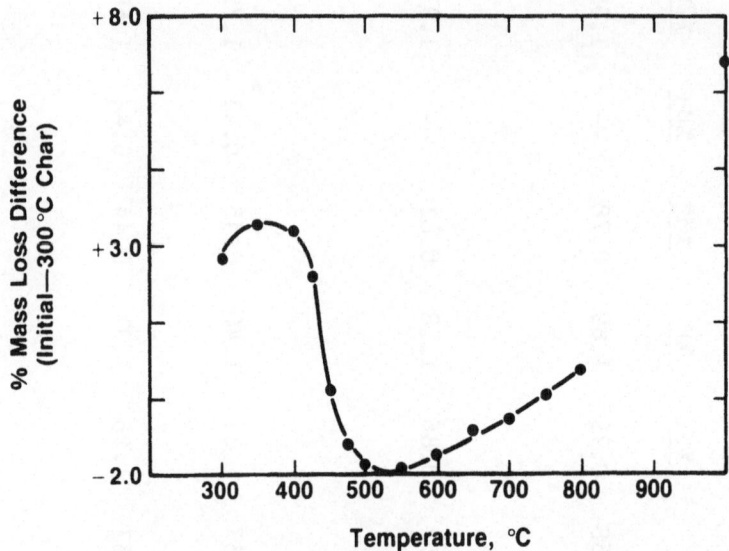

Figure 3. TGA difference profile: Coal - 300°C coal char.

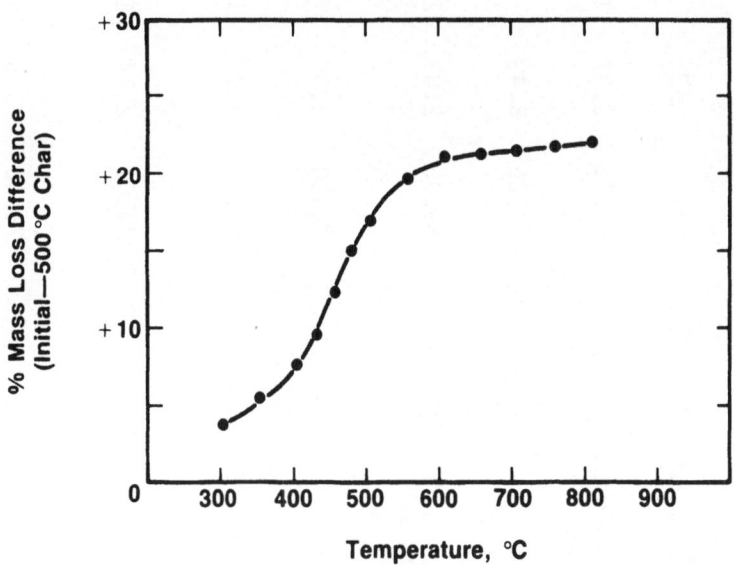

Figure 4. TGA difference profile: Coal - 500°C coal char.

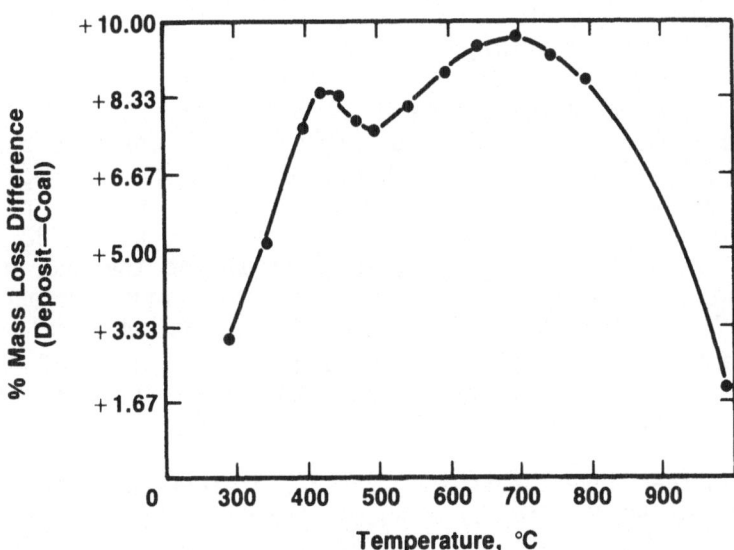

Figure 5. TGA difference profile: deposit (base fuel/
ashless oil) - Illinois #6 coal (Monterey Mine,
69.39% C, 4.80% H on a dry basis).

The reader should note that the deposit loses more
mass than does the high volatile bituminous coal at
all temperatures between 300° and 1000°C.

(The y axis in both cases was taken on a dry
basis, so mass loss due to "free" water is not
included.)

titration. Following the pyrolysis, we recovered 0.622 g deposit residue in the boat, and the microanalysis of this residue appears in Table 3. By difference, we have a mass loss of 0.390 g, corresponding to 38.54% of the initial weight. In part, this lost material appears as a red, oily residue at the end of the pyrolysis tube, and Figure 6 shows this material to have a proton aromaticity of ca. 25%. In part, the lost material appears as gases, with our titration indicating the formation of 2.3 mmol of CO_2, equal to 0.1012 g, which is 10% of the deposit weight. In terms of an oxygen balance on the initial deposit (average % O = 26.48), we have 37% of the oxygen in the residue, 27% of the oxygen as CO_2, and 36% of the oxygen as other volatile material.

To check on the reproducibility of this experiment, we performed a second 500°C/30 minute pyrolysis of the same deposit mixture, utilizing an initial charge of 2.015 g. We recovered 1.284 g in the boat, and the microanalysis of this residue appears in Table 3. The mass loss by difference was 0.732 g, or 36.28%. We again found a red, oily liquid at the end of the tube, and the proton NMR spectrum of this liquid appears in Figure 7. The titration equivalent was 4 mmol CO_2, corresponding to 8.7 wgt.% of the sample. In terms of oxygen balance, 22% remains in the residue, 24% is as CO_2, and 54% is as other volatile material.

To further investigate the oxygen distribution, a third run was made at 500°C for 30 minutes. More material was used (5.151 g), and a U-tube immersed in Dry Ice/acetone was placed in the line to trap condensible gases. A residue of 3.311 g was recovered, and duplicate microanalytical data for this run appear in Table 3. By difference, there is a mass loss of 1.84 g, or 35.72%. Recovery of the red, oily liquid was 0.511 g, and the proton NMR of this appears in Figure 8. The titration equivalent was 10.6 mmol CO_2, equal to 0.466 g, and thus leaving 0.86 g unaccounted for. The U-tube contained a transparent liquid, whose infrared spectrum (3450 cm^{-1} (max), 1638 cm^{-1}, 1200 cm^{-1}, 2080 cm^{-1}) was suggestive of water.

With some assumptions, we can obtain a more refined oxygen distribution. The microanalysis of the red, oil liquid is given in Table 3, with oxygen, by difference, amounting to 13.6%. The proton NMR spectra are consistent with oxygen functionality, specifically phenolic (6.5-7 ppm) and aliphatic alcohol (4.1 ppm. As an aside, our attempts at decoupling the 4.1 ppm doublet were not instructive, as seen in Figure 9). In terms of distribution, 5% of the oxygen of the initial deposit is in the red, oily liquid, 30% is in the thermal residue, and 25% is

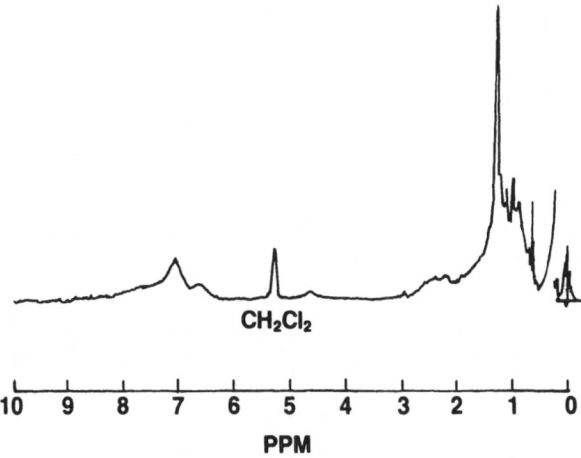

CH$_2$Cl$_2$

10 9 8 7 6 5 4 3 2 1 0

PPM

Figure 6. Proton NMR spectrum of volatiles from run 1.

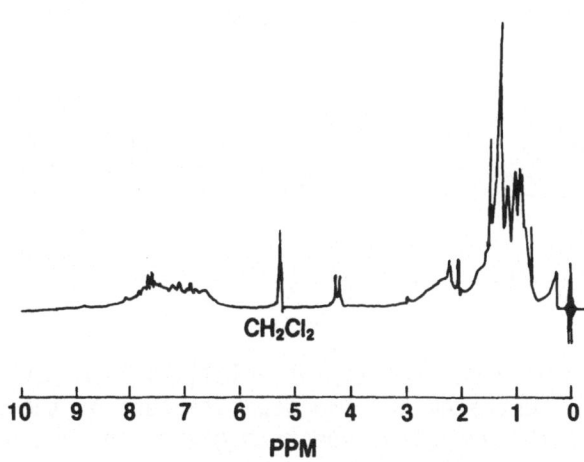

CH$_2$Cl$_2$

10 9 8 7 6 5 4 3 2 1 0

PPM

Figure 7. Proton NMR spectrum of volatiles from run 2.
(Run at 60 MHz in CH$_2$Cl$_2$ at 24°C)

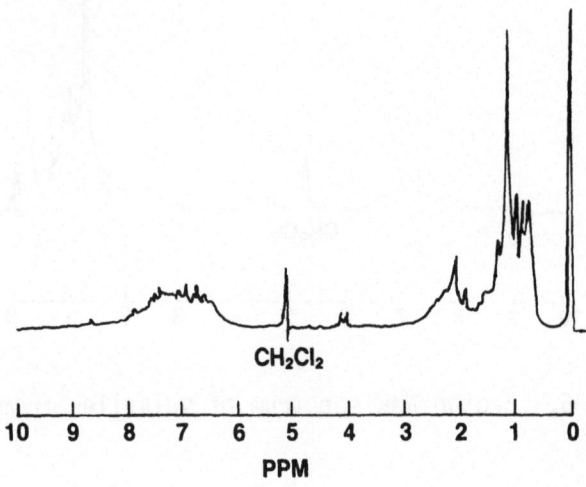

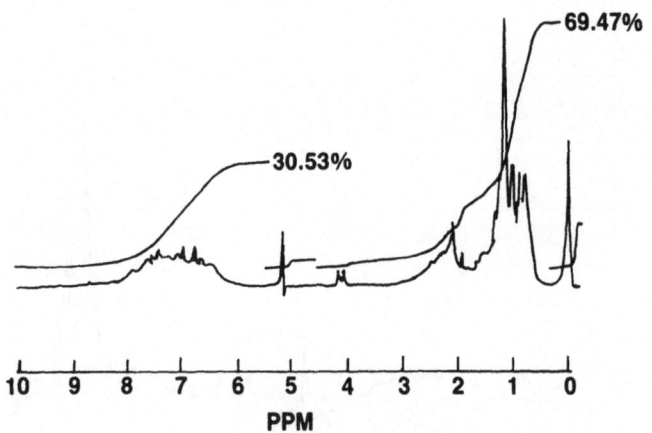

Figure 8. Proton NMR spectra of volatiles from run 3.
Illustrated are two separate runs on the same sample,
with the bottom spectrum giving the output of the
analog integrator.

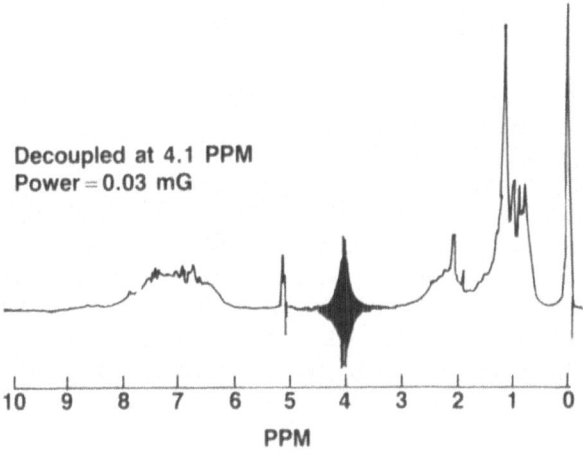

Decoupled at 4.1 PPM
Power = 0.03 mG

PPM

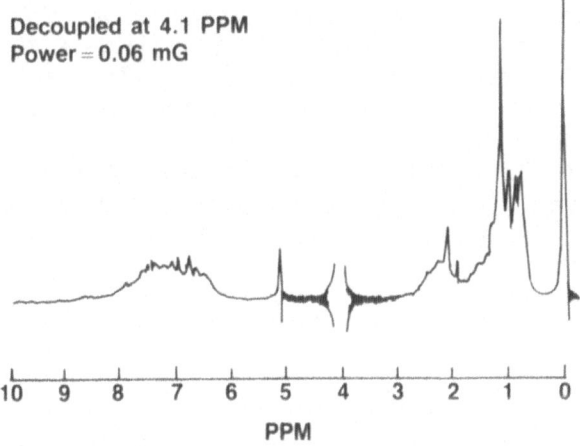

Decoupled at 4.1 PPM
Power = 0.06 mG

PPM

Figure 9.   Proton NMR spectra of volatiles from run 3, with
            decoupling at 4.1 ppm.

recovered as CO_2. If water accounted for the remaining 40% of the oyxgen, we would have evolved 0.12 g H_2O/g, deposit, or 0.618 g H_2O per 5.151 g, deposit.

Such an analysis assumes that all otherwise unaccounted oxygen is as thermally evolved water. To check this, we performed a fourth 500°C/30 minute pyrolysis, with an in-line gas chromatograph, equipped with sampling values, so that we could identify relatively minor amounts of other gas phases. Although our Spherocarb-packed column could, in principle, analyze water, we placed Drierite in the gase line to adsorb water. (The chilled u-tube trap was also in place). Starting with 2.123 g, we recovered 1.408 g of residue, the microanalysis of which appears in Table 3. The mass loss was 0.715 g or 33.68%. Unfortunately, the chromatograph malfunctioned, so a fifth run was necessary. Starting with 2.391 g, we recovered 1.576 g, the analysis for which is in Table 3. In terms of gas analysis, both CO_2 and CO were observed in the evolved gases. Table 4 gives the relative molar ratios during the pyrolysis. Clearly, much more weight is lost as CO_2 than as CO, with the observed mole ratio of 6.5 corresponding to a weight ratio of 10.

To be certain that this CO_2 dominance is characteristic of the deposit, and not the conditions, we altered our temperature program. In the sixth run, there was a 30 minute residence at 400°C, followed by a 30 minute residence at 500°C. Starting with 2.235 g we recovered 1.420 g, the analysis for which is in Table 3. The weight loss observed, 36.47%, was comparable to that observed in run 5, 34.09%. The molecular ratios of CO_2 to CO did depend on temperature, ranging from 3.61 to 4.41 at 400°C and from 5.16 to 5.74 at 500°C. The total amount of CO_2 evolved was 4.34 mmol (8.5 wgt% of deposit) and of CO 0.73 mmol (0.9 wgt% of deposit), corresponding to a molar ratio of 5.9 and a weight ratio fo 9.3.

To verify reproducibility, we repeated these temperature conditions in the seventh run. Starting with 2.141 g of deposit, we recovered 1.378 g of residue, the analysis for which is in Table 3. The molecular ratios of CO_2 to CO were distinct from those found in the sixth run. At 400°C, the ratio ranged from 6.04 to 6.35, but at 500°C, the ratio was between 1.79 and 3.47. The total amount of gas formed still indicated CO_2 predominance; carbon dioxide recovery was 5.65 mmol (11.6 wgt% of deposit) and carbon monoxide recovery was 1.33 mmol (1.74 wgt% of deposit), corresponding to a mole ratio of 4.2 and a weight ratio of 6.7.

In run 7, we noted the presence of two additional GC peaks during the 400°C temperature plateau. To more fully explore this, we ran a more extensive temperature program, involving an 18 minute residence at 250°C, a 15 minute residence at 350°C, a 21 minute residence at 450°C, and a 30 minute residence at 500°C. Starting with 1.188 g, we recovered 0.742 g, the analysis for which is in Table 3. The mass loss observed, 37.54%, is comparable to that observed in the other 500°C runs. The more detailed program of run 8 shows the complexity of this mass loss.

The first evolved molecules, CO_2 and CO, were observed during the 250°C plateau, with the CO_2/CO ratio equal to 0.93. This ratio increased to 3.67 at 291°C, peaked at 6.77 at 393°C, and was only 2.56 on reaching 500°C. The total amount of CO_2 evolved during the program was 4.81 mmol (17.8 wgt% of deposit) and the total amount of CO evolved was 1.38 mmol (3.3 wgt% of deposit). The longer thermal residence time did favor increased gas make. The two additional components proved to be methane and hydrogen. Methane first appeared during the 350°C plateau at a low relative abundance (CO_2/CH_4 = 143), but increased in relative abundance throughout the remainder of the thermal program (at the end of the 500°C plateau, CO_2/CH_4 = 2.14). In terms of absolute abundance, the amount of CH_4 increased by a factor of 52 between the 350°C plateau and the 500°C plateau. Hydrogen first appeared during the 450°C plateau but could not be quantified because we were using helium carrier gas.

The results of the above pyrolyses, which were all carried out on a single deposit mixture, show that deposit oxygen splits into three dominant phases on decomposition:

1. Thermal residue
2. Carbon dioxide
3. Water

To show the applicability of these results to the general issue of deposit chemistry, we have chosen to study one aspect, CO_2 evolution, on a variety of distinct deposit samples. The specific technique selected, coulometric titration, is both fast (allowing time dependent studies) and applicable to relatively small samples.

In table 5, we present results on eight different deposit samples. For each deposit, we have three rows of data: the first giving the weight % carbon formed as CO_2 during nitrogen pyrolysis, the second giving the weight % carbon dioxide formed during nitrogen pyrolysis (related to the corresponding first

Table 4. Molar ratio of CO_2 to CO during
500°C pyrolysis of deposit

Pyrolysis Time	CO_2/CO mole ratio
Reaching 490°C	6.58
At 500°C, 6 minutes	6.07
13 minutes	5.55
20 minutes	5.18
27 minutes	4.78

Interpolating linearly between points and
using the trapezoidal rule for integration,
the total amount of CO_2 is 5.74 mmol and of
CO 0.88 mmol. The fraction of the deposit
evolved as CO_2 is 10.6%, as CO, 1.0%. The
total evolved volatiles constitute 34.09% of
the initial deposit, so the CO_2 amounts to
31% of the volatile phase.

Table 5: Thermal evolution of carbon species at 500°C as a function of time[a]

Sample			Time, (min) = 5	10	15	30	60
249 T2	without O_2	%C=	2.19	2.48	2.58	2.71	2.79
		%CO_2=	8.02	9.09	9.45	9.93	10.22
	with O_2	%C=	12.58	14.01	14.35	14.86	15.05
Chev #3 Phase 1	without O_2		2.21	2.44	2.52	2.65	2.73
			8.10	8.94	9.23	9.71	10.00
	with O_2		13.01	14.09	14.45	14.79	15.09
CLR6-307	without O_2		1.44	1.58	1.69	1.71	
			5.27	5.79	6.19	6.27	
	with O_2		10.25	11.13	11.40	11.67	
Chev 261-13	without O_2		1.95	2.18	2.29	2.37	
			7.14	7.99	8.39	8.68	
	with O_2		11.78	12.83	13.17	13.64	
Car #3	without O_2		1.98	2.17	2.25	2.30	
			7.25	7.95	8.24	8.43	
	with O_2		13.39	14.28	14.63	15.02	
CAR 48	without O_2		2.10	2.32	2.42	2.61	
			7.69	8.50	8.67	9.56	
	with O_2		9.86	10.57	10.81	11.21	
256-2	without O_2		2.40	2.57	2.64	2.78	
			8.79	9.42	9.67	10.16	
	with O_2		10.07	11.41	11.84	12.43	
249 T13	without O_2		2.36	2.47	2.54	2.70	
			8.65	9.05	9.31	9.89	
	with O_2		7.49	8.72	9.14	9.54	

[a]For each specimen we give %C (as CO_2) and % CO_2 evolved under N_2 and %C (as CO_2) evolved under air. The first two numbers are related by the constant 3.66 (= 44/12).

row entry by a factor of 3.66), and the third row giving the weight % carbon formed as CO_2 during <u>combustion</u> of the deposit.

As could be predicted from the tube furnace results, the majority of thermally formed CO_2 evolves during the first five minutes of the experiment. However, one notes continued CO_2 evolution between 30 minutes and 60 minutes, so the attainment of steady state conditions at 500°C will require at least one hour. In terms of absolute amounts of CO_2 formed at 30 minutes, values range from 6.27% to 10.16%, in fair agreement with the tube furnace results.

The choice of 500°C as the temperature for measuring CO_2 evolution is completely arbitrary. In Table 6, we give results for 900°C treatments. As noted for the 500°C treatments, the majority of the CO_2 is evolved in the first five minutes of pyrolysis. The mass of evolved CO_2 constitutes 29 to 31% of the total volatiles.

How does the coulometric technique compare with the tube furnace approach? Two coulometric runs on our deposit mixture at 500°C for 30 minutes yielded the following results:

Evolved CO_2	Total Evolved Mass	CO_2 Fraction
10.32%	34.72%	30%
8.27%	37.30%	22%

At 900°C, 12.84% of the deposit became CO_2, which amounted to 23% of the total volatiles. Additionally, coulometric analyses at both 500°C and 900°C were made on the <u>residue</u> of tube furnace run 2. As might be expected, CO_2 evolution at 500°C was small (0.29 wgt% of the initial residue), although there was 9.44% total mass decrease during this run. At 900°C, there was a 4.17% CO_2 evolution, which amounted to 24.5% of the total mass loss of 17.0%.

C. <u>Infrared Spectroscopy</u>

The infrared spectra of deposits display much detail, and suggest a variety of chemical functionality to be present. A typical spectrum, for a base/Lube A deposit, is given in Figure 10.

In terms of oxygen functionality, we may identify -OH (stretch at 3420 cm^{-1}), aromatic anhydride carbonyl (stretch at 1775 cm^{-1}), other carbonyl (stretch at 1715 cm^{-1}), and C-O (stretch at 1000-1250 cm^{-1}).

Table 6. Thermal evolution of carbon species at 900°C as a function of time

Sample	Time, (min) =	5	10	15	30	% Weight Loss at 30 minutes
CAR 48	without O_2 %C =	3.83	3.86	3.87	3.92	46.87
	%CO_2 =	14.02	14.12	14.16	14.35	
	with O_2 %C =	19.13	19.33	19.45	19.73	53.52
256-2	without O_2 %C =	3.77	3.80	3.84	3.85	48.24
	%CO_2 =	13.80	13.91	14.05	14.09	
	with O_2 %C =	18.44	18.87	19.01	19.29	52.69
249 T3	without O_2 %C =	3.64	3.68	3.70	3.74	46.14
	%CO_2 =	13.32	13.47	13.54	13.69	
	with O_2 %C =	17.20	17.33	17.45	17.61	48.81
Wyodak coal (170-200 mesh)	without O_2 %C =				2.57	49.68
	%CO_2 =				9.41	
Wyodak coal (200-250 mesh)	without O_2 %C =				2.29	45.10
	%CO_2 =				8.38	
Wyodak coal (>250 mesh)	without O_2 %C =				2.37	36.96
	%CO_2 =				8.67	

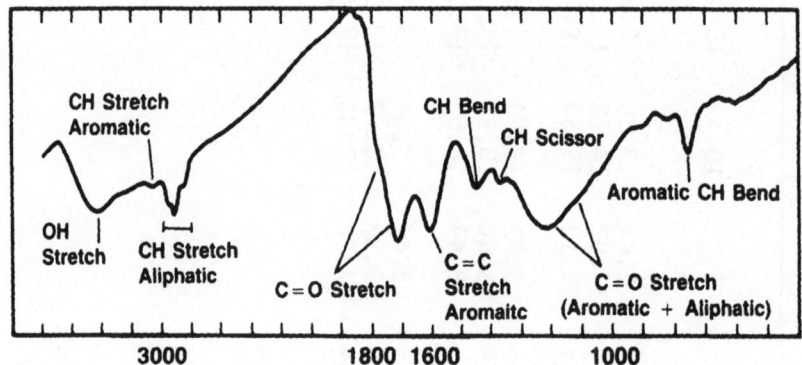

Figure 10. Infrared spectrum of a base fuel/lube A deposit.

Deposits arising from iso-octane/Lube A are somewhat different from base fuel/Lube A deposits. The aliphatic C-H stretch at 2850-3000 cm^{-1} is stronger; there is a peak at 1630 cm^{-1} (rather than at 1605 cm^{-1}), and the C-O stretch is centered at 1100 cm^{-1}.

The infrared spectra of residues following solvent extraction are not significantly different from the initial deposit, as illustrated in Figure 11.

IV. DISCUSSION

The presence of 25 wt% oxygen in the deposits, using either ashless lube or Lube A, suggests that oxygen constitutes a fundamental part of the organic matrix of the deposit. Such an inference is immediately consistent with the infrared spectrum of the solid deposit, which shows -OH, C=O, and C-O functionality.

That we have such high oxygen content is not surprising in terms of the many reactive, oxygen-containing, intermediates formed during combustion.[3] Determining how these intermediates are put together is the key not only to deposit identity but also to potential deposit removal.

There are many organic materials of comparable oxygen content which may serve as models for deposit analysis. Coals, especially those of lower rank, have carbon and oxygen levels comparable to those of the deposit and have been well studied in terms of oxygen functionality.[4,5] Further natural models include humic acids (63%C, 28%O in reference 6), fulvic acids (44%C, 45%O in reference 7), and organic extracts from soils (57%C, 35%O in reference 8). Man-made models include polyaryl ethers[9-11] and poly esters.[12]

One might think that application of techniques useful for the above materials would be suitable for combustion chamber deposits. In the past, Lauer and Friel analyzed deposits of oxygen content between 20 and 36% and inferred the bulk of this oxygen to reside in acid and ester carbonyl groups.[13] To evaluate this hypothesis with respect to our deposits, we first used the calcium acetate titration, explicity as described in reference 4, to analyze for -COOH in a typical deposit (base fuel/Lube A, 60.99%C, 4.26%H, 26.16%O). Although the deposit possessed 16.3 mmol O/(gram deposit), we titrated only 1.2 mmol carboxylic acid/(gram deposit). A total acidity determination, by means of measuring H_2 following exposure to $LiAlH_4$, yielded

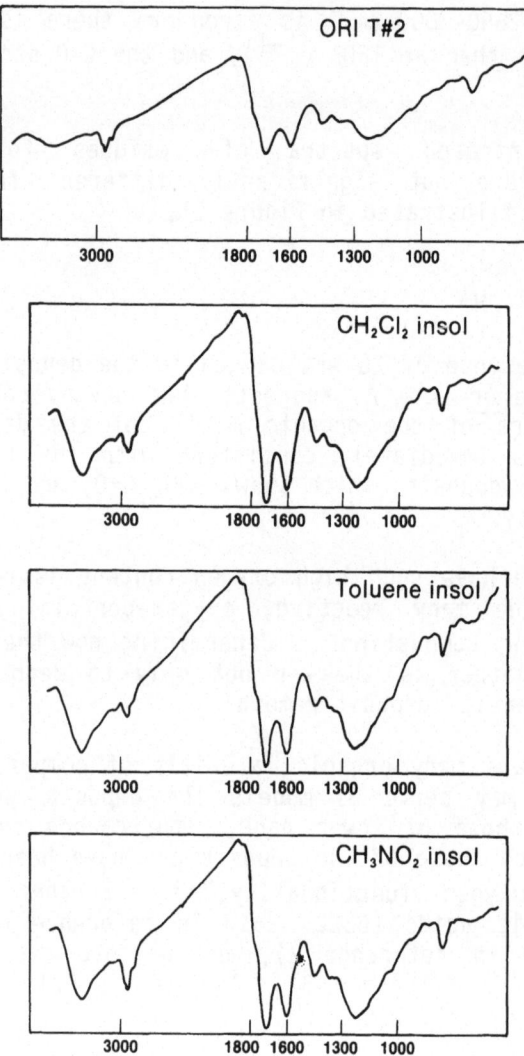

Figure 11. Infrared spectra of a base fuel/ashless oil deposit (top) and residues from that deposit following extraction in various solvents (bottom three).

94

only 1.1 mmol H^+/gram. The absence of evidence for significant amounts of acidic -OH or -COOH functionality is confirmed by the failure of the strong silylating agent, BSTFA, to react with deposit under refluxing tetrahydrofuran, as illustrated in Figure 12. Even the strong reagent n-butyllithium was consumed only to the extent 5.4 mmol/gram.

The difficulty of chemically titrating combustion chamber deposits to determine oxygen functionality is illustrated by a series of chemical reactions, monitored by infrared spectra as illustrated in Figures 13, 14, and 15. Figure 13 shows the infrared spectrum of a base fuel/ashless oil deposit, which is similar to our spectrum in Figure 10. In Figure 14, we see the spectrum of the solid following reduction by n-butyllithium (5.24 mmol consumed per gram; quenched with methyl iodide in diethyl ether). The carbonyl band at 1720 cm^{-1} is absent. Washing such a deposit in 0.1 N DCl in D_2O (138 hours under N_2) not only restores the carbonyl band but also fails to exchange all of the -OH to -OD, as is seen in Figure 15. Whether these difficulties arise from mass transfer limitations (the deposit has a BET surface area of 0.6 m^2/gram), or chemical complexity, remains to be determined.

One additional potential difficulty arises in the area of oxygen microanalysis. As noted in Table 3, the numbers for oxygen for nominally identical samples showed wide variation. In principle, this could be due to chemical heterogeneity or to analytical difficulties. To address this, we submitted three deposit samples to the IRT Corporation for oxygen determination by neutron activation analysis. The results were as follows: 16.6% (Chev#3, phase 1), 18.2% (car 48; RAB#5), and 16.8% (Chev 261-13; CCD; ashless oil). In all cases, these numbers are lower than found by the LECO method.

V. CONCLUSIONS

Combustion chamber deposits are volatile carbonaceous materials, having more in common with bituminous coals than with graphite. A unifying trait among the deposits is the presence of 16-25 wt% oxygen, which is distributed among a variety of chemical functionalities, including -OH, C=O, and C-O. Chemical titration of these deposits is difficult.

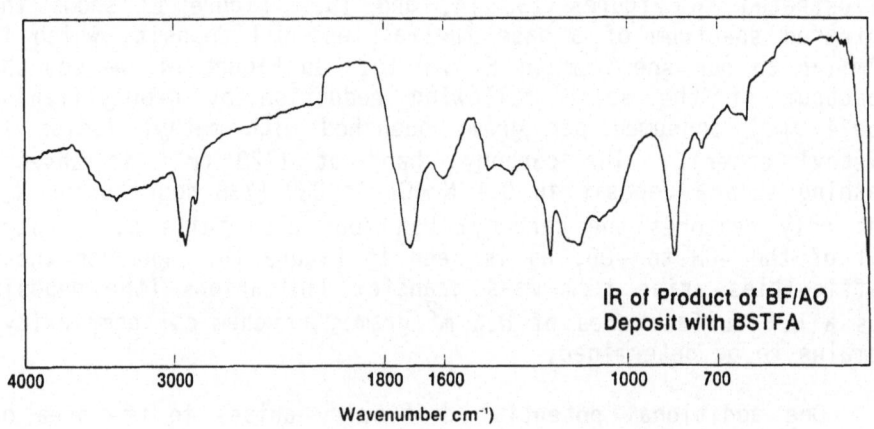

IR of Product of BF/AO
Deposit with BSTFA

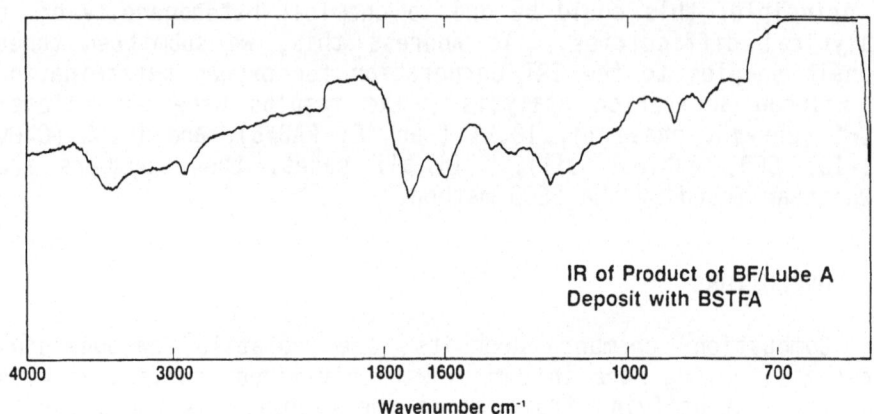

IR of Product of BF/Lube A
Deposit with BSTFA

Figure 12.   Infrared spectra of products of reaction of
BSTFA with two different engine deposits.

96

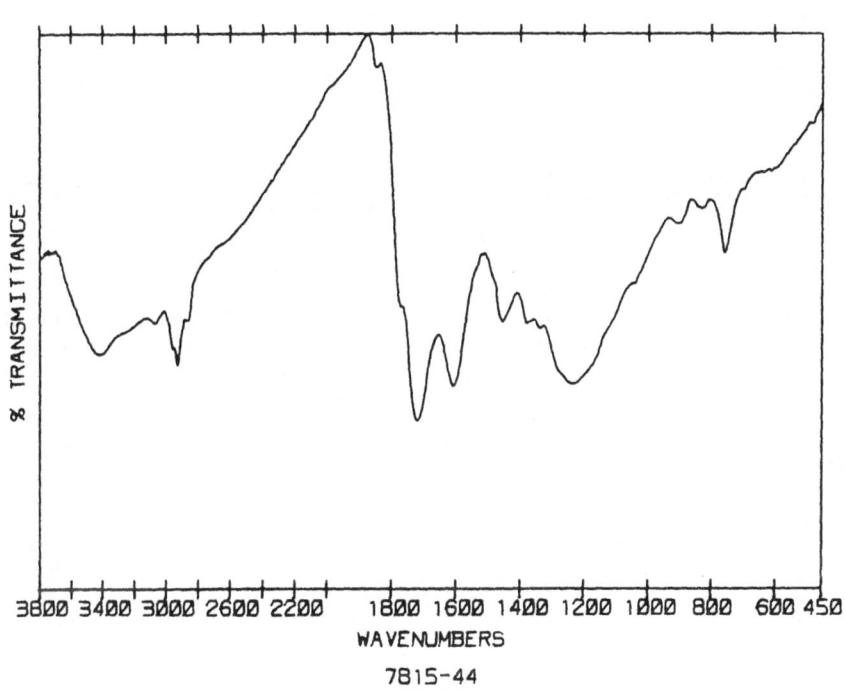

7815-44

Figure 13. Infrared spectrum of base fuel/ashless oil deposit.

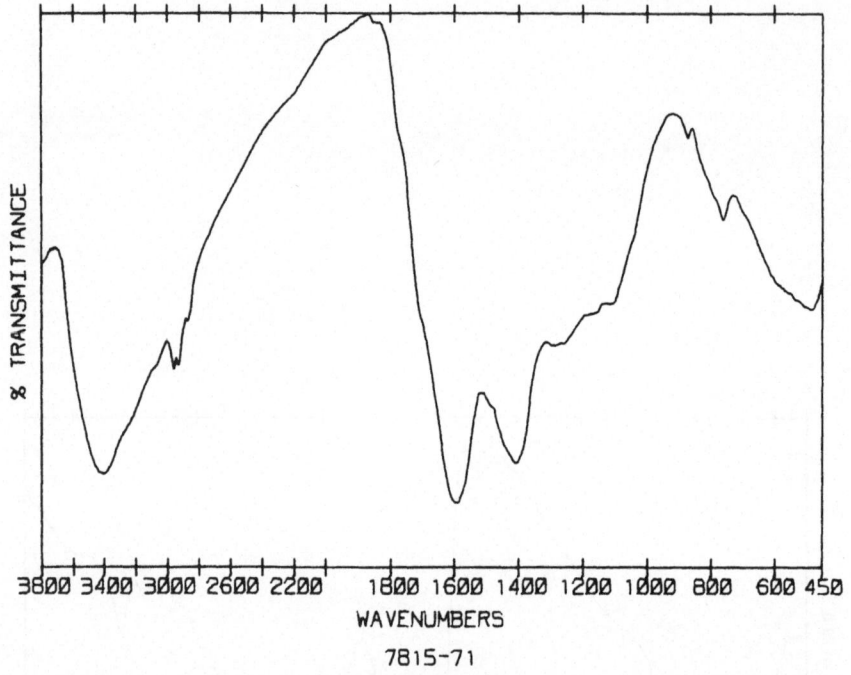

7815-71

Figure 14.   Infrared spectrum of base fuel/ashless oil deposit
following reaction with n-BuLi and quench with CH_3I.

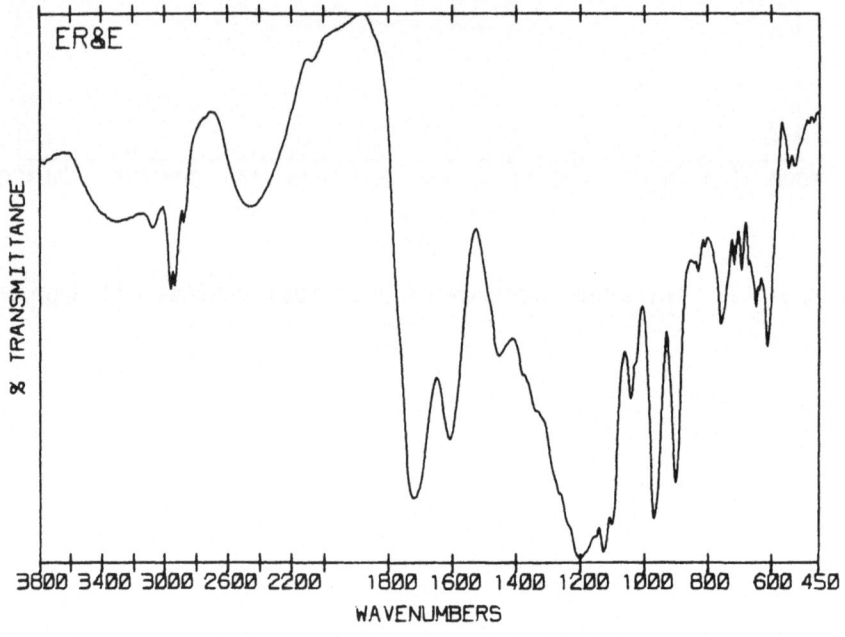

7877-32 SOLID IN FLUOROLUBE MULL

Figure 15.   Infrared spectrum of above following wash with DCl.

VI. ACKNOWLEDLGEMENTS

The work described in this chapter was carried out as part of a collaborative program between Exxon's Corporate Research Laboratory and Products Research Division. We acknowledge the assistance of Bob Lunt, Joe Pecoraro, Jerry Panzer, and Roland Bouffard of PRD. In sorting out analytical problems, we received help from Jim Elliott of Exxon's Analytical and Information Division and from Gail Hutchens of Galbraith and Ed Huffman of Huffman Labs. Finally, we thank Ellen Berglund for the extensive time spent in the preparation of this manuscript.

REFERENCES

1. J. Benson, What Good are Octanes?, Chemtech, 6:16 (1976).
2. L. B. Ebert, L. Matty, D. R. Mills, and J. C. Scanlon, The Interrelationship of Graphite Intercalation Compounds, Ions of Aromatic Hydrocarbons, and Coal Conversion, Mat. Res. Bull., 15:261 (1980).
3. S. W. Benson and P. S. Nangia, Some Unresolved Problems in Oxidation and Combustion, Acc. Chem. Res., 12:223 (1979).
4. L. Blom, L. Edelhausen, and D. W. van Krevelen, Chemical Structure and Properties of Coal XVIII-Oxygen Groups in Coal and Related Products, Fuel, 36:135 (1957).
5. J. D. Brooks and S. Sternall, Chemistry of Brown Coals I. Oxygen-containing Functional Groups in Victorian Brown Coals, Aust. J. Appl. Sci., 8:206 (1957).
6. C. Whitworth and G. K. Pagenkopf, Cadmium Complexation by Coal Humic Acid, J. Inorg. Nucl. Chem., 41:317 (1979).
7. B. Brady and G. K. Pagenkopf, Cadmium Complexation by Soil Fulvic Acid, Can J. Chem., 56:2331 (1978).
8. M. Schnitzer and J. G. Desjardins, Molecular and Equivalent Weights of the Organic Matter of a Podzol, Soil Sci. Soc., Amer. Pro., 26:362 (1962).
9. G. F. Endres, A. S. Hay, and J. W. Eustance, Polymerization by Oxidative Coupling. V. Catalytic Specificity in the Copper-Amine-Catalyzed Oxidation of 2,6-Dimethylphenol, J. Org. Chem., 28:1300 (1963).
10. A. S. Hay, Oxidation of Phenols, U. S. Patent 3,306,874 (February 28, 1967).
11. A. S. Hay, Oxidation of Phenols and Resulting Products, U. S. Patent 3,306,875 (Feruary 28, 1967).
12. R. T. Morrison and R. N. Boyd, "Organic Chemistry", 2nd ed., (Allyn and Bacon: Boston, 1966), pp. 910-911
13. J. L. Lauer and P. J. Friel, Some Properties of Carbonaceous Deposits Accumulated in Internal Combustion Engines, Combustion and Flame, 4:107 (1960).

THE CHEMISTRY OF INTERNAL COMBUSTION ENGINE DEPOSITS —

II. EXTRACTION, MASS SPECTROSCOPY AND NUCLEAR MAGNETIC
RESONANCE

William H. Davis, Jr., Lawrence B. Ebert,
John D. Dennerlein, and Daniel R. Mills

Corporate Research-Science Laboratories
Exxon Research and Engineering Company
P.O. Box 45
Linden, New Jersey 07036

I. INTRODUCTION

The minimum fuel octane required to avoid knocking in an internal combustion engine is not constant, but rather increases as the engine runs. This phenomenon, termed octane requirement increase (ORI), has long been recognized as being attributable to combustion chamber deposits.[1] In spite of many excellent studies which describe the mechanisms by which combustion chamber deposits cause ORI[1-5] and the impact of fuel and lube properties on the magnitude of ORI[1,5-11] little is known about the nature of the deposit.[1,10,11,13] With an understanding of the deposit structure, the chemistry involved in its formation should be more evident. Manipulation of this chemistry could reduce or modify these deposits and thereby achieve the goal of minimizing the octane required for satisfactory engine performance. The initial phase of our work, the characterization of deposit structure, is reported here.

In the preceding chapter, the nature of the deposit as a whole was probed. Since we found that combustion chamber deposits have a substantial volatile component (approximately 35 weight percent of the initial deposit is lost in heating to 500°C), we sought to gain further insight into their structure by studying these volatiles. As reported here, these deposits also can be fractionated by extraction with organic solvents. In this paper we discuss the application of two techniques, high resolution nuclear magnetic resonance and direct insertion probe mass spectroscopy, to the characterization of extractible and volatile fractions of deposits arising from different fuel and lube combinations. The main theme of our findings is a parallel dependence of both extractibles and volatiles on the aromatic content of the fuel: the more aromatic the fuel, the more aromatic the extractibles and volatiles. Our findings are in harmony with the work of Shore and Ockert,[10] who found aromatics to be considerably worse deposit formers than paraffins.

II. EXPERIMENTAL

The combustion chamber deposits examined in this study were generated and collected by the Products Research Division of Exxon Research and Engineering Company. The deposits are described by the fuel and lube from which they were generated, as shown in Table 1. Whole fuels contain complete naphtha and aromatic fractions; lubes are either the complete lube A package or just the ashless fraction.

A. Extractibility

The ability of different solvents to extract a fraction of the deposit was determined in the following manner. A known amount of deposit (from 0.1 to 2 g, preferably about 0.5 g) was placed in the thimble of a mini-Soxhlet extractor, and 7-10 ml of solvent was added to the pot. The extraction was allowed to go to completion (i.e., until the extracting solvent surrounding the thimble was colorless) at a temperature approximately 25-30°C greater than the solvent's boiling point in order to minimize the pot temperature. The non-extractible material remaining in the thimble was washed with cold dichloromethane and dried to constant weight in a vacuum oven at 70°C/100 mm Hg. In addition, the solvent was also removed from the material extracted by drying in the same vacuum oven. The extractibility of a solvent was, therefore, measured by the weight loss from

Table 1. Origin of Combustion Chamber Deposits for Characterization

Fuel	Lube	ORI
Isooctane	Lube A	<1
Isooctane + Alkylbenzenes	Lube A	1-3
Whole	Lube A	5-8
Whole	Ashless	7-8

the deposit remaining in the thimble (non-extractible residue) as well as by the weight of the extracted material (extractibles).

B. Mass Spectroscopy

Using the direct-insertion probe technique, samples of ORI deposits, extraction and devolatilization fractions of ORI deposits, lube, and fuel were continually analyzed on a DuPont 21-491 mass spectrometer as the temperature of the probe was increased from 30° to approximately 450°C.

C. Nuclear Magnetic Resonance Sepctroscopy

Proton NMR spectra were run on a Varian EM-360L, a 60 MHz spectrometer. The samples were prepared by adding about 1 ml of solvent (usually CD_2Cl_2, $CDCl_3$ or CS_2) to about 0.1 g of deposit, mixing well, and filtering the liquid twice through a disposable pipette packed with glass wool. Only by removing most of the insoluble material from the sample (in this way, for example) could satisfactory spectra be obtained. Carbon-13 spectra run on samples prepared in this way, using Varian XL-100 or FT-80 spectrometers, had poor signal-to-noise.

III. RESULTS

A. Extractibility

The ability of various solvents, such as toluene, dichloro-methane (CH_2Cl_2), tetrahydrofuran (THF), pyridine and N-methylpyrrolidione (NMP), to extract a portion of three ORI deposits arising from different fuels and lubes is summarized in Table 2. These data reflect the extractibility of the deposit by the solvent and not the solubility, since the material removed from the porous thimble may only be a suspension of fine particles rather than a true solution. The material in the Soxhlet extraction pot is not a solution either, but rather a mixture of particles and liquid.

Table 2. Extractibility of ORI Deposits by Various Solvents

Deposit	Toluene	CH_2Cl_2	THF	Pyridine	NMP
Isooctane/Lube A	19	21	28	35	-
Whole Fuel/Lube A	16	21	40	57	64
Whole Fuel/Ashless	20	21	40	66	-

Table 3. Elemental Composition of Extractables and Residue From Extraction of a Whole
Fuel/Ashless Oil Deposit by Toluene

Sample	Wt.%	C mmol/g[a]	H mmol/g	N mmol/g	O mmol/g[b]	H/C[c]	O/C[c]
Residue	80	45.43	27.5	1.0	13.3	.61	.29
Extractibles	20	13.32	13.7	0.09	1.6	1.0	.12
Sum	100	58.75	41.2	1.09	14.9	.70	.25
Deposit (AR)	100	57.16	41.6	1.0	15.6	.73	.27

(a) Millimoles of carbon per gram of the as-received (AR)
    (not fractionated) deposit

(b) Oxygen "by difference"

(c) Atomic ratio

The elemental composition of the residue and extractibles from the extraction of a whole fuel/ashless oil deposit by toluene is given in Table 3 in terms of millimoles (mmol) of element per gram of initial as-received deposit, based on the 20% extractibility reported in Table 2. In addition to showing the preferential segregation of oxygen in the residue and hydrogen in the extractibles relative to carbon, Table 3 also demonstrates the mass balance achieved by this extraction. The amount of each element in the residue and extractibles is summed for comparison with the composition of the as-received (AR) deposit given in the last line.

Selected residues from extractions of several deposits with various solvents were also subjected to thermogravimetric analysis, and the data is given in Table 4.

Table 4. Thermogravimetric Analyses of ORI Deposits and Their Extraction Fractions

Sample	TGA Weight Loss[a] %
Whole Fuel/Ashless Oil	
- AR Deposit	46
Whole Fuel/lube A	
- AR Deposit	48
- CH_2Cl_2 Extraction Residue	44
- Toluene Extraction Residue	45
- THF Extraction Residue	45
- Pyridine Extraction Residue	48
- CH_2Cl_2 Extractibles	66

[a]Weight loss on heating to 1000°C at 10°C/min. in Ar.

Both as-received deposits and their extraction residues suffered approximately the same percentage weight loss on heating to 1000°C, 45±3%. As expected, the extractible fraction showed a higher fraction of volatile components; however, not all the extractibles were volatile.

B. Mass Spectroscopy

As the temperature of the direct insertion probe is increased from 30° to 450°C, the material which is volatilized from a whole ORI deposit or an extraction fraction is continually analyzed. Figure 1 shows a typical plot of the relative amount of material detected, embodied as the relative ion current (RIC), versus time during which the temperature is being increased (300°C was achieved at about 3.5 min. and 450°C at about 4.5 min.) for this isooctane/lube A deposit. The maximum ion current, indicating the maximum amount of material evolved, occurs at approximately 440°C, a result in good agreement with the TGA data reported in Part 1. The data handling capability of this mass spectrometer system allows the mass spectra of the

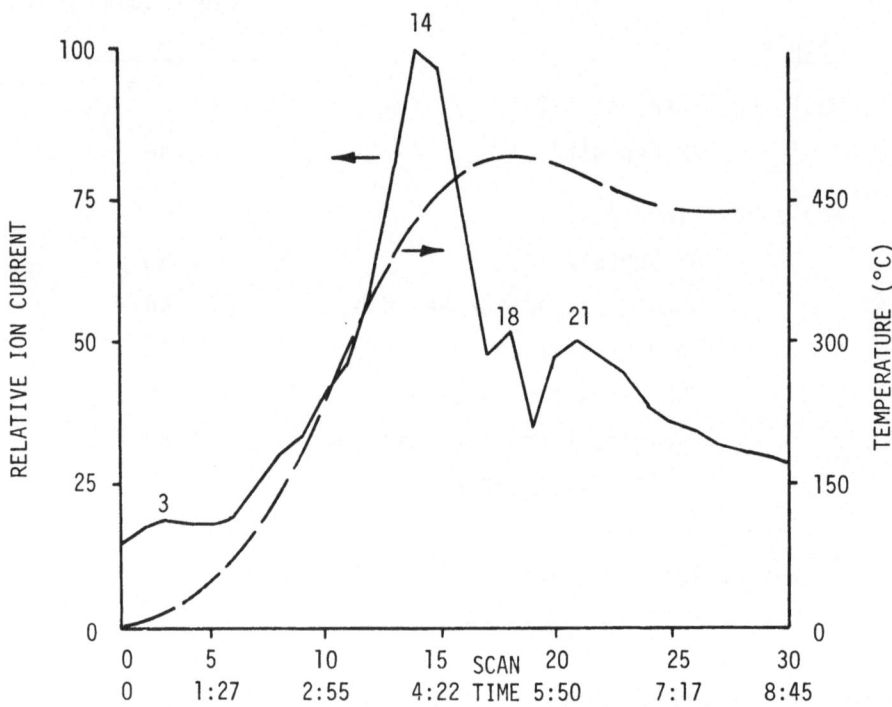

Figure 1. Relative ion current vs. time (temperature) from direct insertion probe mass spectroscopy provides "TGA-like" data.

individual scans to be displayed as well as the sum of all these individual spectra. As we shall see, both of these types of spectra are useful; the individual scans indicate what material is being evolved over a narrow temperature range, and the composite spectrum sums these to show the nature and relative amount of all the material which is evolved.

The composite mass spectrum from the isooctane/lube A deposit is shown in Figure 2. This plot of the relative abundance of an ion versus its mass-to-charge ratio (m/e) consists of peaks which occur in packets with the largest peak generally separated by 14 m/e units (a CH_2 group) from its nearest neighbors. This pattern, typical of the spectrum of high molecular weight alkanes and cycloalkanes, is shown in greater detail in Figure 3, the mass spectrum taken while the maximum amount of material was being evolved from the isooctane/lube A deposit. In neither of these spectra are there peaks at m/e's which are

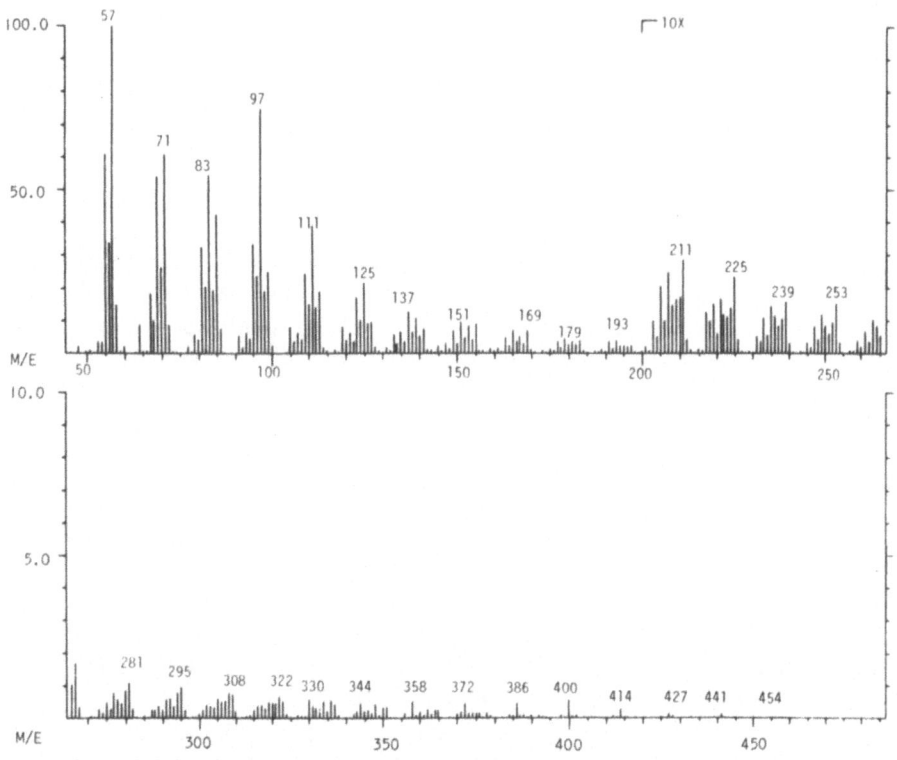

Figure 2. The composite mass spectrum of volatiles from the isooctane/lube A deposit shows aliphatics but no aromatics. Note the scale change, a factor of 10 enhancement, at m/e=200.

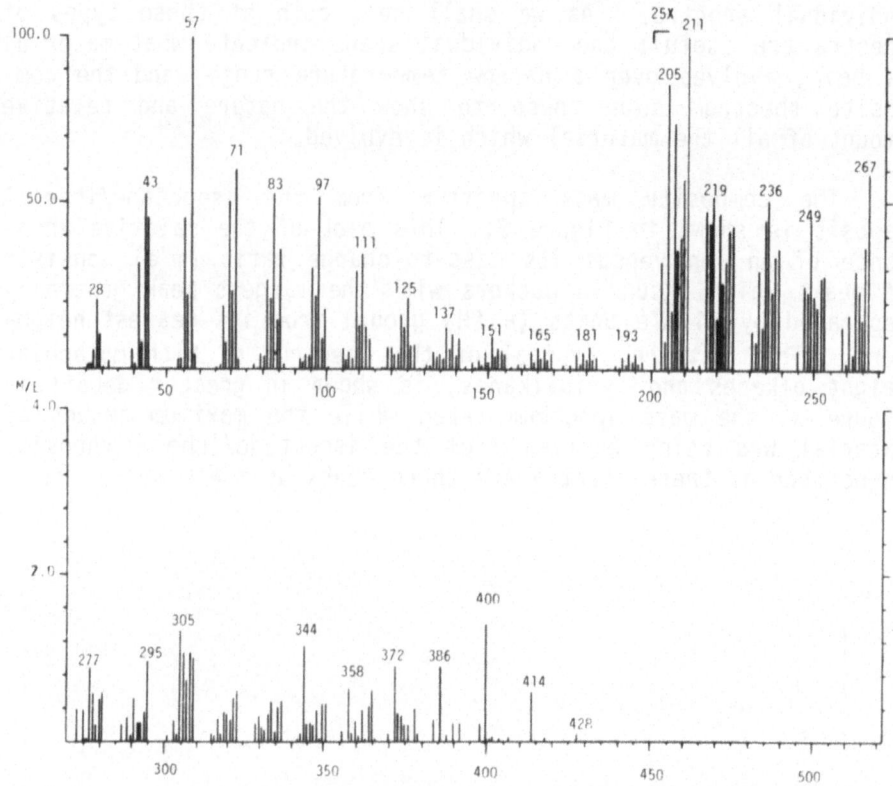

Figure 3. The mass spectrum of the major volatile component (Scan #14) from the isooctane/lube A deposit shows only a homologous series of aliphatics. Note the scale change, a factor of 25 enhancement, at m/e=200.

characteristic of aromatic and polynuclear aromatic (PNA) structures; that is, the peaks at m/e=78 (benzene), 91 (alkylbenzene), 202 (pyrene), 276 (perylene) and 300 (coronene) are either absent or so small that they likely arise from other non-aromatic moieties. However, several homologous series are evident: alkyl groups (C_nH_{2n-1}) at 83, 97, 111, 125, etc.; C_nH_{2n-3} groups at 137, 151, 165, etc. Peaks corresponding to these same groups appear at higher m/e's as well. Since similar non-aromatic hydrocarbon groups are also present in an ashless oil, the mass spectrum of the isooctane/lube A deposit shows a striking resemblance to that of the oil.

When aromatics are present in the fuel, these same aromatics are identified by the mass spectrometer in the volatiles from various deposits. For example, the composite mass spectrum from a whole fuel/ashless oil deposit (Figure 4) indicates the presence of alkyl groups, alkylbenzenes, and polynuclear aromatics (PNA's) ranging from phenanthrene to coronene. In this spectrum, representative structures are drawn over the peaks corresponding to their molecular weights. Since there are many possible isomers, those PNA's shown are just probable examples. Homologous series are seen for the PNA's indicating multiple substitution and/or sidechains of up to five carbon atoms, e.g., homologs of pyrene (m/e=202) are seen at 216, 230, 244, 258, and 272. Although the presence of small alkyl groups is also indicated by peaks at m/e's corresponding to C_nH_{2n-1} ($4 \leqq n \leqq 12$), evidence for larger alkyl groups (m/e = 183, 197, 211, etc.) corresponding to $n \geqq 13$ is not seen.

Mass spectra of the volatiles from deposits produced by 40% alkylbenzenes in isooctane/lube A and 35% alkylbenzenes, 5% naphthalene and 0.2% alkylnaphthalenes in isooctane/lube A are like the fuels from which they arise. That is, they are intermediate between the isooctane and whole fuel spectra shown in Figures 2 and 4. The spectrum from the alkylbenzene-isooctane/lube A deposit shows few aromatics (small alkylbenzene peaks at m/e=91

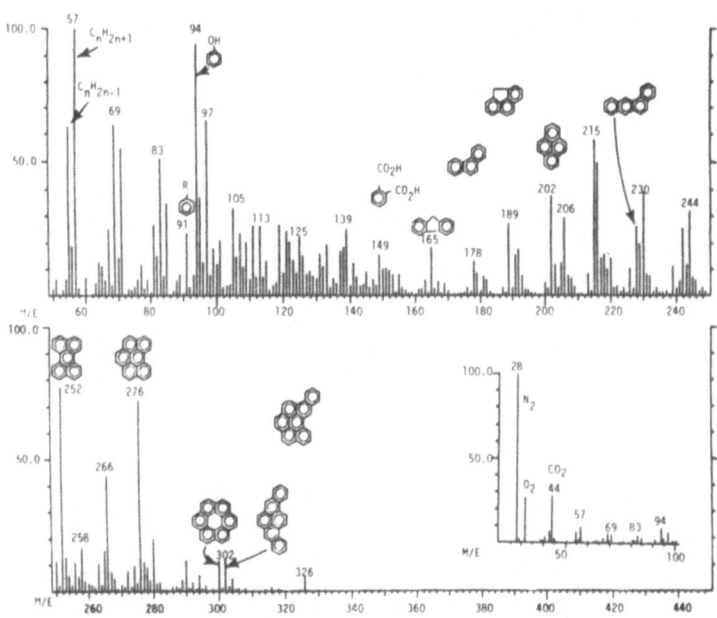

Figure 4. The composite mass spectrum of volatiles from a whole fuel/ashless oil deposit indicates the presence of fuel polynuclear aromatics.

or 105 but large alkyl groups (C_nH_{2n-1}, n > 20). Shorter alkyl groups and more aromatics (larger m/e=91 and 105 peaks and a peak at m/e=128 corresponding to naphthalene) are present in the spectrum from the alkylbenzenes-naphthalene-isooctane deposit.

This mass spectrum (Figure 4) also indicates the presence of oxygenated species such as phenol (m/e=94) and its homologs m/e=108, 122, and 136) and phthalic acid and its esters (m/e=149). However, the major oxygen-containing compound and, indeed, the largest single species detected is carbon dioxide (m/e=44), as shown in the insert in the lower right hand side of Figure 4. Analysis of the individual scans taken as this deposit is volatilized show that CO_2 is not a significant compound in the mass spectra taken at low temperatures. However, CO_2 becomes the major species at high temperatures. Separate mass spectrometry analyses at 23°C and at 150°C also indicate that there is little residual CO_2 absorbed by the deposit. Thus, the observed CO_2 must arise from thermal degradation of the deposit.

The fractions from Soxhlet extractions of the deposits were also submitted for direct insertion probe mass spectral analysis. Figure 5 shows the spectrum of the chloroform extractible fraction from a whole fuel/ashless oil deposit. This spectrum is very similar to that arising from the as-received deposit (Figure 4); the major differences are the solvent ($CHCl_3$) peaks

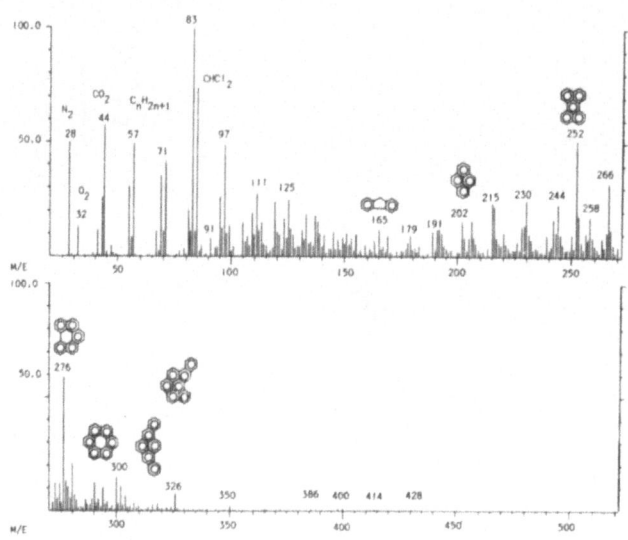

Figure 5. The composite mass spectrum of volatiles from the $CHCl_3$ extractible fraction of a whole fuel/ashless oil deposit. Note the similarity to Figure 4.

in Figure 5 and the decrease in the phenol (m/e=94) peak on going from Figure 4 to Figure 5. The mass spectrum of the fraction which was not extracted by $CDCl_3$ (the residue) is different from the extractible fraction. Only peaks corresponding to chloroform-d (m/e=84, 86, 36, 38, 47, and 49), CO_2 (m/e=44), alkyl groups (m/e=43, 57, and 71) and toluene (m/e=91) are evident and no peaks of higher m/e were observed.

The absence of aromatics in the mass spectrum of the isooctane/lube A deposit encouraged us to probe the ability of this deposit to absorb components from "whole" fuel. The light ends were weathered off a test fuel to obtain a fuel concentrate. Interestingly, after the isooctane deposit had been soaked in this whole fuel concentrate overnight, the fuel washed off the deposit five times with pentane and the deposit dried at

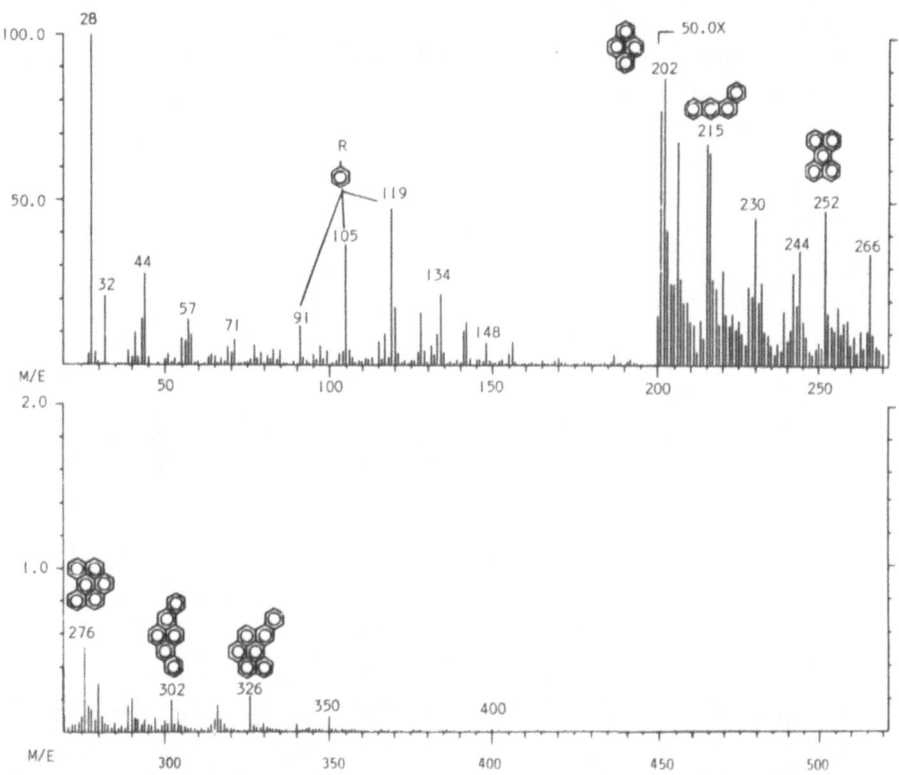

Figure 6. The composite mass spectrum of the volatiles from the isooctane deposit soaked in whole fuel, washed with pentane and dried. Note the scale expansion, 50 times, at m/e = 200.

150°C and 200 mm Hg overnight, the mass spectrum of the volatiles from this isooctane deposit treated with fuel showed (Figure 6) peaks corresponding to aromatics, both alkylbenzenes (m/e=91, 105, 119, etc.) and PNA's (m/e=202, 228, 252, 276, 302) and their homologs.

C. Nuclear Magnetic Resonance

Figure 7 shows a typical high resolution proton magnetic resonance spectrum of the dichloromethane soluble fraction of an ORI deposit arising from whole fuel and ashless oil. Both aliphatic and aromatic protons are present in the solubles, with the aliphatic species to the right of the sharp solvent related peak, the aromatic species to the left.

In order to confirm the mass spectral observation that aromatics in the fuel lead to aromatics in the combustion chamber deposit, proton NMR spectra were obtained on the CD_2Cl_2 soluble fractions of deposits from fuels of varying aromatic content. These results are summarized in Table 5. Not only does the amount of aromatics in the deposit solubles increase

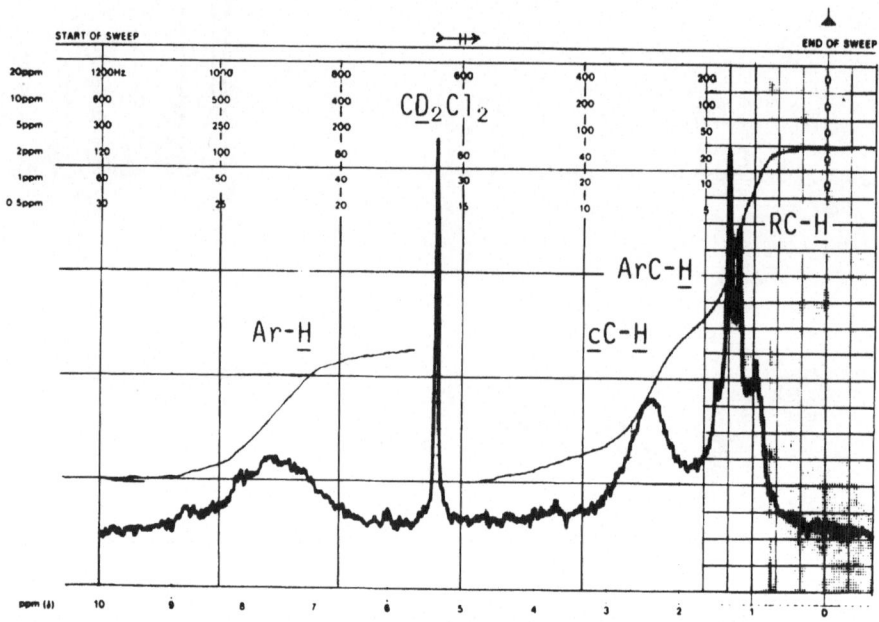

Figure 7. Proton NMR spectrum of the deposit from whole fuel/ ashless oil soluble CD_2Cl_2. The sharp central peak arises from proton impurities in the solvent.

with increasing aromatics in the fuel, but the sizes of the aromatic rings, as judged from the shift of the maximum of the aromatic envelope to higher δ, also increase as the fuel aromatics become larger. Thus, the spectra of deposits from the whole fuels show more material extending further to the left on the left-hand side of the aromatic "hump."

The ability of the isooctane/lube A deposit to adsorb aromatics from a whole fuel was also probed by NMR. The fuel was first concentrated by weathering off the light ends. The proton spectrum of the concentrated test fuel was run and integrated using p-dioxane as an external standard in the inner tube of a

Table 5. NMR Spectra of Solubles Show a Strong Fuel Dependence

Deposit Precursor	Proton Distribution of CD_2Cl_2 Solubles (% Hydrogen)[a]		
	Aromatic	Benzylic/ Naphthenic	Aliphatic
Isooctane/lube A	2	17	80
Isooctane + 40% Alkylbenzenes	12	19	69
Isooctane + 35% Alkylbenzenes + 5% Naphthalene/ lube A	13	26	61
Whole / Ashless Fuel Oil	27	32	39
Whole / lube A Fuel	29	30	40

[a] NMR signals in the 7-9 δ region are termed aromatic, in the 1.9-3 δ region naphthenic/benzylic and in the 0.5-1.9 δ region aliphatic.

coaxial NMR tube. After sitting overnight over the isooctane/lube A deposit (an aromatics free deposit), the same amount of concentrated fuel was placed back in the same NMR tube with the same tube of external standard, and the spectrum was again recorded. After normalizing the p-dioxane areas, comparison of the other peaks shows that contact with the deposit produced the following changes in the fuel: a decrease in aromatic from 14 to 13%, an increase in aliphatic from 69 to 70%, and the fraction of benzylic/naphthenic protons constant at 17%. Although these differences are small, they certainly support the mass spectral observation of selective absorption of aromatics by the isooctane/Uniflo deposit.

IV. DISCUSSION

Combustion chamber deposits are complex materials which are very difficult to characterize. In spite of these difficulties, we have expanded the description of these deposits. We persist in our belief that an understanding of the nature of the deposit will implicate the chemistry involved in its growth and thereby suggests means to modify or reduce its formation.

Our studies of deposits indicate that they are susceptible to fractionation by heating and solvent extraction. This finding is significant because the situation with combustion chamber deposits is analogous to attempting to determine the mechanism of a reaction in which the product (the deposit) is only poorly characterized, the reaction conditions are variable and, therefore, not well defined and the reactants are complex mixtures (fuel, lube and air) which are not well known. By fractionation the deposit via heating and solvent extraction and characterizing these fractions, two of the above unknowns, the nature of the product (deposit) and reactants (deposit precursors) are explored.

A. Extractibility

As Table 2 shows, the extractibility of ORI deposits in different solvents varies widely. In general, extractibility increases with increasing solvent polarity and basicity. These results are in agreement with those of Lauer and Friel[12] who found much higher extractibilities in pyridine than in benzene. However, they did not find the large difference in extractibility, between deposits from isooctane and whole fuel which is shown in Table 2. In fact, Lauer and Friel report that a larger fraction of an isooctane/naphthenic oil deposit than a similar commercial fuel deposit is extracted by pyridine. Orelup and Lee report very different results[1]; they found xylene and chlorobenzene to be better "solvents" than pyridine for cylinder "carbon".

Table 3 shows that the two fractions from extraction are not identical but rather some elemental fractionation has been effected. Since the residue is relatively rich in oxygen and poor in hydrogen compared to the extractibles, the deposit appears bituminous in nature with "heavy" hydrocarbons adsorbed on an oxygenated carbonaceous matrix. The difference in elemental composition of extractibles and residues from various solvents decreases as the solvent extracts more of the deposit. Another way other than extraction to fractionate deposits is volatization as shown in Table 4. Mass spectroscopy was used to analyze these volatiles.

B. Mass Spectroscopy

The volatiles from various deposits which are analyzed by mass spectroscopy reflect the nature of the fuel from which the deposit was formed. That is, as more aromatics occur in the fuel, more aromatics are found in the deposit as evidenced by their presence in the volatiles. In fact, no evidence for aromatization is found. The aromatics observed by mass spectroscopy are the same ones which were present in the fuel. Thus, the deposit growth process apparently occurs by binding small molecules to the existing deposit (probably by oxidative reactions) in a chain growth mechanism (monomer adding to polymer). If growth occurred by a step growth mechanism (small polymers combining to form larger polymers), small adducts of aromatics with oxygen functionalities should be observed.

Although deposits have high oxygen contents (25 wt.% O), very few types of oxygen-containing fragments are seen in these mass spectra. The one oxygenated species which is significant is CO_2. Since CO_2 is a product of pyrolysis of acids and esters at approximately 350°C, its presence in the material which is volatized on heating to 450°C indicates that much of the deposit did not experience temperatures of 350°C in the engine. The presence of long alkyl side-chains on aromatic rings also indicates that the deposit did not see very high temperature (~400°C) because dealkylation to methyl groups occurs rapidly at these temperatures.

C. Nuclear Magnetic Resonance

The extractibles, like the volatiles, from various deposits reflect the nature of the fuel from which the deposit was formed. Thus, the proton distribution of deposits becomes increasingly more aromatic as the fuel becomes more aromatic. Again, no evidence for aromatization is observed since essentially no aromatics appear in the extractibles from the deposit formed from isooctane and no larger aromatics (PNA's are found in the deposits from isooctane and alkylbenzenes or naphthalene).

V. SUMMARY

These studies demonstrate that the deposit is a composite consisting of a refractory skeleton, which is largely unaffected by either extraction or volatilization, and smaller molecules which are adsorbed on this "backbone". Elemental analysis indicates that this intractible matrix is oxygen rich and hydrogen poor relative to the extractible (and volatile) material which is not bound so tightly to the deposit. In fact, these volatiles and extractibles closely resemble the heavy-ends of the fuel (or in the case of isooctane, the lube) from which the deposit was formed. The ability of the deposit to sorb aromatics from the fuel was also shown. Deposit growth probably occurs via a chain-growth mechanism with adsorbed species being oxidatively bound to the existing deposit.

ACKNOWLEDGEMENTS

The mass spectral analyses were carried out by T. R. Ashe and C. S. Hsu.

REFERENCES

1. J. W. Orelup and O. I. Lee, Factors Influencing Carbon Formation in Automobile Engines, Ind. Eng. Chem. 17:731 (1925).
2. L. F. Dumont, Possible Mechanisms by Which Combustion-Chamber Deposits Accumulate and Influence Knock, SAE Quart Trans. 5:565 (1951).
3. J. B. Duckworth, Effect of Combustion-Chamber Deposits on Octane Requirement and Engine Power Output, SAE Quart Trans. 5:577 (1951).
4. J. Warren, Combustion-Chamber Deposits and Octane-Number Requirement, SAE Trans. 62:582 (1954).
5. J. D. Benson, Some Factors Which Affect Octane Requirement Increase, SAE Paper #750933, Detroit (October 13-17, 1975).
6. W. A. Gruse, Summary of Factors Causing Formation of Carbon Deposits in Internal Combustion Engine, Oil and Gas Journal, p. 15 (November 23, 1933).
7. J. D. Bartleson and E. C. Hughes, Combustion Chamber Deposits as Related to Carbon-Forming Properties of Motor Oils, Ind. Eng. Chem. 45:1501 (1953).
8. H. J. Gibson, C. A. Hall and D. A. Hirschler, Combustion-Chamber Deposition and Knock, SAE Trans. 61:361 (1953).

9. J. G. McNab, L. E. Moody, and N. V. Hakala, Effect of Lubricant Composition on Combustion-Chamber Deposits, SAE Trans. 62:228 (1954).

10. L. B. Shore and K. F. Ockert, Radiotracer Study Points Way to Make Clean-Burning Gasoline, SAE Journal 65:18 (1957); L. B. Shore and K. F. Ockert, Combustion-Chamber Deposits - A Radiotracer Study, SAE Trans. 66:285 (1958).

11. C. N. Sechrist and H. H. Hammen, Radiotracer Studies of Engine Deposit Formation, Ind. Eng. Chem. 50:341 (1958).

12. J. L. Lauer and P. J. Friel, Some Properties of Carbonaceous Deposits Accumulated in Internal Combustion Engines, Combustion and Flame 4:107 (1960).

13. T. Nakamura, Y. Yonekawa and N. Okamoto, The Relationship Between Combustion Chamber Deposits and Octane Number Requirement Increase, J. Japan Petrol. Inst. 22:105 (1979).

THE CHEMISTRY OF INTERNAL COMBUSTION ENGINE DEPOSITS —

III. ^{13}C NUCLEAR MAGNETIC RESONANCE EMPLOYING ^{1}H
CROSS-POLARIZATION AND MAGIC ANGLE SPINNING

Lawrence B. Ebert[†], Kenneth D. Rose*, and
Michael T. Melchior*

Corporate Research Labs[†]
Analytical and Information Division*
Exxon Research and Engineering Company
Linden, New Jersey 07036

I. INTRODUCTION

In the previous two chapters, we have discussed the use of relatively classical techniques to characterize engine combustion deposits. Elemental analysis shows the deposits to contain 60 to 69 wgt.% carbon, 20 to 27 wgt.% oxygen, and 3 to 5 wgt.% hydrogen, with varying amounts of nitrogen, sulfur, and lube-derived inorganic elements. Both thermogravimetric analysis and insertion probe mass spectroscopy show the presence of volatile materials, and the mass spectroscopy suggests these volatile materials to contain, in part, polynuclear aromatic hydrocarbons. Infrared spectroscopy shows that the oxygen functionality includes ketones (possibly as anhydrides) and aryl and alkyl ethers. However, one very significant question remains unanswered: What kind of carbon holds the backbone of the deposit matrix together?

To answer this, we have performed experiments to investigate the ^{13}C nuclear magnetic resonance of solid state deposits, using the technique of cross polarization/magic angle spinning (CP/MAS). Different types of carbon resonate at different spectral positions, allowing a distinction between aliphatic (sp^3) and aromatic (sp^2) carbons. We briefly discuss in the next section some of the past work in this field.

II. BACKGROUND

In 1976, the combination of sensitivity-enhanced cross-polarization nuclear magnetic resonance with magic angle sample spinning allowed the generation of liquid-like high-resolution rare-spin nuclear magnetic resonance spectra of solids.[1] Applied most frequently to the ^{13}C (rare spin)/^{1}H (abundant spin) pair of nuclei, this technique has gained widespread popularity in the interpretation of structure of typical organic solids.[2] The isotropic chemical shifts obtained under these conditions are essentially equivalent to those observed in the liquid-state, with little evidence for problems associated with "solid-state effects".[3-5] In fact, magic angle spinning has even been applied to _liquid_ specimens to eliminate bulk susceptibility shifts.[6]

CP/MAS ^{13}C NMR has been especially useful in the characterization of coal and coal-derived materials,[7,8] and one notes that combustion deposits are similar to coals with respect to abundances of carbon and hydrogen. In this context, various studies have been made on the quantitative reliability of the technique, especially in samples of low hydrogen abundance.[9,10] Schaefer has suggested that, if the protons in a sample are more or less uniformly distributed in a rigid organic solid, concentrations as low as 5% will generate cross-polarization spectra of representative carbon intensities.[2] Alemany, Grant, Pugmire, Alger, and Zilm[10] have noted that carbon atoms four or more bonds from the nearest intramolecular proton cannot be fully polarized before $T_{1\rho}$ (H) effects begin to dominate the proton magnetization. Rapid molecular motion aggravates the problem because of motional attenuation of the dipolar interaction.[10] One way of assessing the significance of these potential complexities is to compare CP/MAS results with data from a simple ^{13}C 90° pulse without cross-polarization.[11] Although data accumulation time will be lengthened, the results will not be dependent on the proton $T_{1\rho}$. As it happens, one of the more serious problems in analyzing engine deposits by CP/MAS arises from the presence of ferromagnetic impurities, as discussed in the next section.

III. OBTAINING THE CP/MAS SPECTRUM

Solid-state cross polarization/magic angle spinning experiments were performed on a JEOL FX-60QS solid-state spectrometer.

120

Initial experiments on the material as received were somewhat of a disappointment. Figure 1 shows only a very broad, poorly resolved spectrum which was obtained after long accumulation (67,000 scans)*. Even an aromatic/aliphatic split is difficult in such a spectrum. Based on the premise that the poor quality of the spectrum in Figure 1 was due to the presence of paramagnetic or ferromagnetic impurities, an acid-wash pretreatment was developed.

Dried deposit (<1g of a base fuel/Lube A lube deposit, 249 T13; 65.94% C, 3.56% H) was transferred to a freshly prepared regenerated cellulose dialysis tubing which was sealed and placed in three hundred milliliters of 0.1 N HCl. The ORI sample was dialyzed for one day each against two changes of HCl solution followed by three changes of distilled water. The final two water wash solutions were neutral to pH test paper and all wash solutions were clear and colorless. The dialysis bag was then opened and the ORI sample suction-filtered under nitrogen. The filtered sample was dried in a vacuum oven at 85°C for three days in preparation for solid-state ^{13}C NMR spectroscopy.

Figure 2 shows the dramatic improvement in both resolution and sensitivity using the demetallizing step. Figure 2 was obtained with only 2,000 scans under the conditions used in Figure 1. Figure 3 was obtained by an overnight accumulation to provide optimum signal/noise. Integration of this spectrum, correcting for the spinning sidebands (cf. Figure 3) indicates that this material is 85-90% aromatic carbon. The aromatic carbon peak maximum is 129 δ, the aliphatic at 23.2 δ.[†] While not much can be concluded from the aromatic shift alone, the aliphatic shift suggests a high portion of the aliphatic carbon is in methyl groups.

It is clear that acid washing has had a dramatic effect on the CP/MAS experiment in this case. Paramagnetic impurities provide a source of broadening of the ^{13}C NMR transitions which

---

*Fairly standard conditions (CP time 1 msec., repetition time 1 sec.) are found to give satisfactory results on a variety of carbonaceous materials.

[†]The symbol "δ" denotes shift, in ppm downfield from tetramethylsilane.

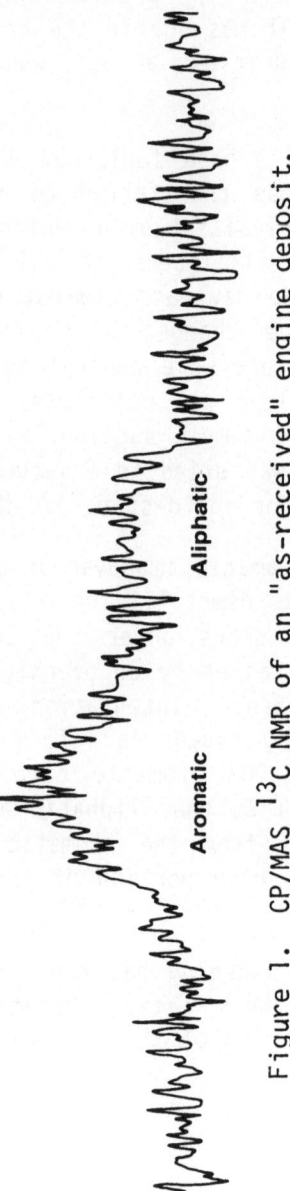

Figure 1. CP/MAS ^{13}C NMR of an "as-received" engine deposit.

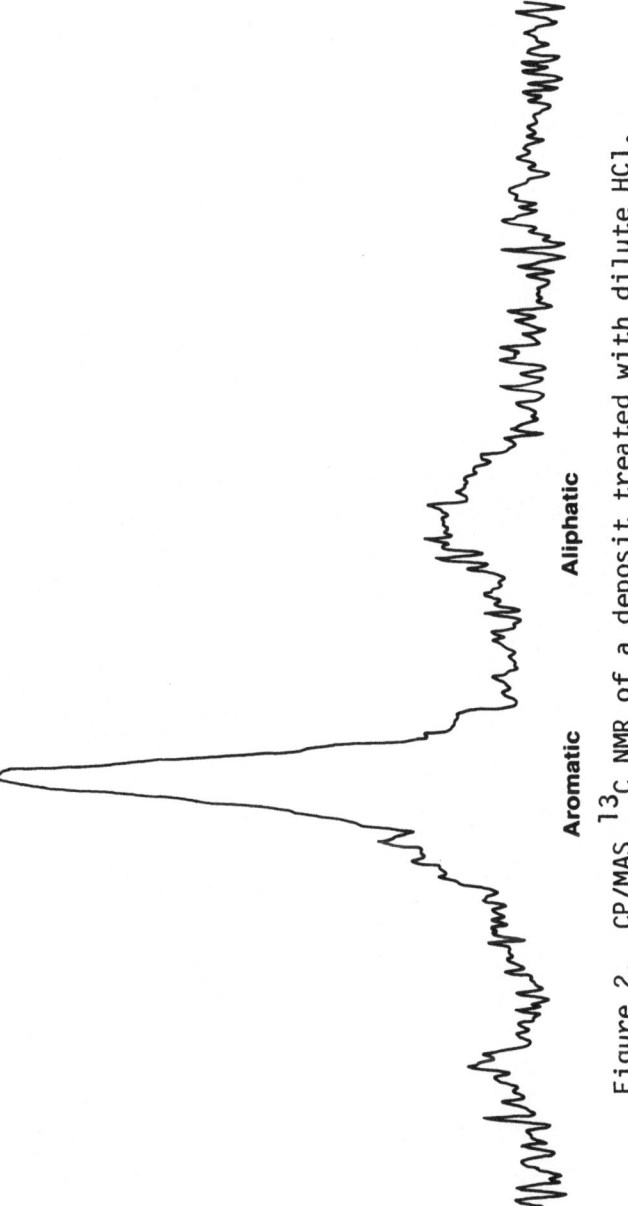

Figure 2. CP/MAS ^{13}C NMR of a deposit treated with dilute HCl.

Aromatic

Aliphatic

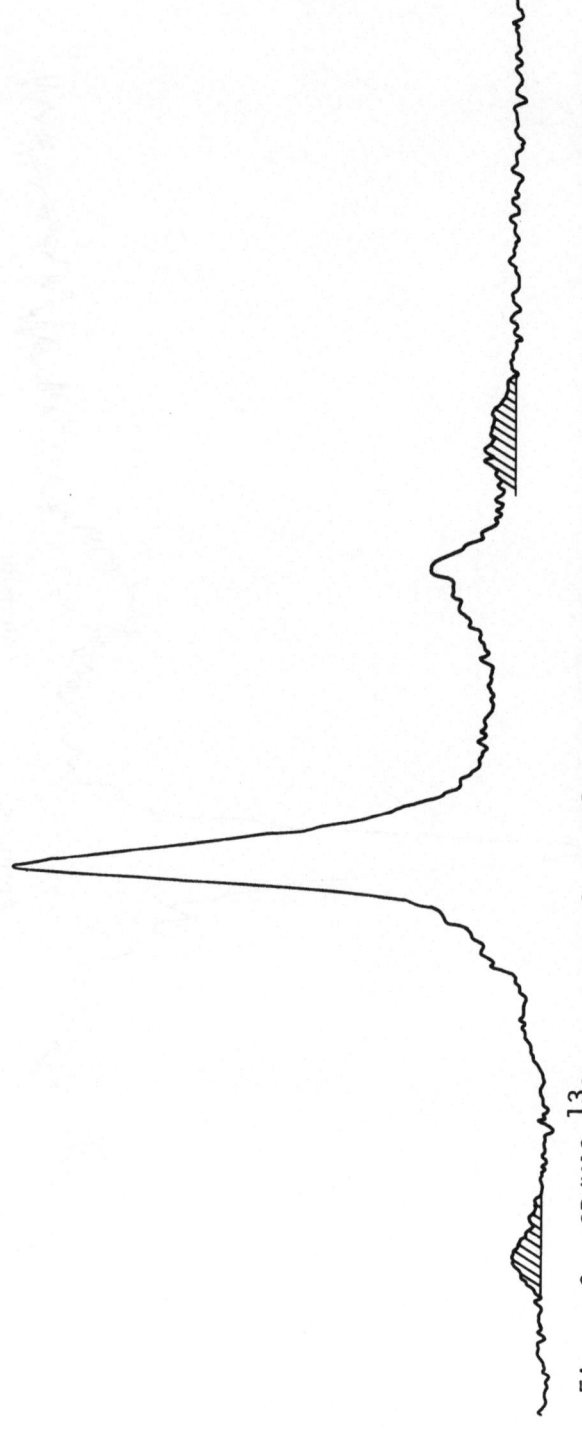

Figure 3.   CP/MAS ^{13}C spectrum of an HCl-washed deposit following overnight accumulation. The spinning sidebands are shaded.

would not be expected to be much reduced by MAS at only 2.4K Hz. It is less obvious why the sensitivity is so dramatically improved. It seems reasonable that this is due a shortening of the proton $T_{1\rho}$ (spin-lattice relaxation time in the rotating frame) in the presence of impurities. Since there is a competition between cross-polarization rate (T_{CP}^{-1}) and proton polarization leakage to the lattice $(T_{1\rho}^{-1})$, shortening $T_{1\rho}$ relative to T_{CP} drastically reduces the realizable ^{13}C polarization. We are currently investigating the general applicability of such pretreatments.

Nevertheless, one must be certain that this dilute acid treatment does not alter the nature of the deposit. Another way to remove the iron-containing impurities is with a small magnet. In Figure 4, we illustrate the ^{13}C CP/MAS spectrum of a base fuel/ashless oil deposit for which the impurities were removed magnetically. The aromaticity is again about 90%, so we conclude that the acid treatment did not distort the aromaticity of the sample. In passing, we note that electron spin resonance offers a useful diagnostic in assessing potential problems with magnetic impurities. Prior to removing the contaminants with the small magnet, the base fuel/ashless oil deposit showed on "unreasonable" g value of 2.0025 and a linewidth of 6.0 G (Figure 5). Impurity removal led to a "reasonable" g value of 2.0031 and a linewidth of 4.8 G (Figure 6). The ^{13}C CP/MAS spectrum of this sample was easily obtained. In the context of engine deposits, we expect an influence of the oxygen spin-orbit coupling on the g value of the radicals, and thus the observed resonance should be in the range 2.0028 to 2.0040.[12]

With respect to possible problems in the ^{13}C CP/MAS spectrum arising from rapid molecular motion, we include in Figure 7 the wideline proton nuclear magnetic resonance spectrum of the base fuel/ashless oil deposit. It has a derivative extremum linewidth of 1.7 G, and is thus in the "rigid lattice" condition. For comparison, we show the wideline proton NMR spectra of coronene (linewidth=1.0 G) and 2,6 dimethylnaphthalene (linewidth=4.9 G) in Figures 8 and 9.

IV. ANALYSIS OF THE SPECTRA

Carbon-13 spectra taken by the CP/MAS technique show that engine combustion deposits are 85-90% aromatic carbon. Whereas

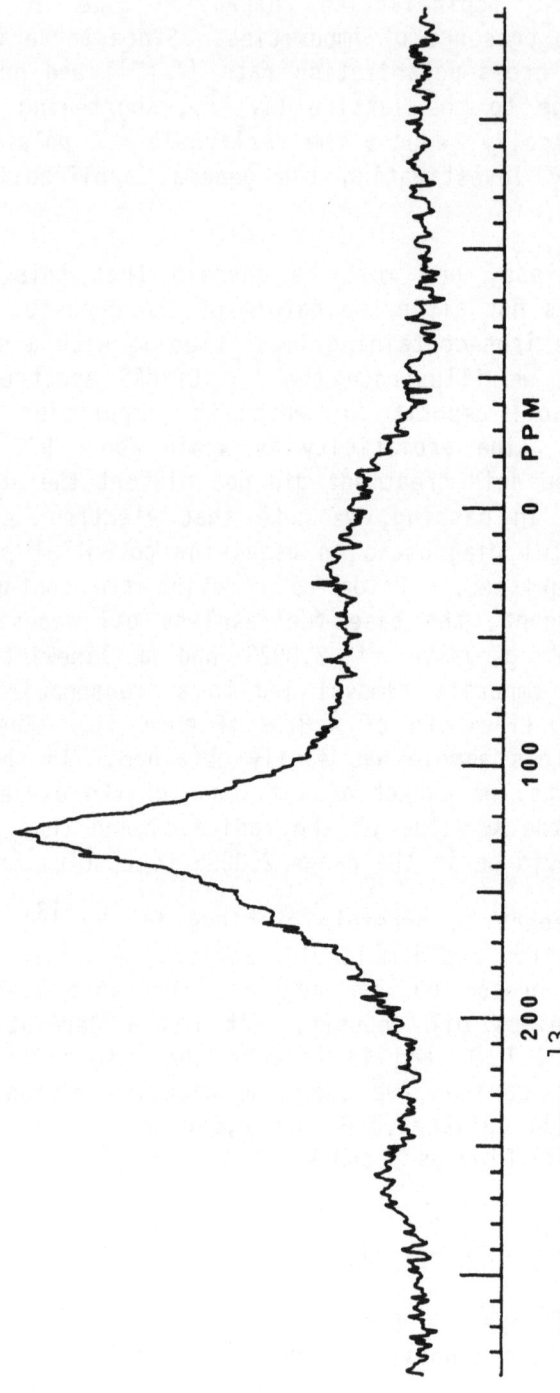

Figure 4. CP/MAS ^{13}C spectrum following removal of deposit impurites with a magnet.

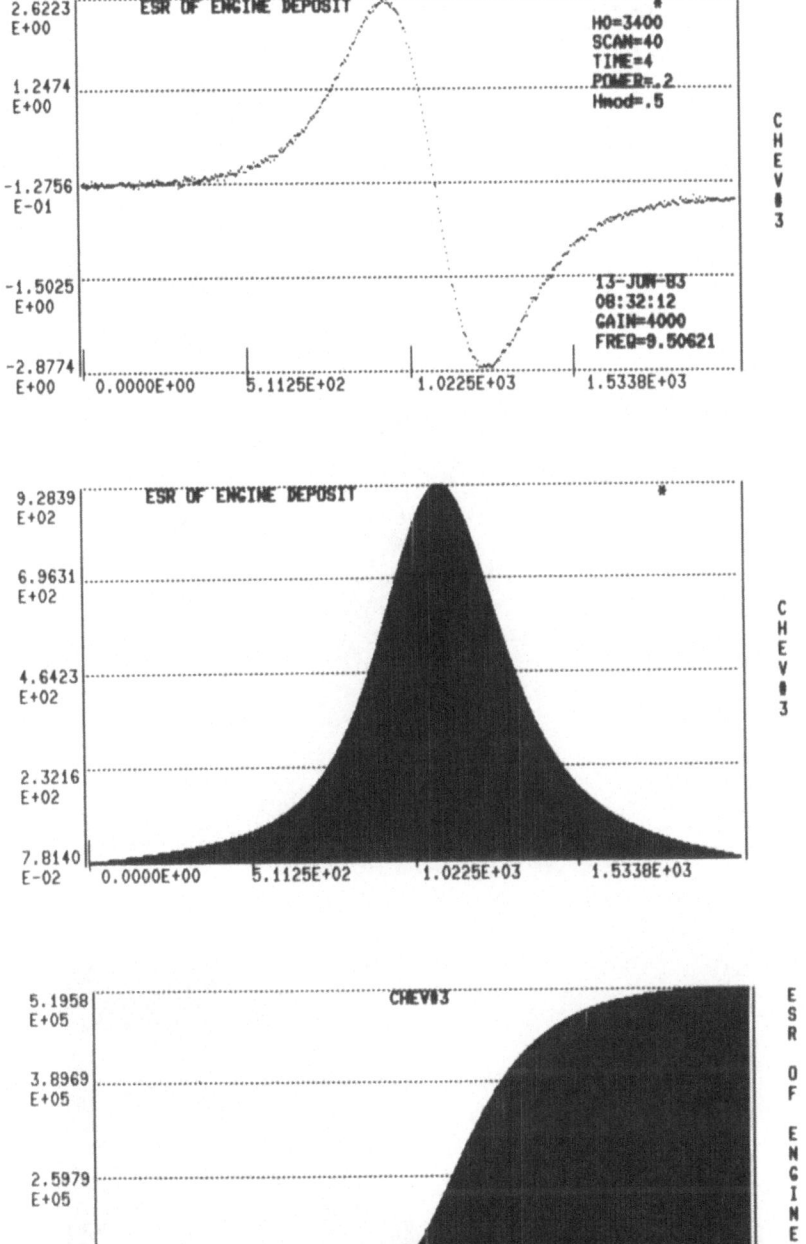

Figure 5. ESR pattern of "as-received" deposit. The derivative
of the absorption mode is at the top, the absorption
mode in the middle and the integral at the bottom.

127

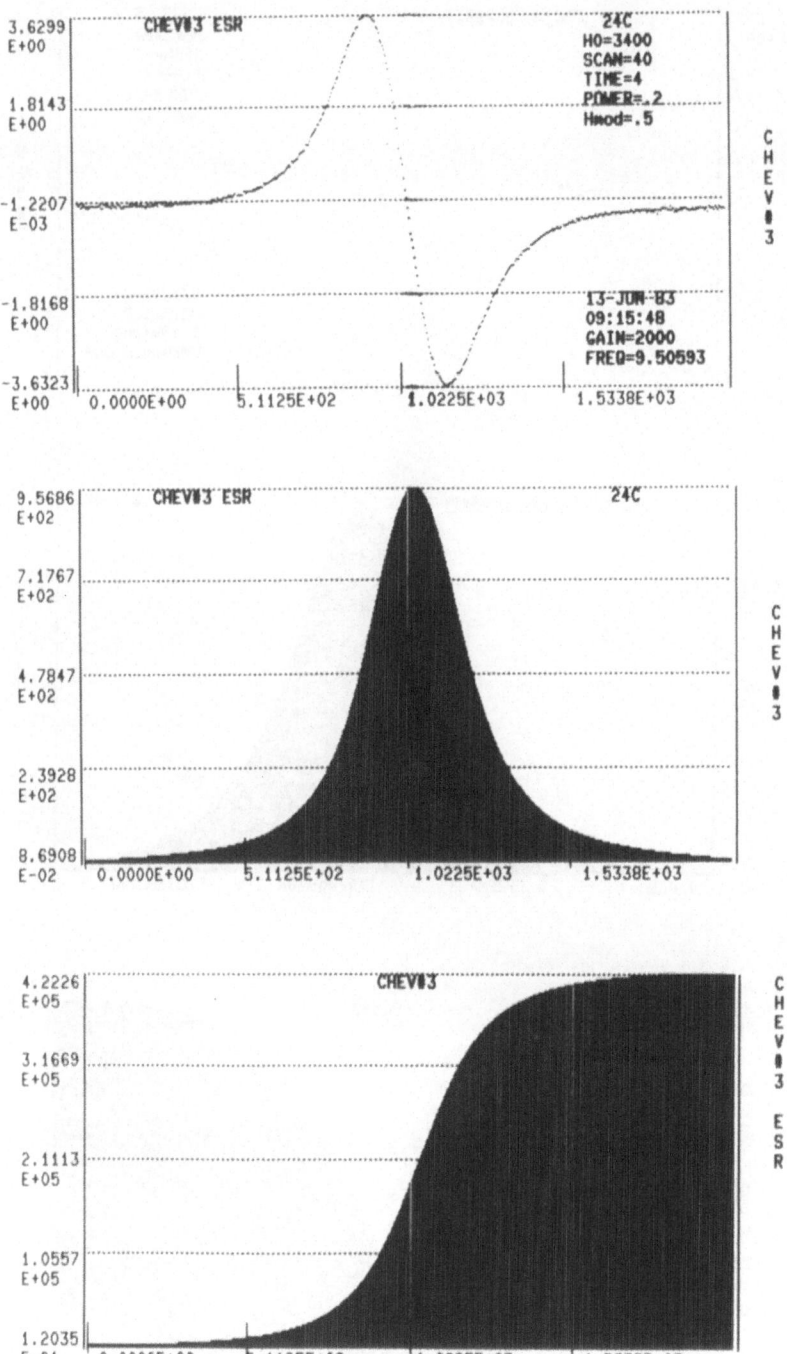

Figure 6. ESR pattern of deposit following removal of impurities with a magnet.

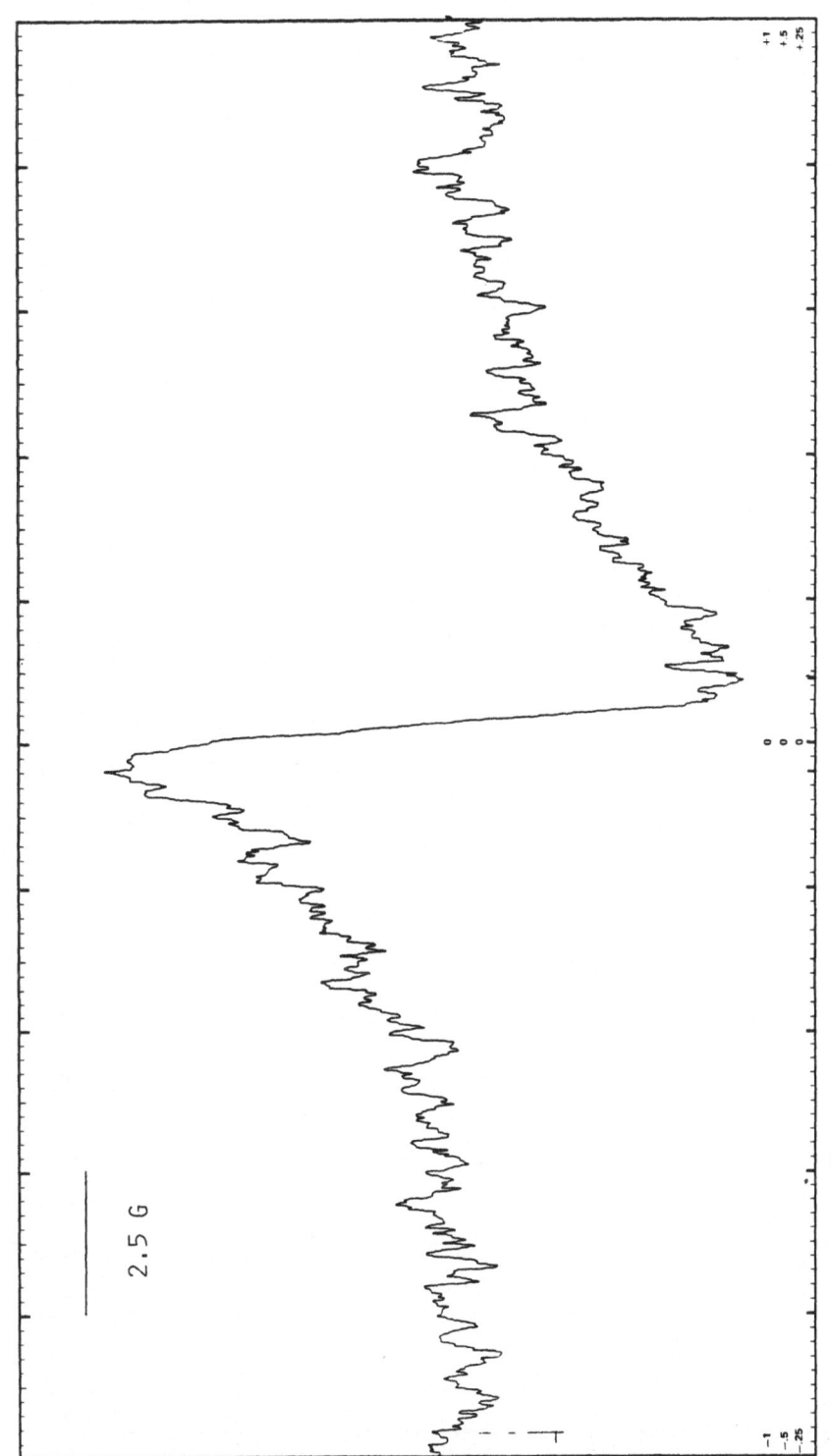

Figure 7. Wideline ^{1}H NMR of base fuel/ashless oil deposit at 15 MHz, 24°C. In separate pulsed NMR experiments at 75 MHz, two T_1 values were observed: 110 ms and 174 ms.

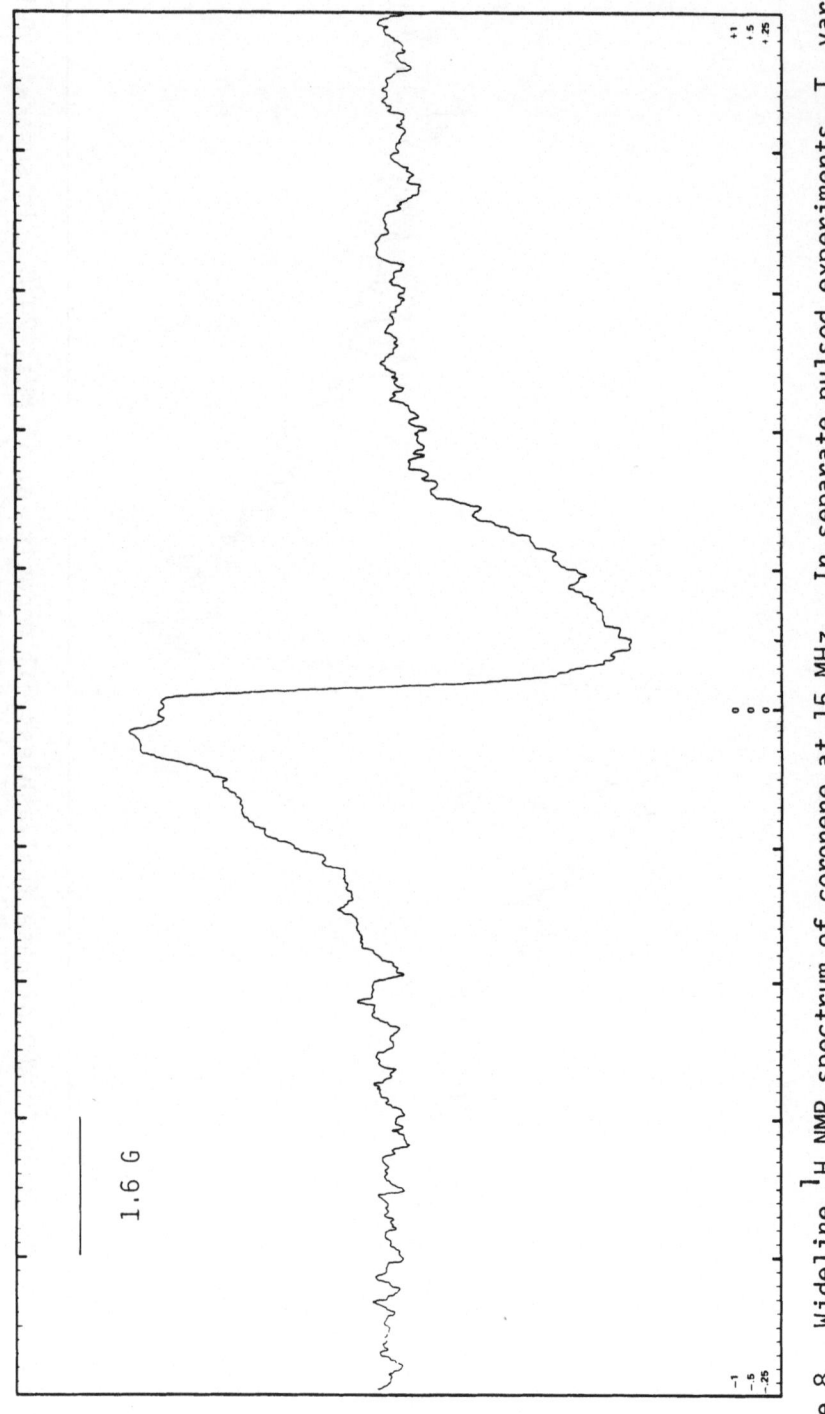

1.6 G

Figure 8. Wideline ^1H NMR spectrum of coronene at 15 MHz. In separate pulsed experiments, T_1 varied with frequency, ranging from 66 ms at 15 MHz to 149 ms at 75 MHz. $T_{1\rho}$ was found to be 40 ms (60 MHz, H_1 = 12.5 G), and T_2 was 52 μs.

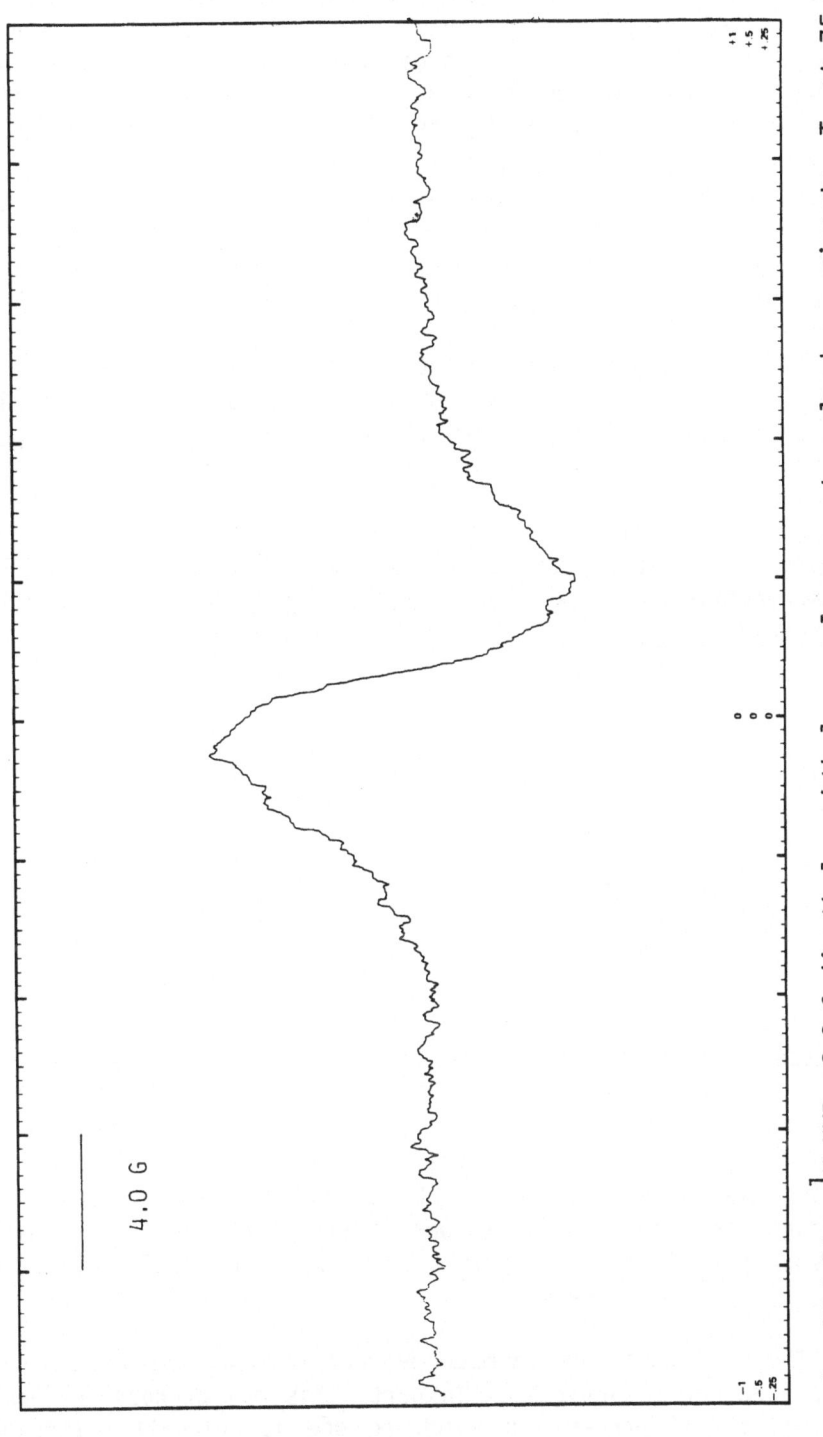

Figure 9. Wideline ^1H NMR of 2,6 dimethyl naphthalene. In separate pulsed experiments, T_1 at 75 MHz was measured to be 36.4 sec! The reader should note that Figures 7 - 9 give the derivative of the absorption mode.

typical Australian and North American bituminous coals show aromaticities between 50% and 80%,[11] one sees that engine deposits are more aromatic than coals of comparable elemental composition. Although it is true that some of this aromaticity arises, in the case of base fuels, from polynuclear aromatic hydrocarbons from the fuel, the low extractibility of these deposits suggests such contribution to be less than 20 wgt.%. If 20% of the total aromaticity arises from entrained aromatic molecules, the "skeleton" is still 81% aromatic ((85-20)/(100-20)). In fact, preliminary work on a deposit derived from iso-octane fuel showed an aromaticity of ca. 50%.

Unfortunately, there is little additional information from the spectra of Figures 1 through 4. The aliphatic peak at 23 δ is reminiscent of benzylic methyl groups, but the aromatic peak at 129 δ is merely that of typical sp^2-hybridized carbon.

As a counterpoint to this, consider the CP/MAS spectra of the simple aromatic hydrocarbon perylene (Figure 10) and of the oxygen-containing material 3,4,9,10-perylenetetracarboxylic dianhydride (Figure 11).

The spectrum of perylene shows typical aromatic lines at 121, 129, and 131 δ. The dianhydride shows not only lines at 117, 124, and 128 δ but also resonances at 134 δ (ipso carbon) and at 161 δ (anhydride carbon). However, these features disappear on heating the dianhydride to 500°C, as illustrated in Figure 12. Although the pyrolysis product still contains 18-25 wgt.% oxygen, the CP/MAS spectrum shows only a single resonance typical of aromatic carbon.

Thus, products of thermal degradation may lose characteristic spectral information. In part, this may proceed through a "randomization" process in which no one functionality strongly

132

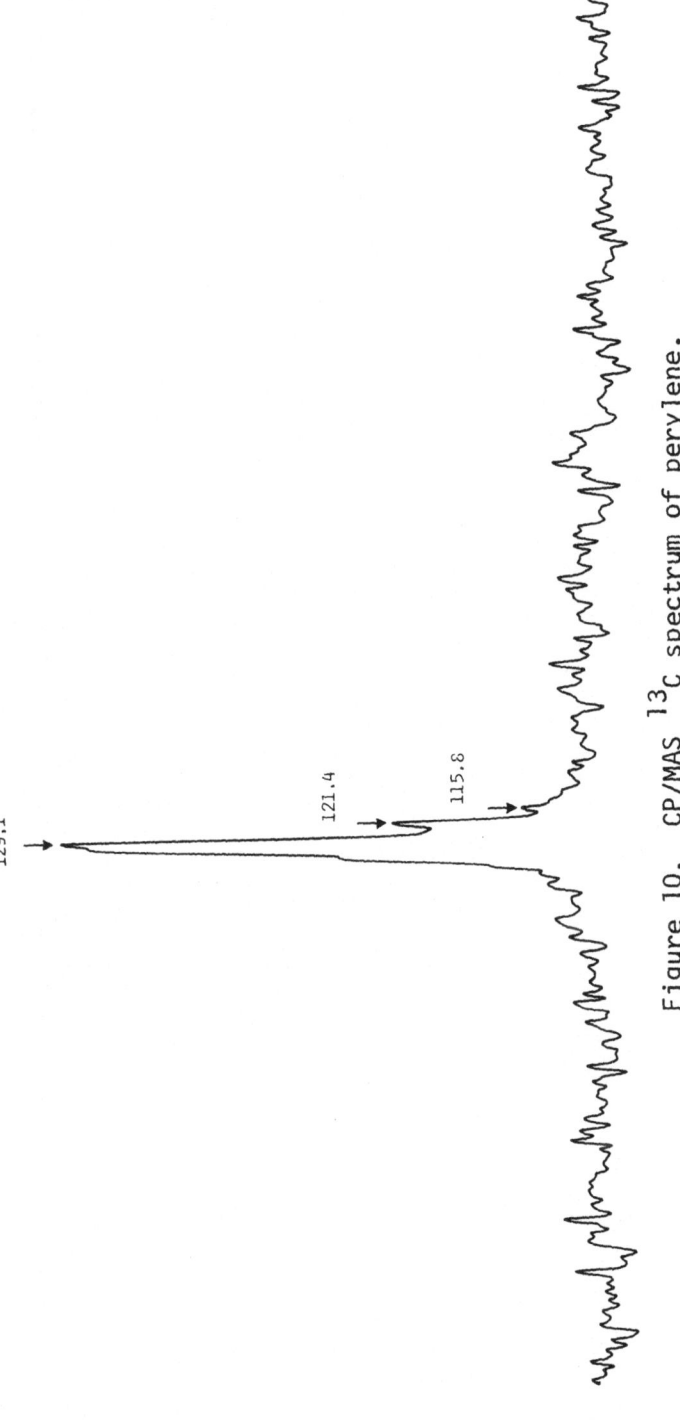

Figure 10. CP/MAS ^{13}C spectrum of perylene.

129.1

121.4

115.8

133

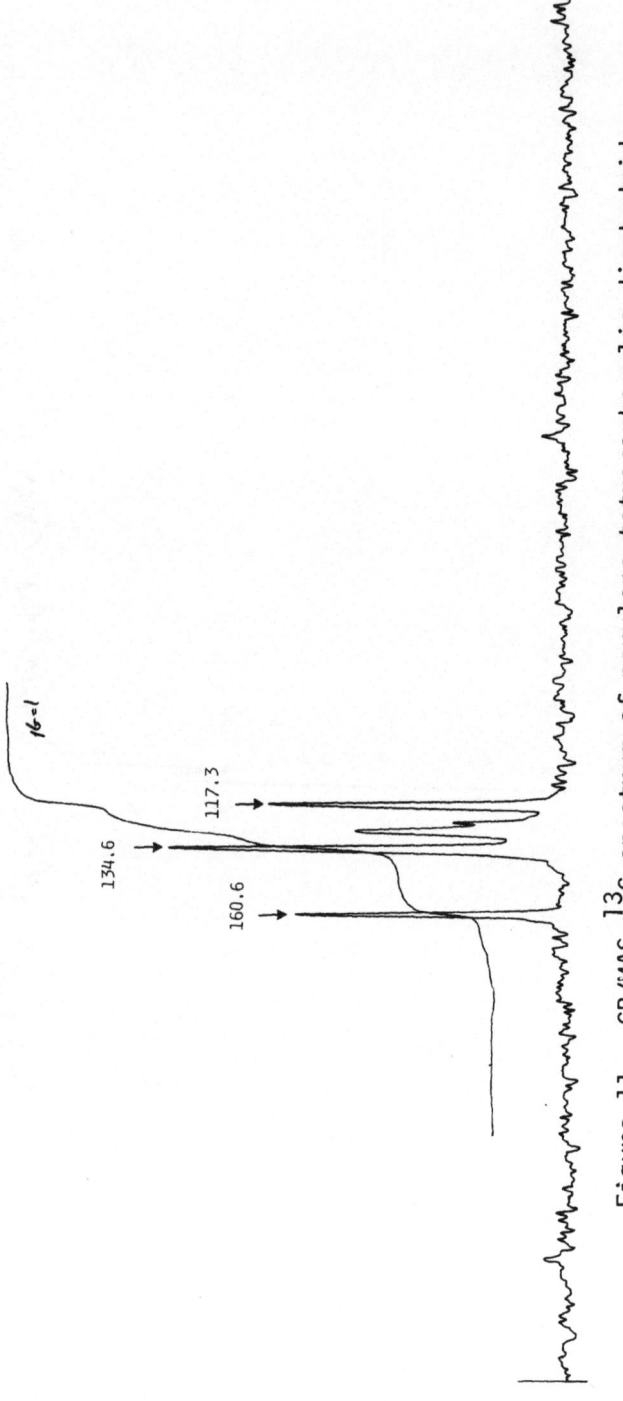

Figure 11. CP/MAS ^{13}C spectrum of perylene tetracarboxylic dianhydride.

134

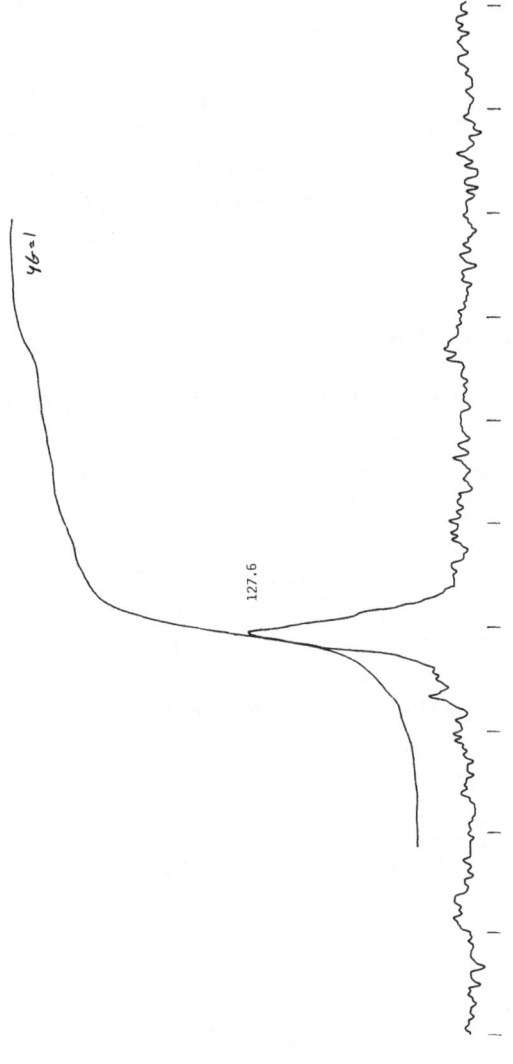

Figure 12. CP/MAS ^{13}C spectrum of perylene tetracarboxylic dianhydride pyrolized at 500°C. Note that this spectrum shows primarily one peak, in contrast to the spectrum of the starting material.

predominates. Additionally, however, there may be selective "interferences" in the CP/MAS experiment, as in oxygen-containing radicals obscuring the ^{13}C resonances of carbon-oxygen functionality. As seen in Figure 13, the ESR pattern of pyrolized 3,4,9,10-perylenetetracarboxylic dianhydride (g=2.0033, linewidth=4.4 G, radical density=1.8×10^{18} spins/gram) is similar to that of the engine deposit (g=2.0031, linewidth=4.8 G, radical density=3.9×10^{18} spins/gram) in suggesting the presence of oxygen-based radicals.

One way to avoid these difficulties is to generate engine deposits under more carefully controlled and less severe conditions. As shown in the next section, this approach can aid in the identification of oxygen-functionality involved in combustion deposits.

V. ANALYSIS OF A MODEL COMBUSTION DEPOSIT

Professor J. P. Longwell (Department of Chemical Engineering, Massachusetts Institute of Technology) submitted to us for analysis a model combustion deposit prepared in the following manner. A stoichiometric ratio of toluene and oxygen with 30 molecular % argon was combusted above a brass frit for 95 hours. A mass of 0.551 g of deposit was recovered, corresponding to 2.24×10^{-4} g deposit/g toluene. Microanalysis of this deposit showed 55.67% C, 4.86% H (done by Galbraith Labs employing Pragl method) and (25.27±0.59) % O (done by IRT Corporation employing neutron activation analysis). The amounts of these three elements sum to only 85.8% and suggest the possible presence of brass scrapings in the deposit.

The CP/MAS ^{13}C spectrum of this deposit (Figure 14) is striking in showing five distinctly resolved peaks. By analogy to known systems,[13-16] the resonances are assigned as follows:

Peak δ	Assignment
205	Ketone or quinone
173	Carboxylic acid or ester
126	Aromatic carbon
76	Ether carbon (such as CHR_2O)
35	Aliphatic carbon

136

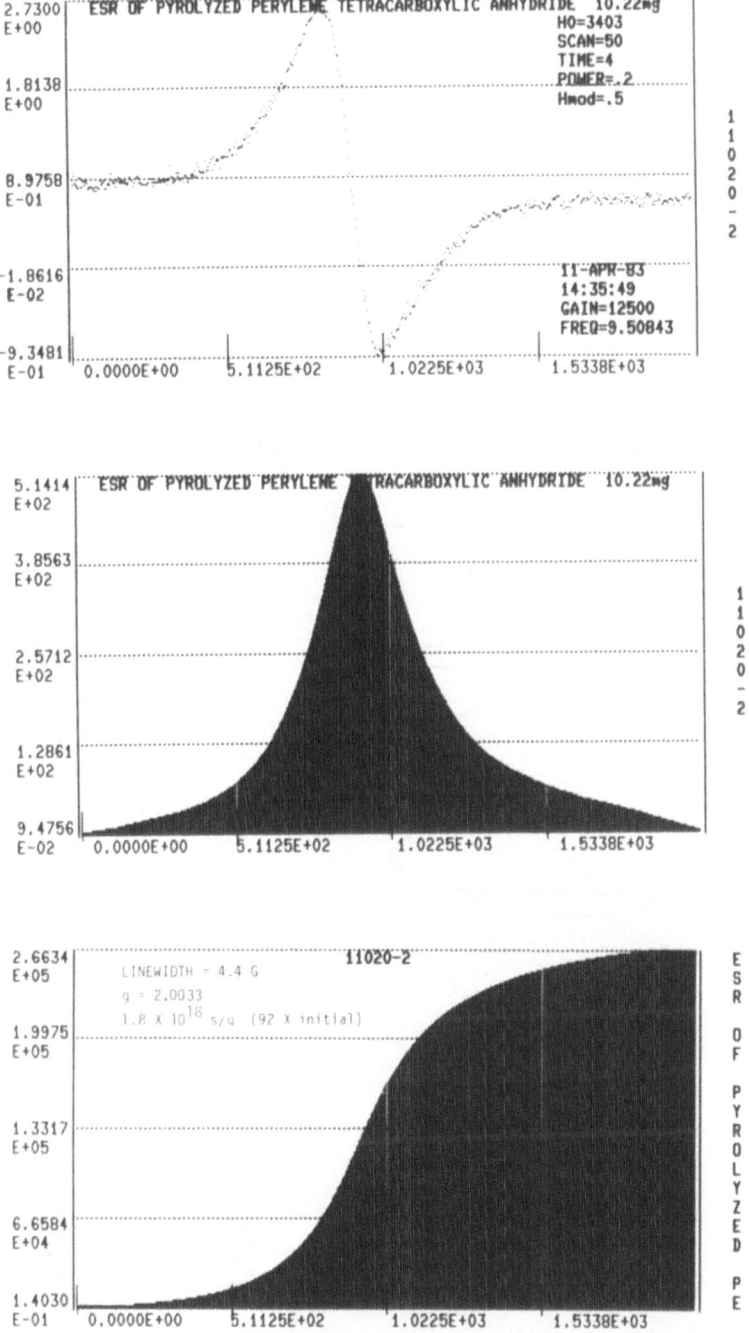

Figure 13. ESR spectrum of pyrolized material described in Figure 12. The top portion is the derivative of the absorption.

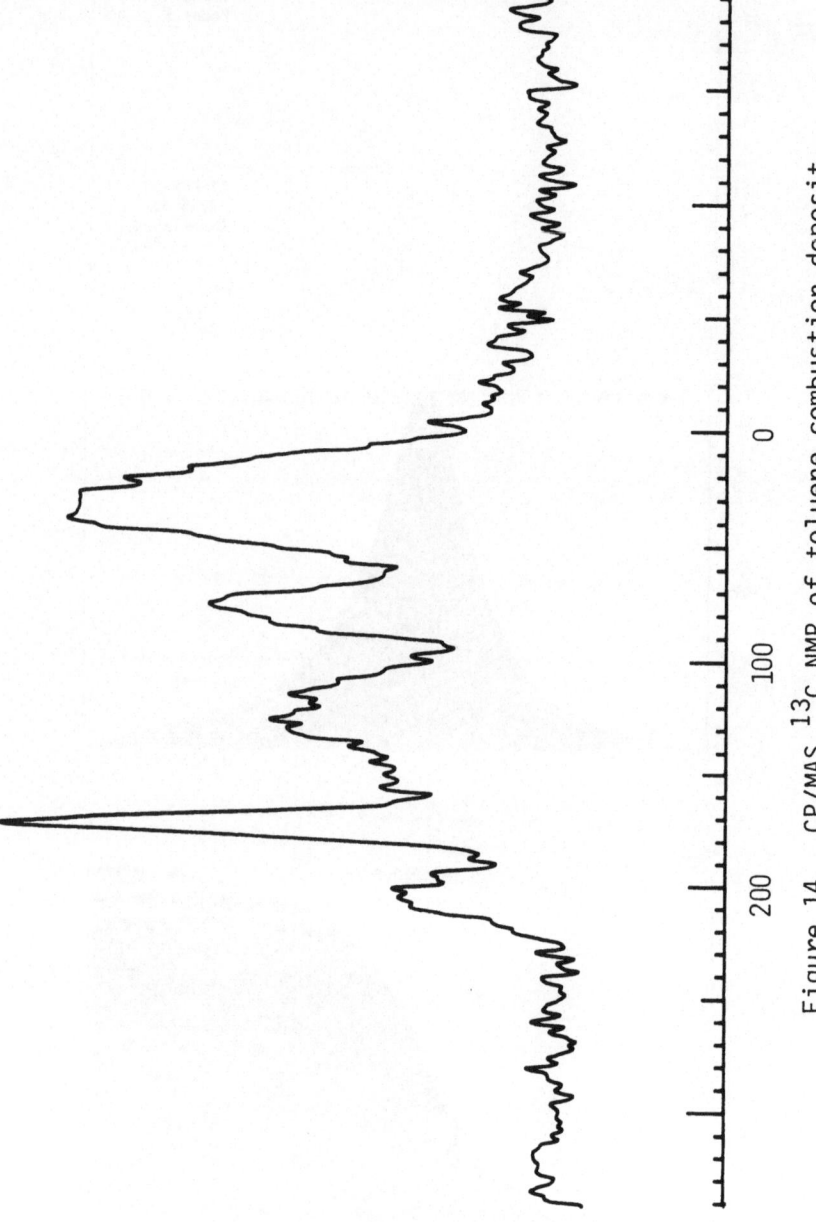

Figure 14.   CP/MAS ^{13}C NMR of toluene combustion deposit.

138

The toluene itself has four aromatic resonances at 138, 130, 129, and 126 δ and a single methyl peak at 21 δ, demonstrating the extent to which the deposit differs from the feed. Furthermore, in addition to the five resonances already mentioned, one observes resonances at 155 and 110 δ, which are consistent with -OH substituted aromatics. There is a methyl carbon shoulder in the range 10-30 δ.

The spectrum of Figure 14 was obtained on a JEOL FX-60QS using a cross-polarization contact time of 2 msec and a pulse repetition time of 2 seconds, with 24,800 transients accumulated. Such conditions produce quantitative ratios of carbon resonances for fossil fuel substrates, so the spectrum was integrated and divided into five chemical shift regions:

Chemical Shift Range	Assignment	Fraction of Total Intensity
220-190 ppm	Carbonyl	8
190-160 ppm	Carboxyl	14
160-100 ppm	Aromatic	25
100- 60 ppm	Oxygen-substituted aliphatic	15
60- 0 ppm	Aliphatic	38

Since the two downfield regions are readily assigned to specific oxygen-substituted carbon types, integration values can be used to estimate the fraction of total oxygen accounted for by these peaks, assuming that all carbon included in the elemental carbon content of 55.67 wgt.% is detected in the NMR experiment. The carbonyl resonance accounts for 5.9 wgt.% oxygen (0.08x0.5567x16/12) and the carboxyl resonance accounts for 20.8 wgt.% oxygen (0.14x0.5567x2x16/12). Since these two amounts sum to 26.7 wgt.% oxygen, which is <u>higher</u> than that measured by neutron activation analysis, <u>and</u> we have not factored in aliphatic ether content, it is possible that not all carbon in the sample was measured by the CP/MAS technique. This problem was not further explored.

Instead, more refined study was made on the available spectrum to extract additional information. In the CP/MAS experiment, one contends with fairly wide resonance envelopes which may obscure detailed structural information. One cross-polarization technique reported in the literature, $^{13}C-^{1}H$ dipolar dephasing,[17,18] has been shown to simplify the NMR spectrum of complex solids in well-defined ways by utilizing differences in the dipolar dephasing time, T_{DD}, for quaternary vs. non-quaternary carbon types. The dephasing delay is added after the ^1H-^{13}C cross-polarization contact, and distinguishes between carbons which have protons as nearest neighbors (-CH and -CH$_2$, for which dephasing is rapid) and carbons which lack protons as nearest neighbors (for which dephasing is slow). Methyl groups are a special case because the -CH$_3$ group generally has motional freedom which averages part of the ^1H-^{13}C dipolar interaction, thereby generating long dipolar dephasing times.

In the present case, a dephasing delay of 60 μsec was chosen to illustrate the principle of the experiment. The resulting spectrum is given in Figure 15-B, beneath the normal CP/MAS spectrum previously given as Figure 14 and now re-illustrated as Figure 15-A. The intensity axis of Figure 15-B was adjusted so that the intensity of the peak at 173 ppm (carboxylic groups) was equal in both Figures 15-A and 15-B. Figure 15-C is the difference spectrum between 15-A and 15-B, which thus approximates the distribution of CH and CH$_2$ carbon types.

Additional features which appear in the dipolar dephasing spectrum include a measureable methyl resonance centered at 17 ppm and a new ether band at 80 ppm probably associated with CR$_3$O-ether types. Intensity between 25 and 70 ppm is almost completely eliminated and substituted aromatic carbon downfield from 130 ppm is favored. The relative intensities of the carbonyl and carboxyl bands are almost unchanged between the regular CP and gated CP results, as would be expected.

With respect to other characterization, it is interesting to note that the infrared spectrum of the model combustion deposit is similar to spectra of engine combustion deposits. As illustrated in Figure 16, one sees relatively broad peaks at 1180, 1730, and 3440 cm^{-1}, as well as shoulders at 1050 and 2600 cm^{-1}. This last feature presumably arises from carboxylic acids, with the 1180 cm^{-1} band a C-O stretch, the 1730 cm^{-1} a carbonyl, and the 3440 cm^{-1} an O-H stretch. The 1050 cm^{-1} shoulder could be associated with a primary alcohol or an aryl ether.

For comparison, Figure 17 is the infrared spectrum of the pyrolysis residue of 3,4,9,10-perylenetetracarboxylic

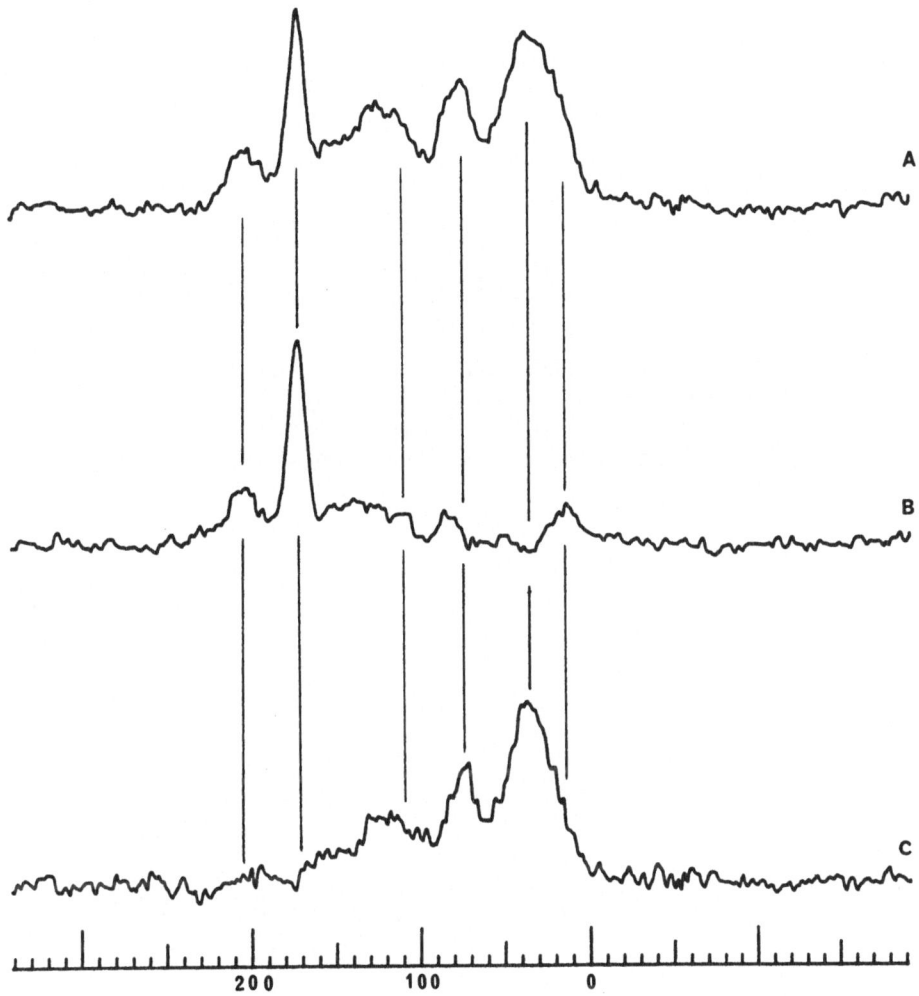

Figure 15. CP/MAS ^{13}C NMR spectra illustrating dipolar dephasing in toluene combustion deposit. Spectrum A is normal CP/MAS spectrum, as previously given in Figure 14; spectrum B is with a dephasing delay of 60 μs, and spectrum C is the difference between A and B.

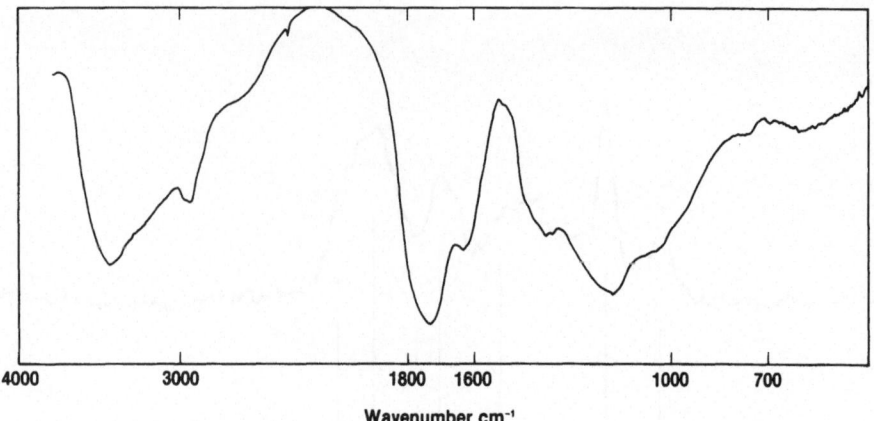

Figure 16. Infrared spectrum of the toluene combustion
deposit. Note the similarity of this "model"
compound spectrum to the spectra previously
given in part I for real deposits.

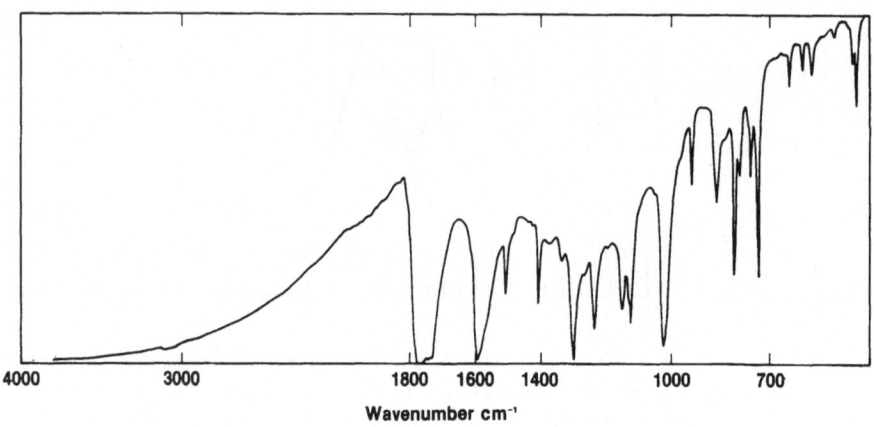

Figure 17. Infrared spectrum of perylene tetracarboxylic
dianhydride following short treatment at 500°C.
Note the number of sharp lines remaining.

perylenetetracarboxylic dianhydride on being heated to 500°C. In contrast to the combustion deposits, this material still shows narrow peaks in the infrared.

VI. CONCLUSIONS

Carbon-13 nuclear magnetic resonance of engine deposits taken with magic angle spinning and cross-polarization to the proton spins shows 85 to 90% of the carbon to be aromatic. This degree of aromaticity is higher than that found for coals of comparable elemental composition. The carbon spectrum of a model deposit formed by the stoichiometric combustion of toluene shows five distinct resonances, of which three correspond to carbon-oxygen functionality (carboxyl, carbonyl, and aliphatic ether). Although there is some evidence for an undercount of carbon spins in this last sample, it is most likely that the missing spins are aromatic carbons not bound to protons (either internal carbons of extended benzenoid networks or substituted peripheral carbons).

REFERENCES

1. J. Schaefer and E. O. Stejskal, Carbon-13 Nuclear Magnetic Resonances of Polymers Spinning at the Magic Angle, J. Am. Chem. Soc. 98:1032 (1976).
2. J. Schaefer, E. O. Stejskal, M. D. Sefcik, and R. A. McKay, Applications of High-Resolution ^{13}C and ^{15}N N.M.R. of Solids, Phil. Trans. R. Soc. Lond A 299:593 (1981).
3. D. L. VanderHart, Influence of Molecular Packing on Solid-State ^{13}C Chemical Shifts: The n-Alkanes, J. Magn. Res. 44:117 (1981).
4. G. E. Balimann, C. J. Groombridge, R. K. Harris, K. J. Packer, B. J. Say, and S. F. Tanner, Chemical Application of High-Resolution ^{13}C n.m.r. Spectra for Solids, Phil. Trans. R. Soc. Lond. A 299:643 (1981).
5. S. J. Opella, Solid State N.M.R. of Biopolymers, Phil. Trans. R. Soc. Lond. A 299:665 (1981).
6. A. N. Garroway, Magic-Angle Spinning of Liquids, J. Magn. Res. 49:168 (1982).
7. D. E. Wemmer, A. Pines, and D. D. Whitehurst, ^{13}C N.M.R. Studies of Coal and Coal Extracts, Phil. Trans. R. Soc. Lond. A 300:15 (1981)

8. C. A. Mims, K. D. Rose, M. T. Melchior, and J. K. Pabst, Characterization of Catalyzed Surfaces by Derivatization: Solid State NMR, J. Am. Chem. Soc. 104:6886 (1982).

9. R. L. Dudley and C. A. Fyfe, Evaluation of the Quantitative Reliability of the ^{13}C CP/MAS Technique for the Analysis of Coals and Related Materials, Fuel 61:651 (1982).

10. L. B. Alemany, D. M. Grant, R. J. Pugmire, T. D. Alger, and K. W. Zilm, Cross Polarization and Magic Angle Sample Spinning NMR of Spectra of Model Organic Compounds. 2. Molecules of Low or Remote Protonation, J. Am. Chem. Soc. 105:2142 (1983).

11. N. J. Russell, M. A. Wilson, R. J. Pugmire, and D. M. Grant, Preliminary Studies on the Aromaticity of Australian Coals. Solid State N.M.R. Techniques, Fuel 62:601 (1983).

12. H. L. Retcofsky, M. R. Hough, M. M. Maguire and R. B. Clarkson, Some Cautionary Notes on ESR and ENDOR Measurements in Coal Research, Appl. Spec. 36:187 (1982).

13. G. C. Levy and G. L. Nelson, "Carbon-13 Nuclear Magnetic Resonance for Organic Chemists", Wiley, New York (1972).

14. K. Müllen and P. S. Pregosin, "Fourier Transform NMR Techniques: A Practical Approach", Academic, New York (1976).

15. F. W. Wehrli and T. Wirthlin, "Interpretation of Carbon-13 NMR Spectra", Heyden, London (1978).

16. R. J. Abraham and P. Loftus, "Proton and Carbon-13 NMR Spectroscopy: An Integrated Approach", Heyden, London (1978).

17. P. D. Murphy, B. C. Gerstein, V. L. Weinberg, and T. F. Yen, Determination of Chemical Functionality in Asphaltenes by High-Resolution Solid-State Carbon-13 Nuclear Magnetic Resonance Spectrometry, Anal. Chem. 54:522 (1982).

18. P. D. Murphy, T. J. Cassaday, and B. C. Gerstein, Determination of the Apparent Ratio of Quaternary to Tertiary Carbon Atoms in Anthracite Coal by ^{13}C-^{1}H Dipolar Dephasing N.M.R., Fuel 61:1233 (1982).

ELECTRON SPIN RESONANCE STUDIES OF INTERNAL COMBUSTION ENGINE DEPOSITS

L. A. Gebhard*, R. S. Lunt[†] and B. G. Silbernagel*

*Corporate Research-Science Laboratories
[†]Products Research Division
Exxon Research and Engineering Company
Annandale, New Jersey 08801

I. INTRODUCTION

The problem of Octane Requirement Increase, or ORI, has emerged as an important consideration in optimizing the use of internal combustion engines for automotive transportation. Studies in our laboratories and elsewhere have linked the increased octane requirement of an engine after ~10,000 miles of operation (see Figure 1a) to the formation of carbonaceous deposits in the engine cylinders. The presence of the deposits causes knock -- i.e. preignition of the gasoline in the end gas region of the engine cylinder prior to the arrival of the combustion flame front.[1] While a totally quantitative description of the "knock" process is still not available, it is believed that poor heat transfer across this carbonaceous layer is primarily responsible for the preignition. This problem has broad implications on engine operation and fuel quality. To reduce knocking, a higher quality, more expensive fuel must be used, the engine's compression ratio must be reduced, or the engine's spark must be retarded. The latter two options reduce the engine's efficiency and therefore carry an associated cost.

A number of basic questions about the nature and function of these carbonaceous deposits have been addressed jointly by researchers and are the subject of the present volume:

● How do these deposits form?

- What are the precursor molecular species, and how are they transformed during the combustion process?

- What are the chemical and physical forces which hold the deposit together?
- What roles do the internal surfaces of the engine cylinder play in the formation of the deposit?

• How do these deposits influence the combustion process?

- How effectively can heat transfer across the deposit be modified?
- Can these properties be related to the combustion process itself?

Engine runs show that combustion is an important component of the deposit formation process. Extended runs (~100 hr) with unleaded fuel in a motored engine (no ignition) produce no deposit. Similarly, low deposit levels occur when "clean" fuels (like isooctane) and lubes (polyisobutene) are used. The ORI does not track deposit rate as variable time runs for a CLR (Cooperative Lubes Research) engine seen in Figures 1b and 1c illustrate. While deposit weight increases linearly (Figure 1c), the ORI rises very steeply at short times (~5 hr - Figure 1b), and then increases much more slowly thereafter. While comparison of ORI with weights of the deposit from the end gas region gives a somewhat better agreement, scatter in the data suggests that it is difficult to spatially isolate this phenomenon. The gross deposit chemistry suggests a substantial metamorphosis of fuel and lube prior to its inclusion in the deposit. Both fuel and lubes have high H/C ratios (~1.7 for fuel and ~1.9 for lube) and few if any heteroatoms (no oxygen and perhaps ~20 ppm nitrogen as additives in the lube). By contrast, the deposit has H/C ~0.6-1.0, ~25 wt.%, O and ~1 wt.% N, suggesting a considerable loss of alkyl components and the inclusion of oxygen in the organic during the combustion process. Solid state ^{13}C NMR reveals the highly aromatic character of the deposit: a 90%/10% aromatic/aliphatic split in deposits generated with unleaded fuel and a 51%/49% split for isooctane as the fuel. The sources of these aromatics are not completely established and can include accretion of polyaromatics from the fuel, reduction of cycloparaffins in the lubricants and formation of aromatic products, presumably by mechanisms like Diels-Alder reactions.

The Electron Spin Resonance (ESR) studies complement elemental and thermogravimetric analyses and selective chemical reaction examinations of deposits because they focus on the unpaired electrons associated with incomplete bonding in the organic. At the carbon radical densities encountered here, ~3.5×10^{18} spins/gm, there is roughly one such electron for each

146

FIGURE 1a: ORI is the increase in octane number for knock-free engine performance which occurs as deposits form in the engine.

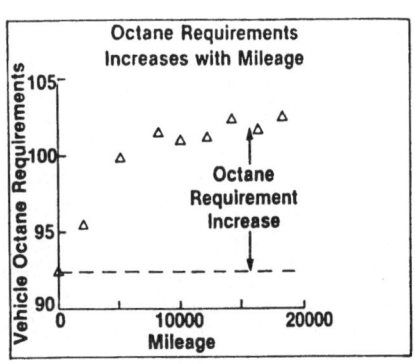

FIGURE 1b: In contrast to the ORI variation the total engine cylinder deposit weight increases linearly with time in CLR runs up to 100 hours.

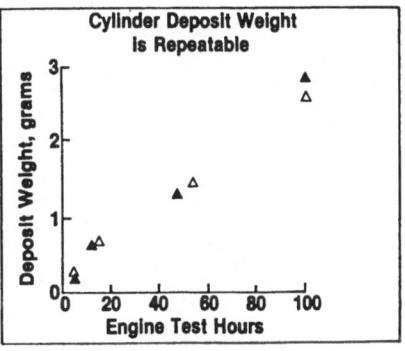

FIGURE 1c: And a rapid initial ORI is observed in single cylinder CLR test engine runs.

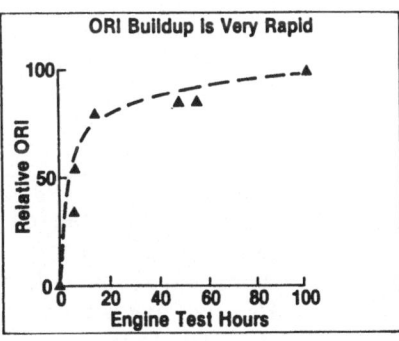

10,000 carbon atoms. Less than 1% of the molecules in the deposit would have such a "dangling bond". By contrast, ^{13}C NMR will observe all carbon atoms in the solid -- at least in principle. Selective chemical reactions can be used to quantify specific chemical functionalities. Radicals are interesting because they reflect some of the more reactive molecular classes in the deposit and because they can provide information about deposit chemistry -- especially in cases where radical types and numbers can be correlated to other physical or chemical properties. Heteroatom chemistry can be determined by studying the resonance position or g-value of the ESR absorption. The radical's environment is reflected in its width, and the intensity of the ESR signal can be integrated to determine the total number of spins. The number and type of radicals in the deposit will be determined by the type of molecular species involved in the combustion process, the character of the reaction, and the history of the deposit. Most radicals observed here are stable by virtue of being buried in the deposits -- the unpaired electrons would be quite reactive if exposed to the proper environment.

The strategy of the present research has been to correlate the ESR and chemical properties of these deposits among themselves and with each other. The result of these studies, which will be discussed at length in later sections, is as follows:

1. A linear correlation between H/C as determined by chemical analysis and V/N (the relative amounts of volatile and non-volatile organics determined by TGA) is found with high H/C deposits yielding larger amounts of volatiles upon TGA heating to 500°C. Recent chemical studies suggest that the volatiles consist of small molecules, CO, CO_2, H_2O and organic fragments from dealkylation.
2. ESR studies indicate that carbon radicals are associated with molecules containing heteroatoms - presumably oxygen. The radical density decreases in deposits with large V/N, H/C.
3. Significant differences are observed for single cylinder (CLR) and six cylinder deposits. The six cylinder deposits have typically lower values of H/C, V/N and lower carbon radical densities, suggesting a more condensed, refractory deposit than seen in the CLR.
4. Short duration CLR runs (5 hr, 26 hr, 50 hr) have lower H/C and oxygen content, suggesting that the first material to form on the engine surface has a more refractory form than subsequent deposits.

148

5. The morphology and properties of <u>deposits in the end gas region of the CLR engine cylinder differ from the rest of the deposit.</u> The material is more "chunky", and has lower H/C and oxygen content but higher radical density. It appears to be associated with condensed heavy molecules in the region.

6. <u>The ESR g-value and line width correlate weakly with O and N levels in CLR deposits and even less so for six cylinder deposits,</u> illustrating the difficulty of comparing a microscopic observation like ESR with macroscopic chemical values.

A discussion of ESR techniques and their applications in the engine deposit problem is included in Section II. Correlation studies are highlighted in Section III. A detailed discussion of individual correlations is presented in Appendix A.

II. EXPERIMENTAL DATA COLLECTION

In Section I, it was mentioned that certain parameters like g-values, lineshapes and widths, and spin densities could furnish microscopic information about the engine deposit. However in materials as heterogeneous as these, it is necessary to exercise considerable care to make certain that the observations are representative of the material. The present section discusses the relevant ESR parameters and their application to engine deposit problems.

A. <u>ESR Parameters and Their Determinations</u>

The <u>g-value</u>, the spectroscopic splitting factor, reflects the chemical form of the molecule bearing an unpaired electron. It is defined as $h\nu_L = g\mu_B H_0$, where ν_L is the frequency of the microwave field, H_0 is the applied external magnetic field strength at resonance, h and μ_B are Planck's constant and the Bohr magneton respectively. Operationally, g can be determined by the relationship $g = 0.714492 \cdot \nu_L$ (in GHz)/H_0 (in kG). For a "free electron", as might be in a molecular beam experiment, $g = 2.0023$. Polynuclear aromatic radicals have $g \approx 2.0026-2.0030$. Molecules containing heteroatoms have higher g values, the shift in g being associated with interaction of the radical electron with the heteroatom. Oxygen-bearing molecules, for example, can have g values ranging from 2.0030-2.0045 and the bulk of the engine deposits observed here have g values in this range.

Measurements at X-band microwave frequencies ($\nu_L \sim 9.5$ GHz, $H_0 = 3380$ G) yield g values which are reproducible to within ± 0.0002, corresponding to variations in field of $\sim \pm 0.3$ G. In principle, g-values can be determined with considerably higher precision for a single radical species but in the present case the distribution of molecular types precludes a more precise analysis. It is interesting to note that the intrinsic widths of these ESR signals are ~5-6 G, so the resonance center is routinely determined to within ±1/20 of the width of the line itself.

The linewidth and lineshape of the ESR adsorption furnish information about the radical's environment. Three factors can come into play in determining the width of an ESR line[2]: (1) the hyperfine interaction with protons on an aromatic host molecule or aliphatic protons on its substituents, (2) the extent of electron delocalization and (3) interactions between radicals which can lead either to dipole broadening or intermolecular exchange. In the present case, hyperfine interactions appear to be the principle source of linewidth although, as discussed in Section IIB of Appendix A, some evidence of broadening appears at high radical densities. As indicated in Figure 2, there are several interrelated ways of characterizing the width: the splitting between derivative maxima, of the absorption spectrum, called ΔH_{pp}, and the width at half intensity of the absorption spectrum itself, commonly called $\Delta H_{1/2}$. Since ΔH_{pp} can be determined directly and with high precision from the "raw" data -- i.e. the absorption derivative curve, it has been used in the correlation analysis of the present report. For convenience, ΔH_{pp} will be referred to as DHH in Section III and Appendix A.

Typical linewidths for the carbon radical ESR of engine deposits are: ΔH_{pp} ~5-6 G, $\Delta H_{1/2}$ = 7-8.5 G. The ratio of $\Delta H_{1/2}/\Delta H_{pp}$ provides a measure of the shape of the ESR absorption line. If the ESR line were a simple Gaussian, this ratio would be 1.18; the ratio for a Lorentzian line would be 1.73 (= $\sqrt{3}$ identically). In the engine deposit samples, this ratio takes an intermediate value of ~1.4 ± 0.05 in almost all cases.

The carbon radical density is defined as the total number of radicals per gram of carbon in the sample. Care must be taken to obtain an accurate spin density determination: (1) The microwave power must be low enough so that the signal is linearly proportional to the number of spins (i.e. no saturation), (2) the modulation for these particular samples, should be one fifth the intrinsic linewidth to optimize sensitivity and (3) the magnetic field scan range must be large enough to include all of the ESR signal and allow convergence of the baseline of the spectrum. All measurements reported here are relative, made with respect to a Varian standard consisting of weak pitch dispersed in KCl salt crystals. The present standard is designated

150

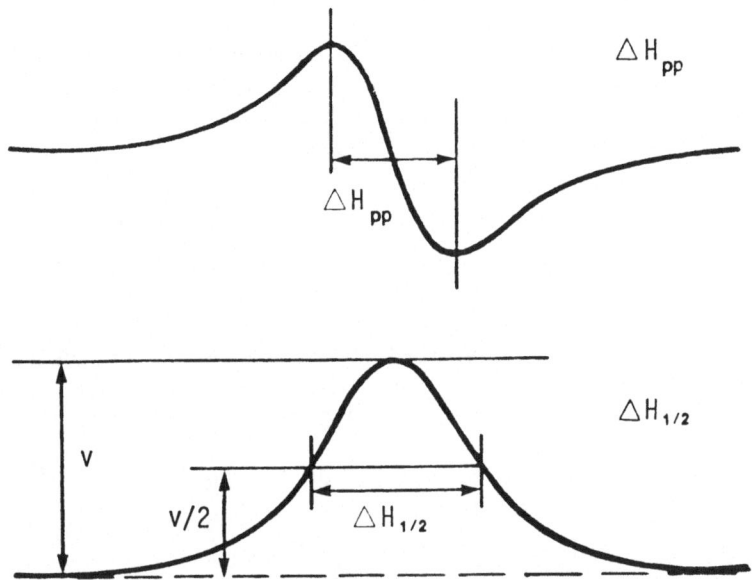

$\triangle H_{pp}$

$\triangle H_{pp}$

V

$V/2$

$\triangle H_{1/2}$

$\triangle H_{1/2}$

FIGURE 2: Schematic of the Derivative and Integral of the Carbon Radical Microwave Absorption, With Critical Parameters Identified.

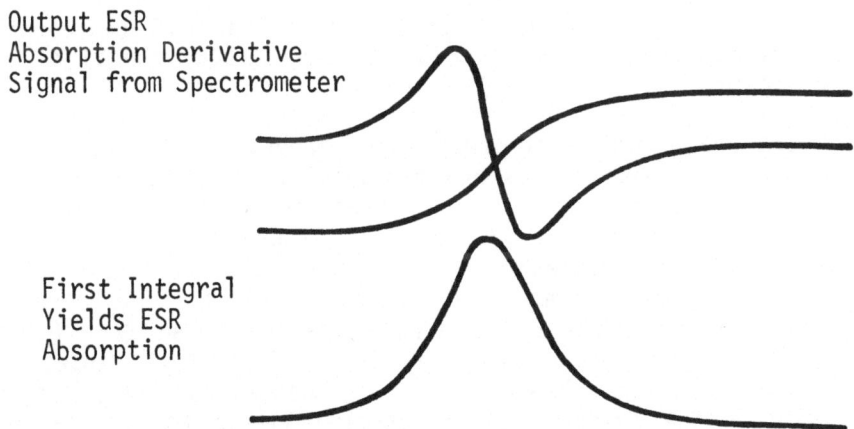

Output ESR
Absorption Derivative
Signal from Spectrometer

First Integral
Yields ESR
Absorption

FIGURE 3: Schematic Showing Derivative, Integral and Second Integral of the Same Trace. Carbon Radical Density is Determined From This Second Integration.

"100% of 0.00033% pitch 032576, with a concentration of 1.13 x 10^{13} spins/cm. The spin concentration of an unknown sample can be expressed as:

$$\#C \text{ spins/g} = \frac{9.393 \times 10^{15} \ I_2 \ (\frac{Scan_x}{40})^2}{G_x M_x P_x^{1/2} N_x Wt_x C_x}$$

where:

9.393×10^{15} = spin concentration of pitch with ESR parameters folded in.

I_2 = integrated absorption area for unknown.

Scan x = scan range in gauss for unknown.

G_x, M_x, P_x = gain, modulation and power in μW of the unknown.

N_x = number of scans.

Wt_x = unknown sample weight, gm.

C_x = %C/100 of unknown.

The I_2 integrated intensities values are obtained on a Nicolet 1180 signal averaging computer system. The data is accumulated on the 1180 as the derivative of the absorption and then a double integration is performed. Figure 3 represents typical computer output of the ESR of the carbon radical of an engine deposit.

There are a number of areas where errors can be introduced in spin density measurements and manipulation of data within the computer is one of them. In an experiment to determine the uncertainties involved in performing a second integral via the computer, various samples were run repeatedly in a situation where the only variations were computer related. The results indicate that computer-related errors are less than 2%. Other sources of error involve sampling and day to day changes in cavity configuration. These day to day changes include position of sample in the cavity, sample rotation within the cavity and sample repacking between runs. The errors associated with sample position and rotation within the cavity amount to 3%, 2% of which are related to the computer. To get a feeling for the overall level of errors related to these day to day changes, four samples were repeatedly examined over the course of several days. The radical densities for these samples were $1.14 \pm .08 \times 10^{18}$; $9.80 \pm 1.35 \times 10^{17}$; $1.26 \pm .20 \times 10^{18}$ and $8.29 \pm .48 \times 10^{17}$ spins/gm. The total variation would then be 7.0%, 13.7%, 15.7% and 5.8%. These levels of uncertainty are directly related to sample properties. When considering only ESR and computer related effects, spin density measurements can be made within 5% uncertainty.

As the microwave power is increased, the signal initially increases in magnitude, then reaches a maximum, and finally falls off at high power levels. This deviation from a linear increase with microwave field strength is known as saturation. Since the absorption is broadened and distorted when it is saturated, lineshape and even position information is lost as well. It is usually wise to be prudent, choosing microwave power levels well below saturation. At low power levels, the signal varies linearly with the microwave field strength or equivalently, varies as the square root of the microwave power. Since microwave power levels are determined with considerable precision, the usual comparison is with $\sqrt{P}$, where P is the microwave power, expressed in microwatts (μW).

The test for saturation involves collecting a series of ESR spectra on a typical sample over the full range of microwave powers (~1μW - 200 mW). All ESR spectrometer parameters except for the amplifier gain are kept constant during the test. The amplitude of the ESR signal divided by the amplifier gain is plotted against $\sqrt{P}$. Two typical deposits from an earlier engine deposit survey were used in the present test and the data are included in Table I. Figure 4 illustrates the systematic variations observed when the intensity as inferred from the height of the ESR derivative maxima is plotted as a function of $P^{1/2}$. The increase in intensity is initially linear, followed by a very broad maximum at high power levels. To quantify the initial deviations from a linear response, the quantity intensity/$G/P^{1/2}$ is included in Table I. While saturation may not be obvious visually until power levels of ~1 mW are used, the initial deviations set in around P~10μW. A convenient means of visualizing this tabulated data is to plot the ratio of intensity/$G/P^{1/2}$ as a function of log P, as shown in Figure 5.

The spectrometer system used in these observations is a Varian E-line Century Series with an E102 microwave bridge operating at ~9-9.5 GHz. Microwave frequencies are determined with a Hewlett-Packard 5245L Electronic Frequency Counter. The data were accumulated on a Nicolet 1180 Computer Data System. The

Table 1

Deposit Type	Power	$p^{1/2}$ μW	Gain	Y' max	Y' max/G × 10^3	Y' max/G/$P^{1/2}$
	100 mW	316	5 × 10^2	10.75	21.5	.068
	64 "	253	=	11.15	22.3	.088
	20 "	141	=	12.0	24.0	.170
	9 "	95	=	12.6	25.2	.265
Base Fuel	1 "	31.6	=	10.95	21.9	.693
Ashless Oil	0.1 "	10	=	5.45	10.9	1.09
	100 μW	10	=	5.45	10.9	1.09
	49 "	7	=	4.0	8.0	1.14
	25 "	5	=	2.95	5.9	1.18
	9 "	3	=	1.85	3.7	1.23
	1 "	1	1.25 × 10^3	1.55	1.24	1.24
	.05 "	.71	"	1.1	.88	1.24
					× 10^2	
	100 mW	316	1.25 × 10^2	17	13.6	.043
	50 "	224	"	18	14.4	.064
Additive in Gas	10 "	100	"	18.5	14.8	.148
Premium B	1 "	31.6	5 × 10^2	54	10.8	.32
	100 μW	10	"	24	4.8	.48
	64 "	8	1.25 × 10^3	49	3.92	.49
	16 "	4	2.5 × 10^3	50	2.0	.50
	10 "	3.1	1.25 × 10^3	20	1.6	.52
	1 "	1	5 × 10^3	25.5	.51	.51

ESR Parameters: 9.5 GH$_z$; 3388 G Field; 40 G Scan;
Hmod = 1.0; 4 min Scan; τ = .25 sec.

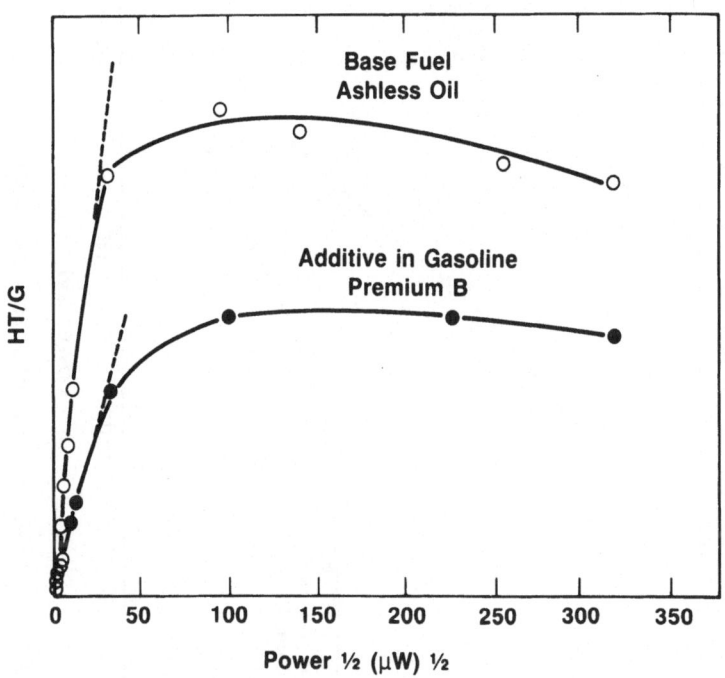

FIGURE 4: Saturation Response of Two Typical Engine Deposits

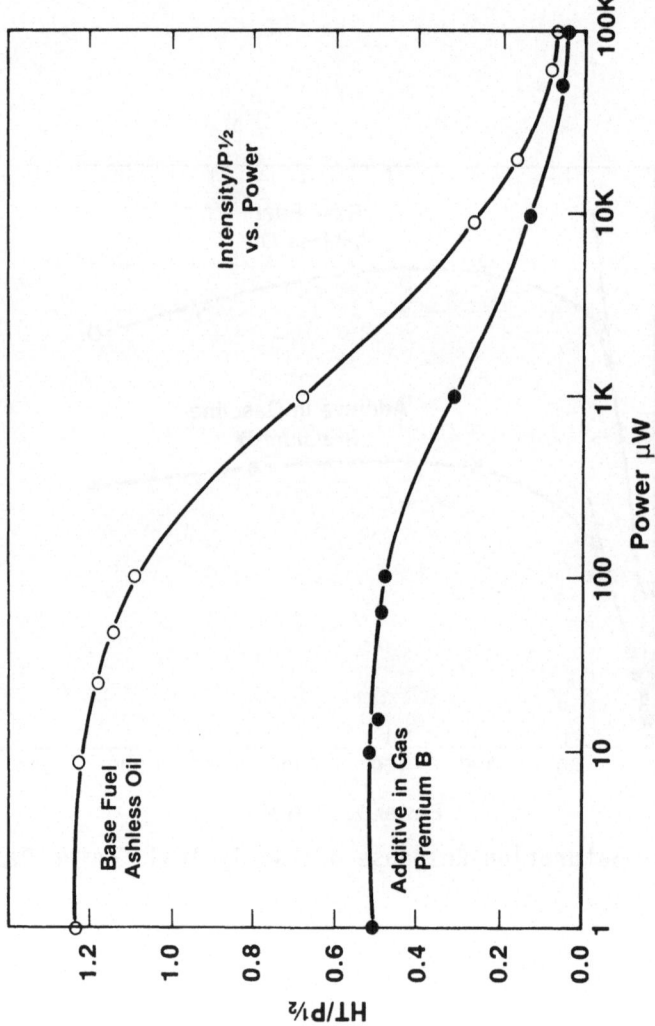

FIGURE 5: An Alternative Representation of Carbon Radical Saturation

156

standard operating conditions for these carbon radical studies have been: central field value 3396 G, a field scan range of 40 G swept in 4 minutes of scan time. The microwave frequency is ~9.5 GHz, and the microwave power level was chosen at 10μW. To employ phase sensitive detection techniques, a 100 KHz magnetic field with an amplitude of 1 G was applied parallel to the static magnetic field. For filtering purposes a 0.25 second time constant was used in the phase sensitive detector.

B. Characteristics of Engine Deposit Samples

1. Radical Density Variations with Sample Type

Visual inspection of the piston head or cylinder of an engine after an ORI run suggests that the deposit is not homogeneous. The character of the deposit often is related to the area of the cylinder or piston from which it originated. For example, the end gas region of the piston head typically produced a hard, chunky surface deposit, whereas the deposit around the outer edge of the piston was more varnish-like (and sometimes even oily) and the spark area produced a flaky, brittle deposit.

Of course when dealing with such a wide variety of deposit types, sampling can become a problem. To get a feel for the scatter in properties, like spin densities, repeated measurements were made on four ORI deposits, Base Fuel/ashless oil, additive in gas/ashless oil, Base Fuel/Premium B and additive in gas/Premium B samples from the earlier survey. Table II includes these data. The measurements were made over a several day period, using standard ESR running conditions. The scatter for the repeated runs ranged from 5.8% to 15.7%. To minimize these errors, a number of sample preparations were considered.

Multiple samples from the same batch did not necessarily increase the observed variations. Two samples of the Base Fuel ashless oil sample were each measured repeatedly under similar conditions over the course of several days. It is presumed that a sample that is run daily is removed from the ESR and probably repacked just from normal handling. The less than 7% error for repeated measurements on the same sample is greater than the difference between samples from the same batch. This indicates that the errors are associated with particle size, packing, sample conductivity, or other factors related to sample inhomogeneity.

Errors associated with repacking the same sample between ESR measurements can be as large as 14%. Four ORI deposits were randomly chosen and samples were inverted and repacked between

Table II

Errors Associated with Repeated Runs: Samples As Received

Base Fuel Ashless Oil	Additive in Gas Ashless Oil	Base Fuel Premium B	Additive in Gas Premium B
1.1588×10^{18}	1.0769×10^{18}	1.1364×10^{18}	8.558×10^{17}
1.0145	.7808	1.2305	8.378
1.1008	1.1518	1.1668	8.193
1.0112	.9029	1.1821	9.201
1.1014	.9279	1.2045	8.423
1.1737	1.0366	1.4515	7.738
1.2147		1.4190	7.526
1.1806		1.3954	8.580
1.1214		1.2342	7.913
1.1390		1.4806	8.352
1.0787		1.4870	
1.2495		1.1217	
1.2919		1.5150	
1.1645		.7153	
		1.1084	
		1.2602	
		1.2260	
$1.1429 \pm .080$ $\times 10^{18}$	9.795 ± 1.345 $\times 10^{17}$	$1.2550 \pm .197$ $\times 10^{18}$	$8.286 \pm .478$ $\times 10^{17}$
7.0%	13.7%	15.7%	5.8%

Table III

Base Fuel Ashless Oil

Sample 1:	7.33, 6.42, 6.96, 6.39 6.97, 7.42, 7.66, 7.47	$7.08 \pm .48$ 6.82%
Sample 2:	7.09, 7.20, 6.82, 7.90 8.17, 7.37	$7.43 \pm .51$ 6.90%

each ESR measurement. The variations for a minimum of 6 measurements on each sample were 6.82%, 13.73%, 4.09% and 13.03%.

Particle size is apparently contributing substantially to the errors. A base fuel ashless oil deposit was sieved through 230 mesh and the repacking experiment was repeated on no less than 8 times for each fraction; the variations were 10.4% for the coarse particles and 27.8% on the fines.

In addition to sieving, samples were ground in KBr and suspended in paraffin to try to eliminate some of the errors associated with particle size. Two samples were dispersed in melted paraffin in ESR tubes, cooled and repeated spin densities measured. The variations for the two samples were 12.3% and 25.8%.

Since particle size distribution is crucial, a number of extensive sieving experiments were performed on ORI deposits prior to making ESR measurements. Specifically, samples that were generically different, i.e. either from the cylinder or the piston of the six cylinder test engine, were sieved into meshes. The results are shown in Table IV. Sieving has virtually no effect on the piston area deposits but there are obviously fines in the cylinder head deposits with low g-values, lower radical densities and very broad lines. These fines contain magnetic particles which radically alter the ESR measurement. For samples that have no magnetic contamination, the precision is much better.

A number of repeated runs were made on noncontaminated samples, in which the samples were repacked, rotated in the cavity or not disturbed at all. Table V is a summary of that data. Errors associated with the computer only, involving no change in the position or packing of the sample, are quite small, less than 2.5%. When the sample is rotated in the cavity, the error is up to 3.5%. But the errors increase to 8% when the sample is repacked. Generally then, the level of errors for a noncontaminated deposit might be as high as 8%. Three to four percent of that error is related to the ESR cavity and the computer, the remainder due to the inhomogeneity of the sample, which can vary depending upon the sample's morphology, size, and packing configuration in the tube.

It was decided that sample preparation of ORI deposits should only involve taking a representative portion of the total deposit from each area sampled. Since magnetic contamination is not intrinsic to the deposit and not present in the majority of deposits, no standard procedure was included in the sample prep to remove the fines. Data that is known to have magnetic

Table IV

Sieving Experiment Results

Sample	> 40 Mesh			40 > 100 Mesh			< 100 Mesh		
	g	°C x 10^{18}	ΔH	g	°C x 10^{18}	ΔH	g	°C x 10^{18}	ΔH
Head 1	2.0037	1.8	4.6	2.0036	1.9	5.2	2.0014	.45	12.8
Head 2	2.0033	1.9	5.0	2.0032	1.7	5.2	2.0014	.33	17.5
Piston 6	2.0033	1.3	6.1	2.0033	1.1	5.8	No Fines		

Table V

Percent Variation on Repeated ESR Runs

Repacked	Rotated	No Change
6.35	3.34	1.25
4.90	1.98	1.02
1.69	2.24	2.32
7.78		1.75
		2.31
		.89

contaminants is duly noted; otherwise the level of uncertainty for repeated measurements of carbon radical density should be less than 10%.

2. Magnetic Contamination

The initial ESR observations of g-values and linewidths for six cylinder engine deposits presented some unusual results. Carbon radical absorptions had linewidths ranging from 6 to 14 Gauss and g-values from 2.0020 to 2.0040. A linear correlation existed between the g-values and $\Delta H_{1/2}$ as Figure 6 shows. Some deposits from the six cylinder engine heads have no observable ESR signal. A similar linear variation of g and $\Delta H_{1/2}$ was observed in the isooctane engine deposits as well.

A second observation further implicated magnetic contaminants. In several experiments, separate samples from a given engine cylinder were recombined after ESR measurements for subsequent chemical analyses. Samples from one six cylinder run, an example of such a case, are shown in Table VI. The piston sample from cylinder 3 has a radical behavior very typical of organics from the six cylinder runs: g=2.00331, ΔH_{pp}=4.8G and radical density = 0.48×10^{18} spins/gm. By contrast, the head sample from the same cylinder has an anomalously low g value (g=1.9915), a very large linewidth (~35G), and low radical density (0.19×10^{18}). Mixing roughly equal quantities of the cylinder and head deposits leads to abnormal results for the whole sample. The ESR spectra for these three samples are shown in Figure 7. This change of piston deposit properties upon mixing suggests that a broadening agent is present in the head sample.

A series of observations and experiments provide strong evidence to support the theory that broadening and shifting of the ESR line in these deposits is caused by fine magnetic particles which are collected along with the engine deposit during the engine scraping process. These observations follow:

(1) The pistons in the six cylinder (Chrysler) engine used for these experiments are aluminum, while the heads are made of steel. Since the effect is confined almost exclusively to head samples, it appears that small magnetic particles are removed from the wall of the head during the scraping process.

(2) Such broadening and shifting of the line could be caused if the carbonaceous deposits had a high electrical conductivity -- as is sometimes found in

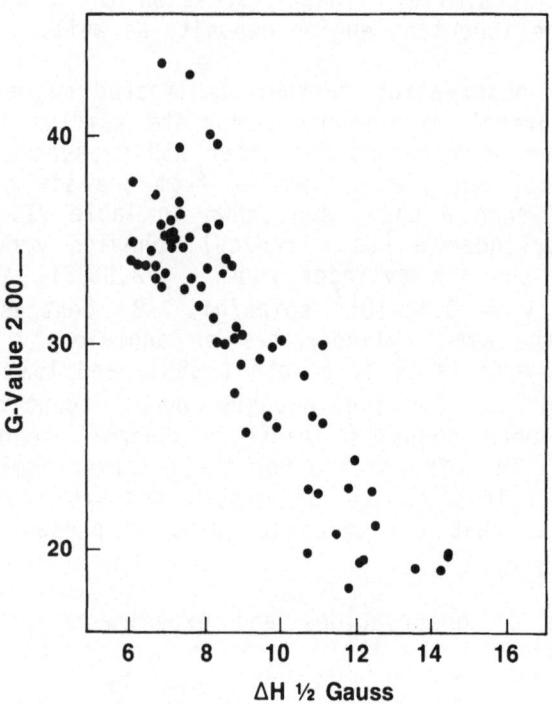

FIGURE 6:   The Dramatic G-Vaue Lindwidth Correlation for
Six Cylinder Engine Deposits Indicates Magnetic
Broadening of the Radical Signal

Table VI

JMP-602	g-Value	ΔH_{pp}	°C Spins/gm x 10^{18}
Piston 3	2.00331	4.8	0.48
Head 3	1.9915	~35	0.1
Mixed Sample	2.0000	13.48	0.73

Table VII

Effect of KBr Dilution on Magnetic Contamination

Sample	ESR Parameter	As Received	grd in KBr
JMP-607	g-value	2.00261	2.00357
Head 3	ΔH_{pp}	10.35	5.5
JMP-607	g-value	2.00379	2.00389
Piston 3	ΔH_{pp}	4.4	4.5
JMP-604	g-value	1.9984	2.00261
Head 3	ΔH_{pp}	15.4	8.4

materials like graphite and high rank coals. However, ESR lines from such conductive materials have a characteristic asymmetry which is not observed here. The mixed sample in Figure 7 still shows a highly symmetric line.

(3) g-values from 1.999 to 2.0020 are not appropriate for aromatic or aliphatic carbon radicals or for such species containing the common heteroatoms O, N or S. Furthermore, the elemental analyses and TGA results from piston and head deposits were very similar, precluding the possibility that the head deposits might be a different carbon type -- perhaps "graphitized".

(4) Dilution of samples with broad ESR lines disperses the broadening agent and narrows the line. Table VII shows the effect of diluting three such samples by mixing them in KBr. While the piston sample with high g and low ΔH_{pp} (2.00379, 4.4 G) is unaffected by dilution, a dramatic loss of width and increase in g values toward the 2.0030-2.0040 range is observed in the KBr mixed samples. Had the low g and high ΔH_{pp} values been intrinsic to the organic, they would have been unaffected by dilution. The ESR lineshapes before and after dilution are shown in Figure 8.

(5) Treatment of head deposit with HCl dissolves the magnetic particles and the residual organic has a narrow ΔH_{pp}, high g value, and high radical density. A composite deposit was treated with HCl. The ESR parameters before and after treatment and the ESR traces are included in Table VIII.

(6) Particle size separation and extraction of magnetic particles also removes or concentrates the broadening agent. Two head samples were separated into three fractions: >40 mesh, 40 mesh >100 mesh, <100 mesh. The results on Table IX and in Figure 9 show that the contaminants concentrate in the deposit fines (<100 mesh). The contaminants can also be extracted by passing a bar magnet over the deposit. The organics remaining after these separations have ESR properties typical of the uncontaminated material.

Understanding these contamination effects requires the explanation of three observations:

(1) The reduction in g-value.

Piston Deposit - 40 G Scan Range

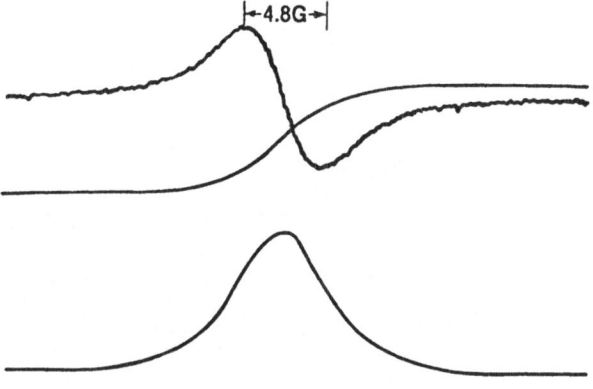

Head Deposit 100 G Scan Range

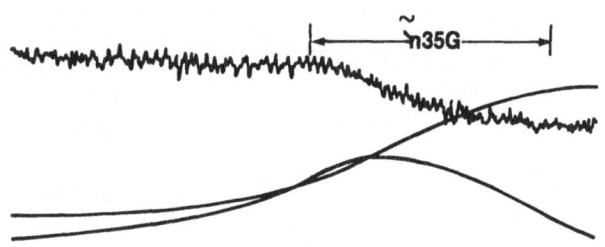

Combined Deposit 100 G Scan Range

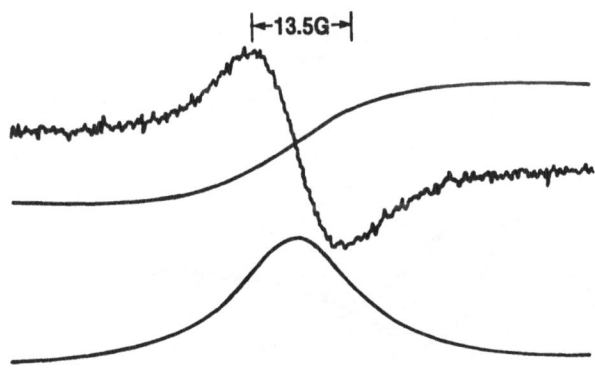

FIGURE 7: An Illustration of Magnetic Broadening in Six Cylinder Engine Head Deposits

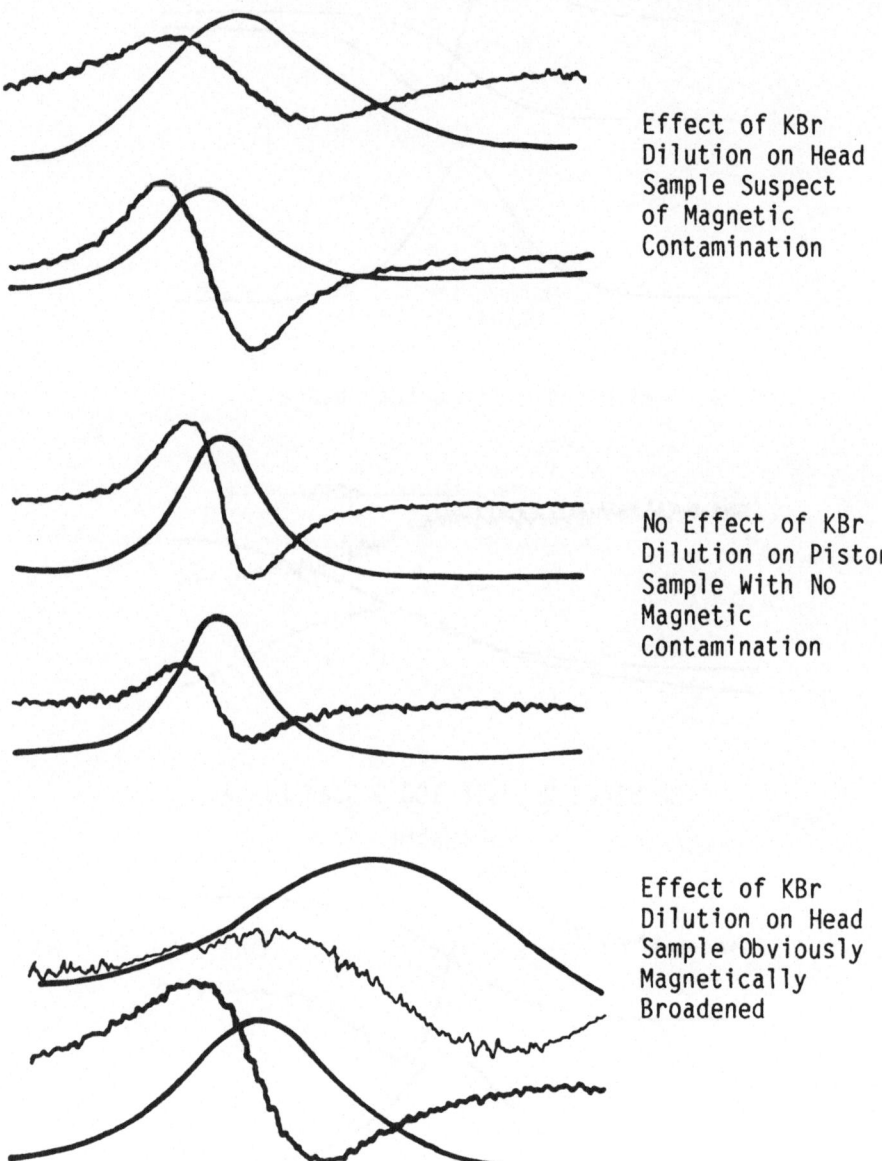

Effect of KBr
Dilution on Head
Sample Suspect
of Magnetic
Contamination

No Effect of KBr
Dilution on Piston
Sample With No
Magnetic
Contamination

Effect of KBr
Dilution on Head
Sample Obviously
Magnetically
Broadened

FIGURE 8:   Dilution of Samples With KBr Suggests That Broadening
            Comes From Isolated Magnetic Species

TABLE VIII

Effect of HCl Treatment on Iron Contamination

Sample	ESR Parameter	Before HCl Treat.	After HCl Treat.
JMP-605	g-value	2.00224	2.00330
Entire Deposit Mixed	ΔH_{pp} °C spin/gm x 10^{18}	8.22 .54	4.7 1.25

|←8.22g→|

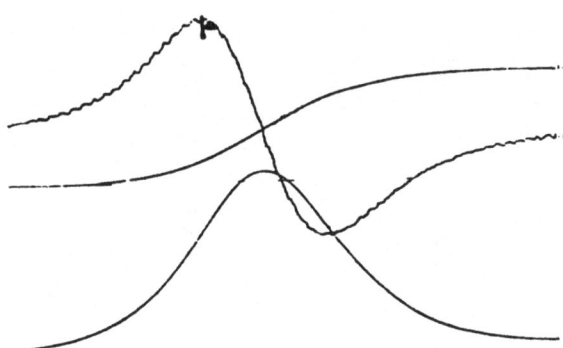

Deposit Before HCl Treatment

|←4.7g→|

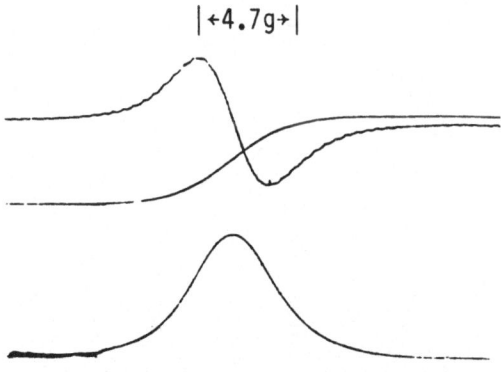

Deposit After HCl Treatment

TABLE IX

Results of Sieving Experiment

Sample	ESR Parameters	Original Deposit	>40	40>100	<100	Magnetic Separation on <100 Mesh
JMP-608	g-value	2.00166	2.00328	2.00317	1.9999	2.00359
Head 2	°C spins/gm x 10^{18}	0.30	1.22	1.09	0.21	--
	ΔH_{pp} Gauss	10.3	5.0	5.2	17.5	5.8
JMP-608	g-value	2.00274	2.00366	2.00356	2.00143	2.00378
Head 3	°C spins/gm x 10^{18}	0.57	1.20	1.25	0.30	0.47
	ΔH_{pp} Gauss	7.8	4.6	5.2	12.8	5.15

(2) The linear correlation in g-value and linewidth.

(3) The linear correlation of ESR intensity with both linewidth and g-value.

These data can be explained by means of the following picture. The sample contains a mixture of carbon deposits and small magnetic particles. These particles polarize in the presence of an applied magnetic field, as shown in Figure 10, with the "south" pole of the particle pointing toward the "north" pole of the magnet. This opposite polarization reduces the effective applied field in the sample, causing the lowering of the g-value which is observed. If the particle has a large magnetization, molecules near that magnetic particle will experience large field shifts and be excluded from the ESR absorption observed. Further away from these particles, the local field will consist of the sum of fields from neighboring particles. Broadening of the line by statistically distributed moments has been investigated previously.[3] For a dilute array of randomly arranged moments, the lineshape should be Lorentzian with a width proportional to the moment density (N). Shifts in the local field should also be proportional to the number of these magnetic particles. Thus, $\Delta g \propto \Delta H_{1/2}$ as observed in item 2. Loss of ESR intensity comes from the exclusion of carbon radicals that lie adjacent to the magnetic particles. As a first approximation at low density, radicals within some fixed instance of any particle would not be seen. Since $\Delta g \propto N$ and $\Delta H_{1/2} \propto N$, this implies that $D = D_0(1-AN) = D_0(1-A \propto \Delta H_{1/2})$. At higher density these simple linear relationships break down and a sharp fall-off would be expected, indicated schematically in Figure 11. These details will be discussed at length in Appendix B.

III. CORRELATIONS

Faced with at least sixteen variables in the present engine test results, we have divided them into classes as indicated in Figure 11, which deals with engine run conditions, engine run results, chemical and physical parameters, and lastly the ESR parameters: g, ΔH_{pp} and the carbon radical density (D). The rationale, systematics, and detailed analyses of 18 different correlations as a function of engine type, lubricant, and run duration are examined in Appendix A. In the present section, we will briefly summarize the conclusions.

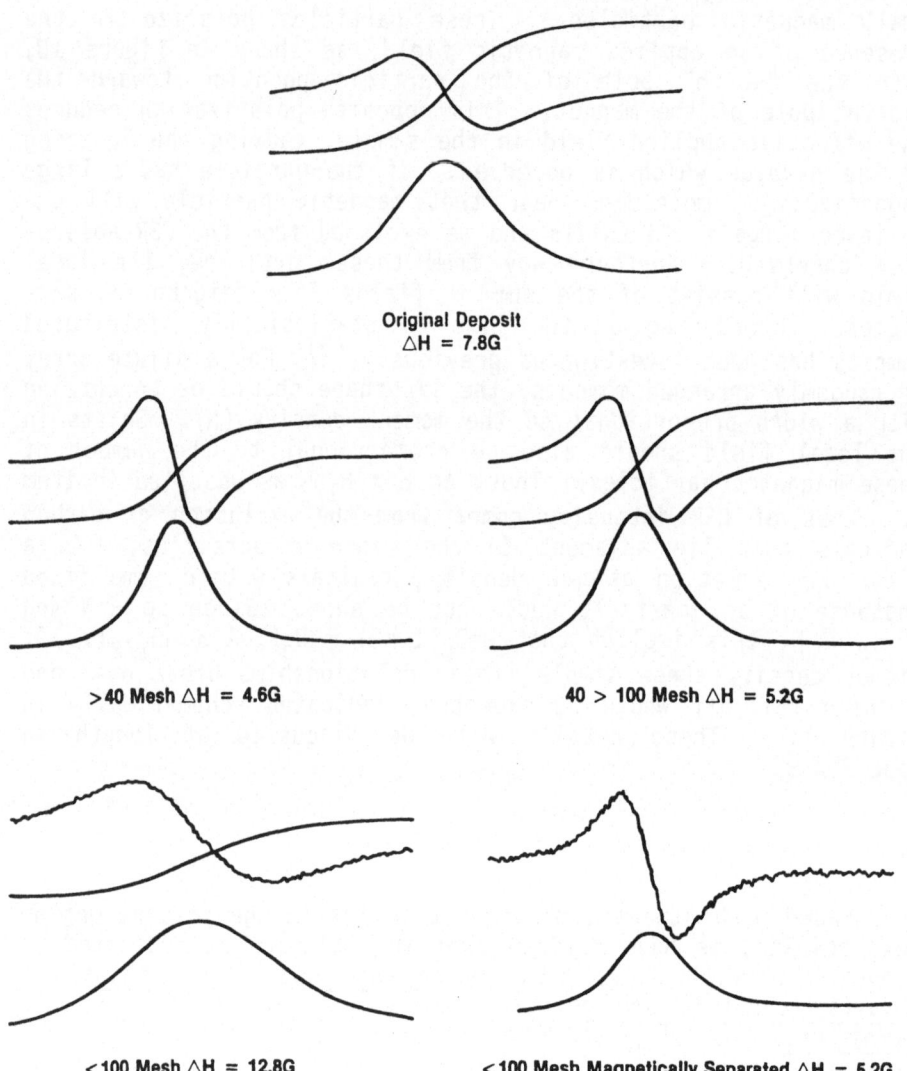

Original Deposit
$\triangle H = 7.8G$

>40 Mesh $\triangle H = 4.6G$

40 > 100 Mesh $\triangle H = 5.2G$

<100 Mesh $\triangle H = 12.8G$

<100 Mesh Magnetically Separated $\triangle H = 5.2G$

FIGURE 9: Sieving Experiments of a Typical Head Deposit Sample Establish the Small Dimensions of the Magnetic Species.

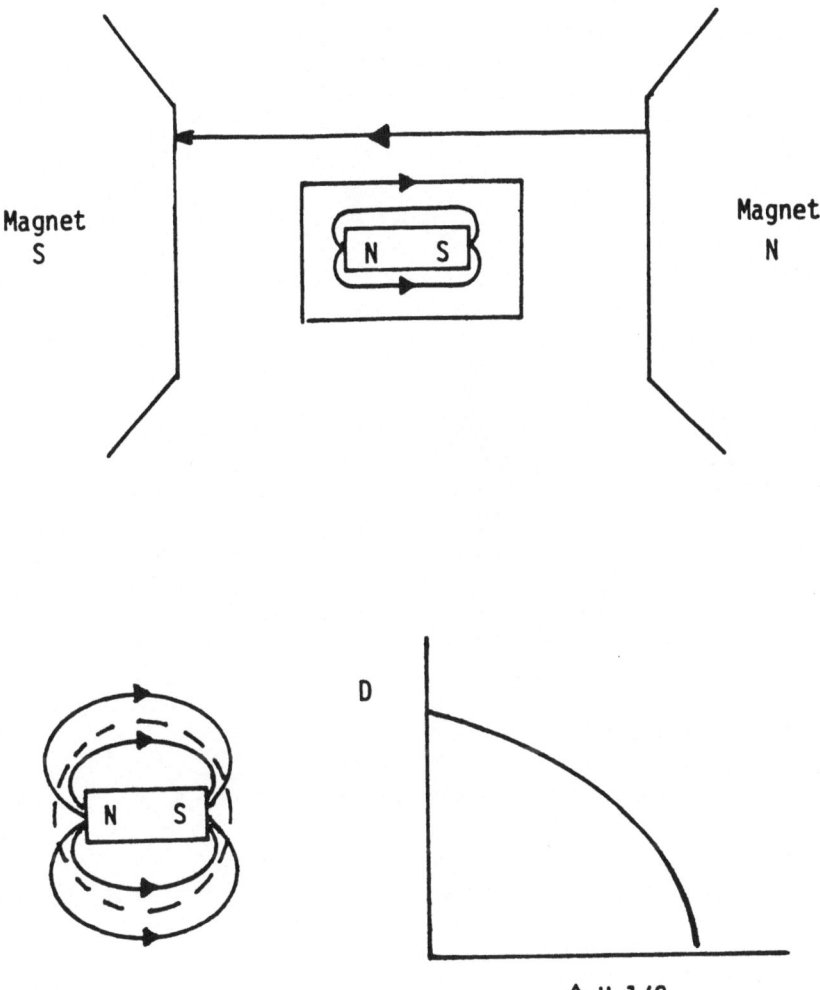

Magnet
S

Magnet
N

D

Δ H 1/2

FIGURE 10: Physical Model for Discussing Demagnetization
Effects

172

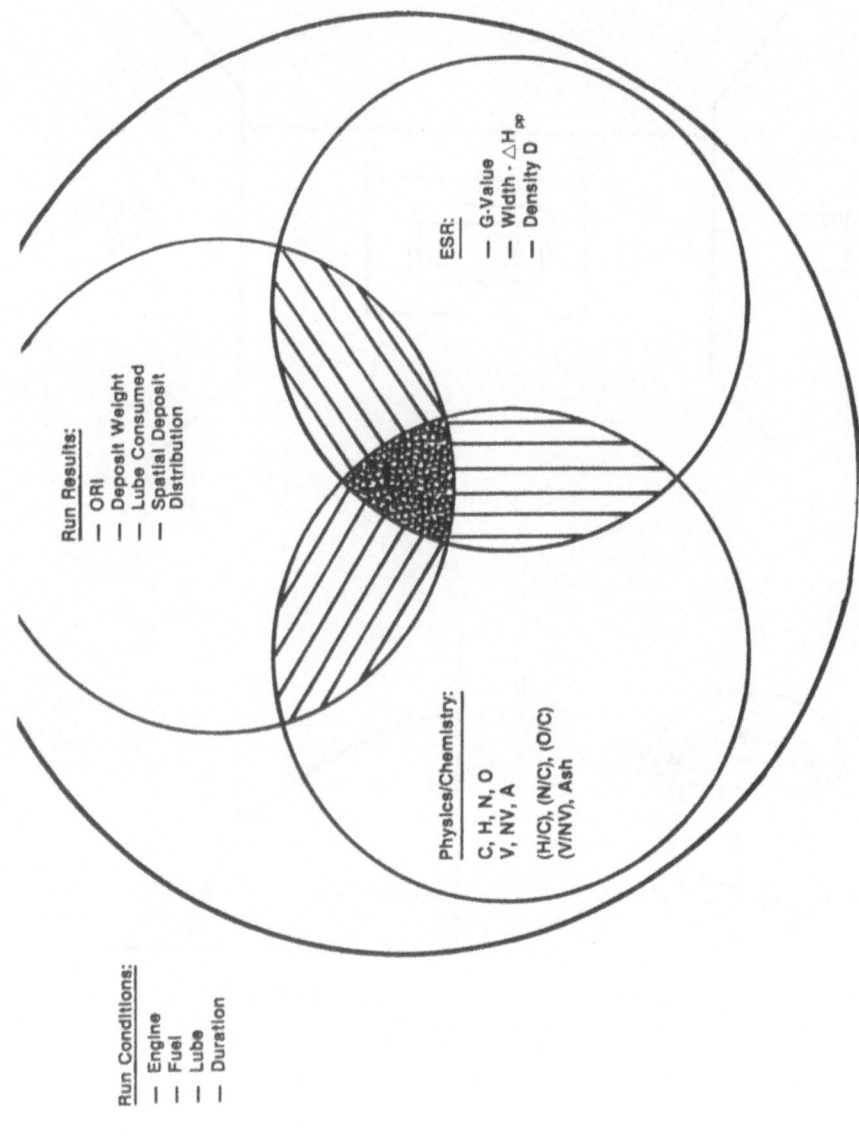

Run Conditions:

— Engine
— Fuel
— Lube
— Duration

Run Results:

— ORI
— Deposit Weight
— Lube Consumed
— Spatial Deposit
  Distribution

ESR:

— G-Value
— Width - $\triangle H_{pp}$
— Density D

Physics/Chemistry:

C, H, N, O
V, NV, A

(H/C), (N/C), (O/C)
(V/NV), Ash

FIGURE 11: Schematic Diagram Identifying the ESR, Chemical and Engine Run
Variables Associated With Deposit Formation

A. Chemical/Physical Correlations

- There is a universal correlation of volatile to non-volatile organic fractions (V/N) to hydrogen to carbon ratio (H/C) for six and single cylinder engines and for both lubricants employed (Premium B and Premium A).

- There are significant differences in six cylinder and single cylinder deposits. The six cylinder deposits are more spatially homogeneous. They have a tighter distribution of H/C and V/N values and these values tend to be lower. The ash component of the deposit is high as are O/C ratios. All these data suggest a more complete oxidation of organic in the fuel.

- Short duration CLR engine test deposits have low V/N, H/C, O/C, and Ash. These data suggest that the initial deposits are more refractory organics which may serve as "anchors" for the deposition of later material.

B. ESR Correlations

- The range of g values for the organic (~2.0030-2.0040) suggests oxygen inclusion in the molecular species containing the free electron.

- The ESR linewidth is nearly independent of carbon radical density at all but the highest concentrations. Local proton hyperfine fields are responsible for the observed width.

- CLR runs with Uniflo lubricant show a narrow distribution of g-values (2.0037±0.0002) and a narrow linewidth suggesting that a molecular form in the lube may be a major carrier of unpaired spins.

- Carbon radical densities are 5x-10x lower for six cylinder engine runs than they are for CLR runs, suggesting a more complete reaction of the molecular precursors in the six cylinder engine case.

C. Comparison of ESR and Physical/Chemical Correlations

- Weak correlation of radical densities with H/C, V/N and O/C are seen in these engine runs, suggesting that there is little relationship between radical properties and the bulk of the deposit.

- g-values and linewidths do not correlate with H/C, N/C, O/C, reinforcing the fact that bulk chemical values do not relate to radical chemistry. The only exception to this rule is a clustering of narrow linewidths of CLR deposits at higher O/C, which may reflect the formation of peroxyradical species.

D. Spatial and Fuel Effects

- In the CLR engine, there is a notable difference in end gas region deposits, which have low H/C, low V/N, and high radical densities. The condensation of heavier aromatic molecules from the unleaded fuel is indicated.

- There is little difference in deposit properties outside of the end gas region.

- Runs with isooctane fuel have high V/N and H/C values, low radical densities, and relatively high ESR linewidth, all of which suggest a deposit formed primarily from the lubricant.

- Runs containing isooctane and 4 wt.% 2-methylnaphthalene give a typical broad resonance line. There is no evidence for a methylnaphthalene radical in the deposit.

IV. SUMMARY AND CONCLUSION

The present data provide the following picture of deposit formation. The initial material deposited on the piston top and cylinder head of the engine is very refractory and highly metamorphosized. This picture is consistent with scanning electron microscope observations of refractory looking material collecting near pointed projections of the engine cylinder surface. The latter deposits appear to form by condensation at less stringent conditions. There is a good deal of metamorphosis of the initial deposits as compared to the fuel and lubes, the presence of 50% aromatic carbon atoms in the isooctane deposits, and the absence of 2Me-naphthalene radicals in the deposits formed from the isooctane + 4 wt.% 2Me-naphthalene fuel. The prominent role of oxygen is reflected in the high O/C values and oxygen signature in the carbon radicals. Spatial variations in the CLR runs testify to the inhomogeneity of the combustion processing the cylinder. These effects are much less pronounced in the six cylinder configuration.

REFERENCES

1. Jack D. Benson, "Some Factors which affect Octane Requirement Increase", SAE Paper #750933, Oct., 1975.

2. For a review, see C. P. Poole, "Electron Spin Resonance: A Comprehensive Treatise on Experimental Techniques", (Interscience NY, 1967).

3. P. W. Anderson, Phys. Rev. $\underline{82}$, 342 (1951).

4. J. P. Helwig and K. A. Council, eds., SAS Users Guide 1979 Editors, (SAS Institute, 1979).

APPENDIX A

I. INTRODUCTION

A complex, multiparameter problem of the present type requires a systematic approach for its analysis. As Figure 11 shows, there are at least 16 variables that are being examined in the engine tests that have been performed. For the purposes of the present discussion, these can be divided into four classes:

Run Conditions - of which four are noted, including engine type, fuel lubricant, and run duration. This does not totally describe the runs since many other parameters must also be specified, such as the programmed engine cycle (idle, acceleration, high speed operation), cooling jacket temperature, and carburetion technique.

Run Results - such as Octane Requirement Increase, weight and spatial distribution of the deposit, and the amount of lubricant consumed.

Chemical and Physical Parameters - deduced from elemental analyses and thermogravimetric analyses (TGA) of the deposit samples. The elemental analyses of H, C, and N content of these samples were performed by the Analytical and Information Division (AID), while TGA analyses were performed in both AID and the Products Research Division (PRD) of Exxon Research and Engineering Company. The organic volatiles were determined by weight loss upon heating the samples up to ~500°C in the presence of an N_2 atmosphere. The ash was determined by burning the residue of the volatilization test in an O_2/N_2 mixture. The organically bound oxygen content of the deposit was estimated by difference, subtracting %C, %H, %N, and % Ash from the total. In the following correlations, five Chemical/Physical variables have been used: the ratio of volatile to nonvolatile organics (V/N), the hydrogen to carbon ratio (H/C), the oxygen and nitrogen to carbon ratios (O/C and N/C), and the percent ash in the deposit (Ash). Ratios have been chosen in each case because they have potentially microscopic significance. For example, H/C is an approximate measure of the deposits' "aromaticity".

ESR Parameters - have been chosen using the motivation outlined in Section II of the body of the report. The g-value (G), equivalent to a chemical shift, indicates the chemical form of the molecule containing the unpaired electron. The linewidth (ΔH_{pp} or DH) provides information about the local environment of the carbon radical i.e. the number of protons and other radicals in its vicinity. The radical density (D) is a measure of the stability of the radicals and reflects the chemistry of molecules and how they were deposited.

In this Appendix, correlations within a class will be explored first, followed by cross correlations between classes. The major component of the analysis in the next section will be to examine variations of the eight parameters: V/N, Ash, H/C, O/C, N/C, G, DH and D for varying engine type (six cylinder and single cylinder CLR), and lubricant (a Premium A lube compared with Premium B). The fuel chosen is from batches of standard unleaded blend. All run conditions, except for run duration, are held constant. Extensions to other fuels and detailed analyses of spatial variations within the engine chamber are discussed in Section IID. The nature of these correlations can be presented in matrix form, as shown in Figure A1.

The correlation and statistical analysis of these parameters was performed using a TSO timesharing computer system. Data accumulated from 19 six-cylinder engine runs and 13 CLR runs was stored in the computer memory. Variables were extracted by pairs from this matrix and plotted. Existence of linear correlations between parameters was examined by using the PROC-GLM program of the SAS statistical software package available on the TSO system.

In the correlation analyses which follow, several issues will be examined.

1. Is there any obvious qualitative difference in behavior between data from different types of experiments?

2. Do subsets of the data tend to cluster?

3. Is there any obvious functional dependence to the correlation, linear or otherwise?

The first two questions can be answered by inspection of the distribution of data points on the correlation plot. The latter point is best addressed statistically. The PROC-GLM regression procedure mentioned above yields slope and intercept values for the optimized linear fit as well as useful statistical diagnostics, including the correlation parameter R^2. We will use R^2 as a figure of merit in subsequent discussions. To provide an approximate measure of the quality of fits implied by values of R^2, several examples are shown in Figure A2. An R^2 value of 0.70 reflects a strong linear correlation. For $R^2 = 0.23$, a good linear correlation is seen but there is considerably greater scatter in the points. For R^2 values near 0.1, correlations are often weak. The regression line which minimizes the

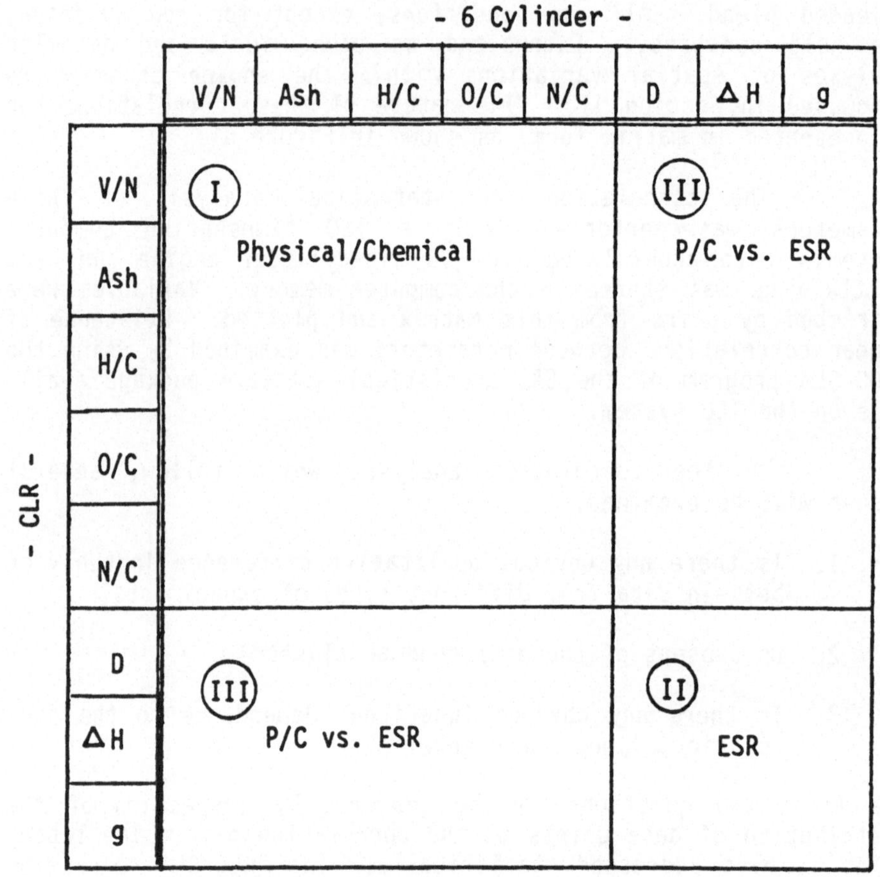
- 6 Cylinder -
V/N
Ash
H/C
O/C
N/C
D
ΔH
g
- CLR -
V/N
Ash
H/C
O/C
N/C
D
ΔH
g
I
Physical/Chemical
III
P/C vs. ESR
III
P/C vs. ESR
II
ESR

FIGURE A1: Correlation Results for Both Single Cylinder (CLR) and 6-Cylinder Engines are Conveniently Presented in Matrix Form.

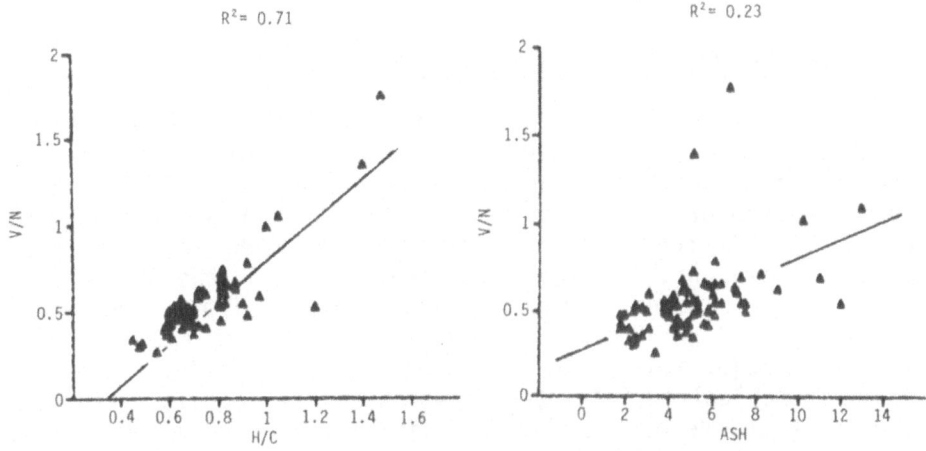

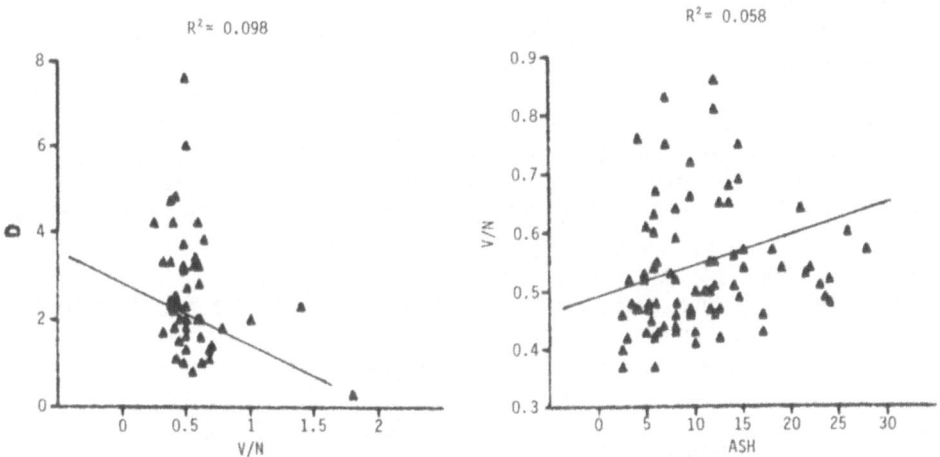

FIGURE A2:   Examples of Correlation Quality

residuals may not necessarily be the most obvious fit to the data based on visual inspection. Lower values, $R^2 = 0.05$, imply a large amount of scatter to the data and no obvious correlation. In the following discussions, the strong, good, and weak nomenclature will be applied.

II. CORRELATION ANALYSIS

A. Chemical/Physical Correlations

The results of the TGA and elemental analyses are of particular importance because of the way in which they bear on the ESR interpretation. As mentioned above, the H/C ratio is a rough measure of the aromaticity of the deposit and can suggest whether the source of the deposited carbon comes from the fuel or the lubricant. Oxygen represents 25% of the deposit by weight and was added to the organics of the fuel and lube during the combustion process. Carbon-oxygen bonds are also believed to play a major role in establishing the intermolecular bonding, forming the "polymeric" material which is the deposit. Nitrogen can enter the deposit either by additives or during combustion. Since the nitrogen levels are much lower (~1%), it is more difficult to trace its history in detail, but we retain N/C as a potential diagnostic.

The results of the TGA studies reflect both physics and chemistry. Some of the species observed in the mass spectra of deposits come from molecules which became entrained during the deposition process. Others, particularly alkyl fragments, are the product of the pyrolysis of the deposit, which is well underway at 500°C. An analysis of the results can clarify the picture of what the deposit is on a microscopic scale. Ash levels are also important. The concentrations in the deposits are usually significantly higher than in the lubricants from which they come, suggesting an enrichment process which occurs as the organics are volatilized or burned away. It is also possible that the amount of ash occurring in a particular position in the cylinder is proportional to the amount of lubricant which has found its way to that spot, a fact which might be reflected in higher H/C values.

1. V/N vs. H/C

If the level of volatiles was determined by pyrolytic dealkylation, then the V/N ratio should be proportional to H/C. This conjecture is supported by the strong linear correlation seen in a composite plot of 6 cylinder and CLR data for both Premium A and Premium B lubricants. R-square values of ~0.7 in all cases support a strong linear correlation. The CLR data shows a strong tendency of Premium B deposits to have low H/C and V/N values, perhaps because there is less deposition of Premium B lube into the engine chamber. Conversely, extremely high H/C values (~1.7) are seen for certain Premium A samples and may reflect lubricant leakage into the engine cylinder. By contrast, the range of V/N and H/C is much smaller for 6 cylinder deposits and Premium A and Premium B values are nearly uniformly dispersed. Deposits from the short time runs have lower H/C and V/N values, suggesting that the initial material deposited is more refractory.

2. V/N vs. Ash

An initial inspection of the composite data set shows no obvious V/N vs. Ash correlation. However, separate analyses for CLR and 6 cylinder runs suggests a considerably stronger correlation within a given data set. The CLR data has quite linear variation (R^2 = 0.23) and relatively low ash values, mostly <8-10%. The Premium B values lie at the extreme low H/C, low V/N end of this plot. Ash levels are much higher in the 6-cylinder deposits, exceeding 20% in some cases, but V/N values lie in a fairly narrow range. Premium B and Premium A lube values are uniformly dispersed for the 6 cylinder deposits. There is also a significant difference between the dispersion of values for head and piston top samples from the 6 cylinder engine. The head samples have typically low V/N values and exhibit some linearity. There is a much greater scatter in the 6 cylinder piston top data. Short time CLR runs typically have low ash levels.

3. V/N vs. O/C

Since significant levels of oxygen occur in the deposit and since they have been implicated in the intermolecular bonding process in the deposit matrix, one might expect a correlation between oxygen level and relative volatility of the deposit. Examining the V/N vs. O/C data for CLR deposits we find that the samples divide into two separate distributions, depending upon the lube used. The deposits formed with runs using Premium B have low V/N values and high O/C values. The remaining deposits generated in runs with Premium A lubes exhibit a weakly linear correlation of V/N with O/C (R^2 = 0.15), in contradiction to the bonding hypothesis suggested above. The six cylinder data exhibits a good deal of scatter and the clustering of Premium B data points is not obvious. The mean value for O/C is somewhat higher for the 6 cylinder engine runs: ~0.29 vs. ~0.24 for the CLR runs. A weak linear correlation is also observed for the time dependent engine runs employing unleaded fuel and Premium A lube. Short time deposits (5-10 hr) show low V/N and O/C values - again suggesting a different chemistry for these initial deposits.

4. V/N vs. N/C

There is no evidence for a correlation between the fraction of volatile organics and the nitrogen content. There is evidence in the CLR data for a high nitrogen level in Premium B deposits, which may reflect the chemistry of the Premium B additives.

5. H/C vs. O/C

If oxidation were selectively influencing one molecular class which occurred in the deposit, a correlation between H/C and O/C might be anticipated. Analyses of the CLR data shows two specific classes of data, depending on lube used - as in the V/N vs. O/C correlation. CLR runs with the Premium A lube show an increase of H/C with increasing O/C. The 6 cylinder deposits show the opposite correlation: H/C decreases as the O/C increases, which might suggest an association of oxygen with the aromatic constituents of the fuel. There is no obvious lube effect on this correlation since the Premium B samples are scattered uniformly throughout the deposits generated with a primol lubricant.

6. Ash vs. O/C

The significance of the deposit ash is still not completely understood. As suggested in Sec. A.II.2, it might reflect the amount of lube which was decomposed in a given region by the cylinder. The ash also might be playing a role in the deposit chemistry - serving perhaps as a potential catalyst to oxidation of the deposit. There is a lube dependence in the O/C vs. ash correlation. Premium B samples have high O/C and low ash levels and the Premium A samples have lower O/C and higher ash levels. The time dependent CLR runs show a progression, with increasing O/C and ash levels with increasing run time. The only exception to this observation are the high ash levels for the 5 hr runs which may reflect contamination problems associated with these very small deposit samples.

- Summary of Chemical/Physical Correlations

  - There is a universal linear correlation of V/N with H/C which holds for both types of engines and for both lubricants employed. The volatile fraction appears to reflect the amount of aliphatic material in the deposits which is volatilized by a dealkyla-

182

tion process at ~500°C, the temperatures employed in the TGA studies. While the relative magnitudes of V/N and H/C for a particular deposit class may cluster (e.g. CLR runs with Premium B have low V/N and H/C) they still lie on the curve.

• There are significant differences in deposit properties for the CLR and 6 cylinder engines. While there are differences in CLR deposit properties for different lubes employed, these distinctions are not nearly as obvious for the 6 cylinder data. The 6 cylinder deposits have a much tighter distribution of H/C and V/N values, considerably higher ash values in the deposit and higher O/C atomic ratios. These data suggest more homogeneous conditions of combustion within the 6 cylinder engine and a more complete oxidation of the organic. The average H/C values, however, are comparable for both engines suggesting similar constituents.

• Short-time runs in the CLR engine reveal different deposit properties; particularly low values of V/N, H/C, O/C and ash. These data suggest that the first deposits are refractory organics which may serve as "anchors" for subsequent deposition of more aliphatic material from the lubricants.

• Correlation of other properties with heteroatom levels, either O/C or N/C are not obvious and weak at best.

B. ESR Correlations

As mentioned above, three ESR parameters have been continuously monitored during these studies: the g-value or chemical shift of the carbon radical, the linewidth which reflects the interaction of the carbon radical with proton spins or other radicals in its vicinity, and the radical density, the number of such radicals per gram of carbon in the sample. There is an underlying chemical relationship between these parameters since the chemistry of a molecule will determine the stability of an additional electron in its structure, the width of the ESR line coming from interactions with its protons and the chemical shift. In this brief section we explore the interrelation of ESR parameters, while the direct connection with chemistry will be discussed in A.II.C.

    7. D vs. DH

The correlation of linewidth with radical density establishes the source of line broadening. If the width of the ESR signal, DH, results from radical-radical interactions, it should increase linearly with radical density in this dilute limit. This is not the case. Except for the variation seen in the magnetically contaminated 6-cylinder head samples, DH is nearly independent of D for D values which vary over more than one order of magnitude. The observed linewidths, ~5.5 G, are consistent with proton hyperfine broadening of the line. This lack of radical broadening is also consistent with approximate estimates of the radical-radical linewidth. From Anderson's statistical theory of line broadening, $\Delta H_{pp} \simeq 4.39 \, \gamma \, hn$, where γ is the free electron gyromagnetic factor, h is Planck's constant and n is the spin density in spins/cm^3. Assuming a density of ~1.3 gm C/cm^3, typical of anthracene, pyrene, and coronene, and the radical density of 5×10^{18} spins/gm C, one finds $\Delta H_{pp} \simeq$ 0.54 G, roughly one order of magnitude smaller than observed. Radical broadening should not contribute significantly to the linewidth until the radical densities approach $2\text{-}3 \times 10^{19}$ spins/gm C, well above the range of densities observed here.

The CLR runs show a significant difference in line-widths for Premium B and Premium A based deposits, even for comparable radical densities. The Premium B lube deposits have narrower lines, ΔH_{pp} as low as 4.8 G, while the lowest Premium A values are ~5.2 G. This indicates that the Premium B deposits may be proton deficient compared to the Premium A ones. While the Premium A data points are scattered, the Premium B line-widths show an approximately linear increase in DH with increasing D, the magnitude of which is consistent with the calculations of the previous paragraph. The 6 cylinder radical densities are significantly lower, on average, than the CLR ones, another indication that the combustion-deposit formation process is not the same in these two engines. Time dependence studies do not reveal any obvious trend in linewidth or radical density with run time except for the anomalously high DH values seen at 5 hrs and attributed to magnetic contamination.

8. DH vs. G

The 6 cylinder head deposits show the classic linear correlation of G value and linewidth associated with magnetic contamination. The rest of the data clusters tightly between G values of 2.0030 and 2.0040, suggesting chemical shifts associated with nearby heteroatoms - particularly nitrogen and sulfur. The CLR data again shows the unique character of the Premium B deposits. The G values cluster near 2.00375 and the linewidths are low suggesting that these radicals are associated with a particular class of molecules. Interestingly, these systematics do not appear in the 6 cylinder data, where Premium B deposits have larger DH values than Premium A deposits and where Premium B G values scatter from 2.0030-2.0040 with a mean value of ~2.0035. There is a weak trend to lower G values for the short time CLR deposits.

- Summary of ESR Correlations

  ● Magnetically contaminated samples show linear correlations between DH and D, and DH and G.

  ● In uncontaminated samples the width of the ESR line does not depend on D and presumably reflects the local proton distribution in radical bearing molecules.

  ● The range of G values suggests radical species in close proximity to oxygen or nitrogen heteroatoms.

  ● CLR runs with Premium B lubricant show a narrow distribution of G values, low DH, and radical densities comparable to Premium A runs. A specific class of molecular radical appears to dominate in this case. Analogous runs with 6 cylinder engine do not produce this class of radicals.

  ● There is no correlation of run time and ESR properties in the short duration CLR tests.

C. ESR and Chemical/Physical Cross Correlations

We wish to see if some of these distinctions in ESR properties will correlate with chemical and physical deposit properties. A connection here would clarify the types of molecular species which condense to form the deposit. As an initial step, the variation of carbon radical density on H/C, V/N, O/C, and N/C will be examined. These correlations are difficult to predict a priori because, at spin densities of ~5 x 10^{18} spins/gm C, there is only 1 unpaired electron for every 10,000 carbon atoms. Furthermore, the survival of a radical depends upon several factors including the intrinsic stability of the added electron on the host molecule, the chemical pathway by which it is formed, and the access of other reactive species to the radical site in the deposit. Certain general principles can guide this correlation effort. Large polynuclear aromatic species tend to have stable radicals because the additional electron can delocalize quite extensively over the molecule.

This would suggest an increasing radical density with decreasing H/C, all other factors being equal. Since large numbers of heteroatoms exist in the deposit and the g-values show a positive shift which appears associated with heteroatoms, it would be interesting to see if a linear correlation exists between g-values and macroscopic heteroatom content. Since linewidths are attributed to proton densities, it would be useful to see if DH varies with H/C. Variations of D, G, and DH will be examined in turn.

9. D vs. H/C

There is a weak correlation at best between D and H/C. Premium B and Premium A values appear fairly evenly distributed and no lube related differences are indicated. There are also no obvious time dependent correlations. Thus, macroscopic H/C ratios do not adequately reflect radical properties.

10. D vs. V/N

There is a loose linear correlation of D with V/N, the R^2 value being ~0.1. The 6 cylinder data lie at the low D high V/N range of the plot. The decrease of D with increasing V/N is plausible since the radicals would be expected to occur mostly in the nonvolatile large aromatics of the deposit.

11. D vs. O/C

The CLR data do exhibit a correlation between D and O/C while no similar relationship is seen in the 6 cylinder deposits. Deposits with Premium B lubes are clustered at lower D and high H/C. The decreasing radical density with increasing O/C coupled with the low radical density observed for the 6 cylinder deposits suggests that more effective oxidation during the combustion process will lower the total radical density. Shorter time CLR runs have lower O/C and higher D, but there is a great deal of scatter between data sets from different duration engine runs.

12. D vs. N/C

High R^2 values (=0.32) for the D vs. N/C correlations of the CLR runs suggest that there may be a relationship between N/C and the stability of the radicals. The density drops with increasing N/C, suggesting that nitrogen compounds may be serving as radical scavengers. The Premium B lube deposits appear fairly uniformly scattered throughout the data set suggesting no strong lube effect. A looser, but still linear, relationship is also seen for the time dependent CLR runs. By contrast almost no correlation is seen for the 6 cylinder data, even though similar N/C ratios are observed.

13. G vs. H/C

While there is no obvious G vs. H/C correlation for the 6 cylinder data points, the CLR correlation appears somewhat linear. As mentioned above, the Premium B values cluster quite tightly near 2.00375. The significance of this correlation is chemically unclear, since G increases as H/C decreases. If the ESR signal were to come from aromatic radicals with peripherally included heteroatoms, the opposite variation would be expected. Larger polynuclear aromatics allow the unpaired electron of the free radical to be more delocalized, spending less time at any given position on the molecule. The effect of a single hetero-atom at one position would be smaller for a larger molecule and the shift from the free electron g value (g=2.0023) should be smaller. The possible discrepancy seen here is not presently understood.

14. G vs. O/C

Simple considerations would suggest that the g values should increase with increasing O/C if the macroscopic elemental analysis reflects the microscopic inclusion of oxygen at the molecular sites. There are no obvious correlations for either CLR or 6 cylinder engine runs, suggesting that application of macroscopic parameters must be viewed with some caution.

15. G vs. N/C

As in the G vs. O/C calculations, the data suggest no obvious correlation between the chemical shift of the carbon radicals and the macroscopic chemical composition of the deposit.

16. DH vs. H/C

Simple considerations would suggest that the linewidth of the carbon radical would increase with increasing proton density, i.e. that DH should be proportional to H/C. Excluding the magnetically contaminated 6 cylinder head data, we find little evidence for the effect.

17. DH vs. O/C

While there is no correlation of DH with O/C for the 6 cylinder data, a reasonably linear variation is seen for the CLR runs with R^2 = 0.21. These data suggest a decrease in linewidth with increasing O/C, which is particularly marked for deposits from Premium B lube runs.   One explanation is that peroxy-radicals occur in greater abundance at high O/C levels and that their linewidths are less because of the concentration of free radical density at the oxygen sites - somewhat removed from the protons.

18.   DH vs. N/C

There is no dependence of DH on N/C values.

- Summary of ESR vs. Chemical/Physical Correlations

  ● Radical densities appear to reflect the conditions of deposition more than the chemistry of the con-stituents.   There are weak correlations with H/C, V/N for the CLR.   Decreasing D with increasing O/C suggests that sustained combustion is a natural radical killer.

  ● The 6 cylinder results show very little chemical correlation.   The radical densities are low and nearly independent of lube and chemical parameters of the deposit, suggesting significant metamor-phosis of the components prior to the deposition process.

  ● g-values do not correlate with H/C, N/C, O/C, sug-gesting that bulk chemical values do not reflect the form of the radical bearing molecular species.

  ● Linewidths do not depend on H/C or N/C but do depend on O/C for the CLR data.   This may suggest the formation of peroxy radical species.

D.   Overview of Correlations with Unleaded Fuel

The combined results of these correlations can be conveniently presented in the matrix form shown in Figure A3. Following the conventions suggested in Section I of this Appendix, strong correlations are indicated in the matrix by circles, good correlations by large squares, and weak correla-tions by smaller squares.   The strong H/C, V/N correlation appears for both engines used and the magnetic effects on the ESR parameters for the 6 cylinder engine are very pronounced. There is a weak correlation between D and V/N for both engine types.   No other evidence for correlations are observed for six cylinder engines.

188

- CLR -

	V/N	Ash	H/C	O/C	N/C	D	ΔH	g
V/N		.06	.79	.05	.03	.08		
Ash	(.47) .23			.01				
H/C	.71			.01		.10	.01	.03
O/C	(.15) .002	.09	.01			.02	.01	.01
N/C	.003					.01	.001	.002
D	(.10) .10		.01	.17	.32		.22	
ΔH			.01	.21	.01	.10		.83
g			.03	.002	.09		.20	

FIGURE A3: Correlation Coefficients for Linear Regression Analysis.

By contrast, several correlations appear for CLR deposits, notably variations of ash content with V/N and also O/C. Heteroatom effects also influence the ESR parameters. One explanation of these phenomena, which arose in the earlier discussions, is that molecules are less completely combusted in the CLR chamber and retain more of their original character. The V/N, Ash correlation suggests that the lubricant is the source of much of these CLR deposits. By virtue of its position on the cylinder wall, it would be less susceptible to the intense combustion process than molecular species in the fuel.

One last comment on the sensitivity of these correlations is in order. Throughout the analysis, correlation plots have been characterized by several "figures of merit". The correlation coefficient, shown in Figure A3 has been examined because it provides a measure of the relationship between variables. Another measure of "scatter" in the data is the standard error of estimate of the slope and intercept parameters in the data. These are furnished directly as a part of the output of the PROC-GLM Program of the SAS statistical package employed here. One concern has been with the effect of several widely disparate points in an ensemble on R^2 and the error estimates. These effects can be quantified as in Table A1. The D vs. V/N correlation was not strong initially ($R^2=0.098$) and does not change greatly after the data are extracted. However, the slope of the best linear fit changes by a factor of two. This strongly suggests that values for the slope obtained from regressions are not well defined for R^2 = ca. ~0.1. By contrast, removal of the Premium B data yields a respectable correlation of V/N vs. O/C for the CLR data. The large change in slope is expected. Large error ratios, defined as the ratio of the standard error of estimate of slope (or intercept) to its actual value, reflect the scatter of data points. In the V/N vs. Ash case, R^2 is already high (0.23) and increases dramatically with the removal of three divergent points. However, the slope of the linear fit to the data does not vary significantly - reflecting the strong correlation which already exists.

E. Spatial Effects

The position of the deposit within the engine cylinder has not entered into the previous discussions, although, as mentioned in Section II, there are strong noticeable differences in the appearance of the deposits from different sections of the cylinder. These differences are considerably more pronounced for the CLR engine and, as a consequence, efforts were made during the CLR sample collection process to isolate materials of a given morphology. The results of this effort for a single engine run (CLR engine, unleaded fuel, Premium A lubricant) are

Table A1

Effect of Selective Data Removal

Correlation	R^2		Slope		$\frac{\Delta \text{Slope}}{\text{Slope}}$		$\frac{\Delta \text{Intercept}}{\text{Intercept}}$	
	Before	After	Before	After	Before	After	Before	After
D vs. V/N CLR	0.098	0.077	-1.7	-3.1		0.52		0.19
V/N vs. O/C CLR	0.02	0.15	0.16	1.73	3.14	0.40	0.255	1.40
V/N vs. Ash CLR	0.23	0.47	0.047	0.044	0.43	0.13	0.046	0.095

schematically indicated in Figure A4. Deposits from the end gas region of the cylinder, which are of a thick, flaky type, have low H/C and V/N values and high carbon radical densities. The more varnish-like deposits from other regions of the cylinder have higher H/C and V/N values and lower carbon radical densities.

Any such comments are necessarily of a "statistical" nature, since there is overlap in these parameters for different deposit types. Furthermore, the region of the cylinder covered by a particular type of deposit will vary somewhat from one engine run to the next. Finally, the deposit type and amount varies in the cylinder during an engine run. Inspection of the deposits shows that the chunky deposit material in the end gas region "flakes off" during the run and more forms in its place. To illustrate some of the major correlations, we will focus on a subset of deposits, collected from the first ten runs on the CLR, where the engine heads and piston tops were scrapped on the basis of deposit appearance. End gas region deposits tend to have high radical densities and low g-values. Attempts to distinguish between other spatial regions in the engine cylinder, near the spark plug or opposite the intake or exhaust ports, did not produce unique values for D or G. Note also that there are some end gas deposit points with low radical densities. This segregation is also seen in chemical parameters as the V/N vs. H/C correlation. As in one CLR engine run, end gas points tend to cluster at low H/C and V/N values. They also tend to have lower O/C and Ash values.

F. Fuel Effects

For comparison, several single cylinder and 6 cylinder engine runs were conducted with special test fuels: isooctane and an isooctane + 4 wt.% 2Methylnaphthalene blend. Solid state ^{13}C NMR analyses of the resulting deposits showed great differences in the character of the carbon species: for unleaded fuel - 90% aromatic carbon/10% aliphatic carbon; for isooctane + 2Methylnaphthalene 71% aromatic/29% aliphatic; for isooctane - 51% aromatic/49% aliphatic. The isooctane deposits are much smaller and tend to have high V/N and H/C values compared to unleaded fuels. They also tend to have low carbon radical densities (< 10^{18} spin/gmC) and relatively high ESR linewidths (~6 G).

192

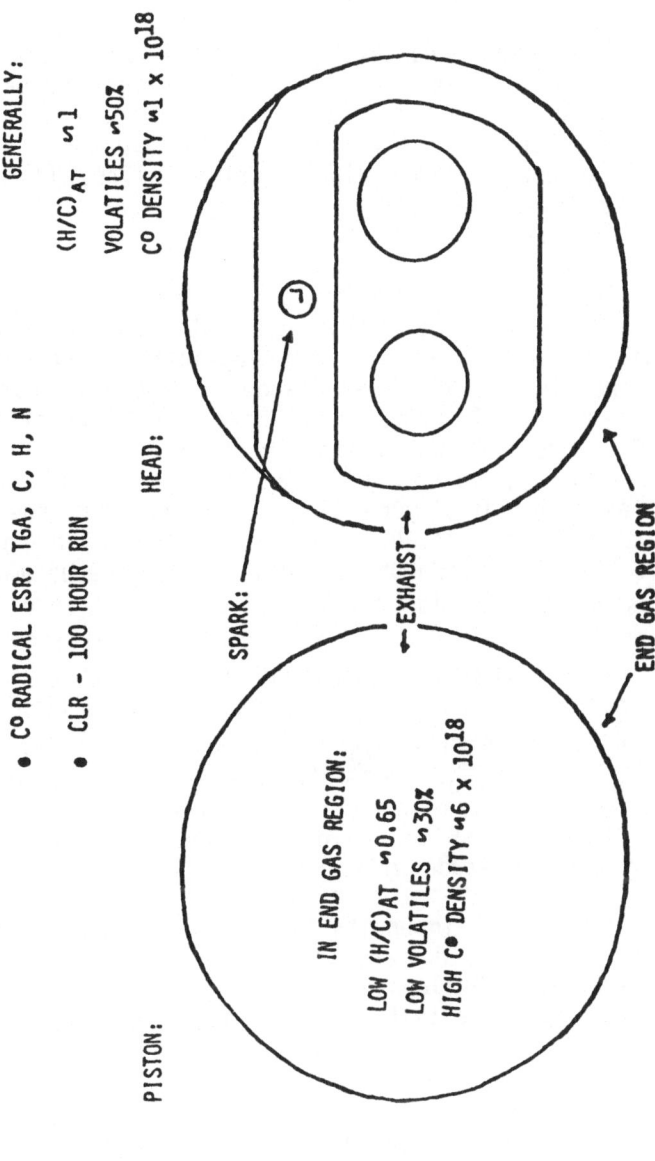

SIGNIFICANT SPATIAL VARIATION OF DEPOSIT TYPE IS OBSERVED

- C^o RADICAL ESR, TGA, C, H, N
- CLR - 100 HOUR RUN

GENERALLY:

$(H/C)_{AT}$ ~1

VOLATILES ~50%

C^o DENSITY ~1 x 10^{18}

HEAD:

SPARK:

← EXHAUST →

END GAS REGION

PISTON:

IN END GAS REGION:

LOW $(H/C)_{AT}$ ~0.65

LOW VOLATILES ~30%

HIGH $C^{\bullet}$ DENSITY ~6 x 10^{18}

FIGURE A4: Summary of Spatial Variation of Engine Deposits

193

APPENDIX B

MAGNETIC EFFECTS ON CARBON RADICAL ESR PROPERTIES

The phenomena associated with contamination of the internal combustion engine deposits with magnetic particles have been outlined in section II.B.2 of the paper. The present appendix is intended to provide a justification of the phenomenological arguments concerning g value shifts, linewidth variations, and intensity loss by setting the discussion on a more mathematical basis.

Variation of the field for resonance by polarization of a sample is very common in both NMR and ESR. It serves as the basis for the chemical shift applications commonly employed in high resolution NMR spectroscopy. Magnetic particles in a deposit sample will cause it to have a net magnetization which will affect the internal fields in the sample. We can visualize this directly by choosing an extremely simple model for the sample, as shown in Figure B1. We assume that the deposit can be visualized as a slab oriented perpendicular to an applied magnetic field of strength H_O, with a homogeneous magnetization, M. Outside of the slab the magnetic field H_O and the magnetic induction B_{OUT} are equal. The boundary conditions of magnetostatics require that the component of B normal to the slab must be continuous, so $B_{in} = B_{out}$. But, by definition of the magnetic induction itself, $H_{in} = B_{out} - 4\pi M$. Thus, since $B_{in} = B_{out} = H_O \rightarrow H_{in} = H_O - 4\pi M$, reflecting a reduction in magnetic field strength inside the slab due to the magnetization of the sample. The sign of the effect does not depend on sample geometry but the magnitude will.

Width and intensity loss effects can be visualized by examining local field variations in the vicinity of an individual particle. While the details will depend on the particle shape, the basic effect can be seen by choosing a simple geometry - for example a magnetized sphere of radius a. As shown in Figure B2, the sphere will be assured to be magnetized along a direction we will call z. The field at any point outside of this sphere, which we designate by the coordinates $(r,0)$, is defined as

$$\vec{H} = \vec{B} = -\nabla\Phi_m, \quad \Phi_m = \frac{4\pi}{3} M_o a^3 \frac{\cos\Theta}{r^2} \qquad (B-1)$$

194

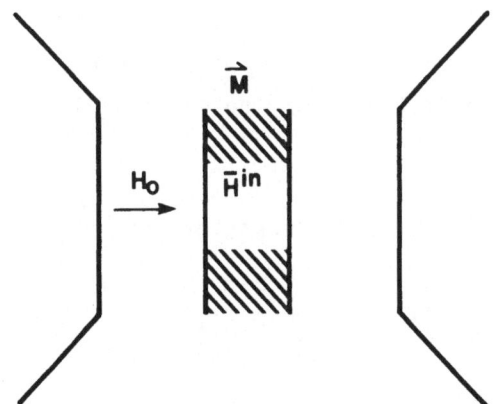

FIGURE B1: Geometry for Calculations of Internal Field Values in a Magnetized Slab.

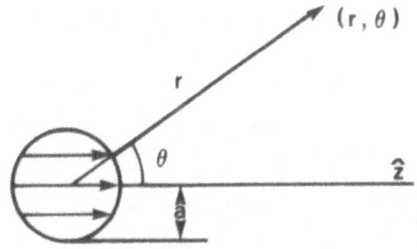

FIGURE B2: Definition of Variables for the Calculation of Local Fields From a Magnetized Sphere.

so that the field associated with the sphere becomes

$$\vec{H}sp = \frac{4\pi}{3} M_0 \left(\frac{a}{r}\right)^3 (2 \cos \Theta \hat{r} + \sin \Theta \hat{\Theta}) \qquad (B-2)$$

If we superimpose an external magnetic filed H_0 along the z direction, the total field at any position will be

$$\vec{H}_{tot} = \vec{H}_0 + \vec{H}_{sp}.$$ The z component of this field becomes:

$$H^z_{tot} = H_0 + \frac{4\pi}{3} M_0 \left(\frac{a}{r}\right) (3 \cos^2 \Theta - 1) \qquad (B-3)$$

the radicals will thus experience a distribution of local fields which will depend on their position (involving both r and Θ) with respect to the particle.

For comparison with the experimental ESR spectra, we would like to obtain a profile of local field values throughout the sample from the dispersed magnetic particles. We begin by examining the profile of local fields at a fixed distance, r, from a single spherical particle. Then

$$H^z_{Loc} = H_0 + \frac{H_L}{2} (3 \cos^2 \Theta - 1) \qquad (B-4)$$

where we define $H_L = \frac{8\pi}{3} M_0 \left(\frac{a}{r}\right)^3$. We wish to determine

$P(H_{Loc})$. The probability that a given point a distance r from the sphere makes an angle Θ with respect to H_0 is $\alpha \sin\Theta \, d\Theta$. We can invert equation B-4 to express this probability in terms of H_{Loc}:

$$P(H_{Loc}) \, d \, H_{Loc} = \frac{1}{3 \, H_L} \cdot \frac{d \, H_{Loc}}{1 + 2 \frac{H_{Loc}}{H_L}} \qquad (B-5)$$

This field distribution is plotted explicitly in Figure B3. The local field can vary from $H_0 + H_L$ to $H_0 - (H_L/2)$. As expected

196

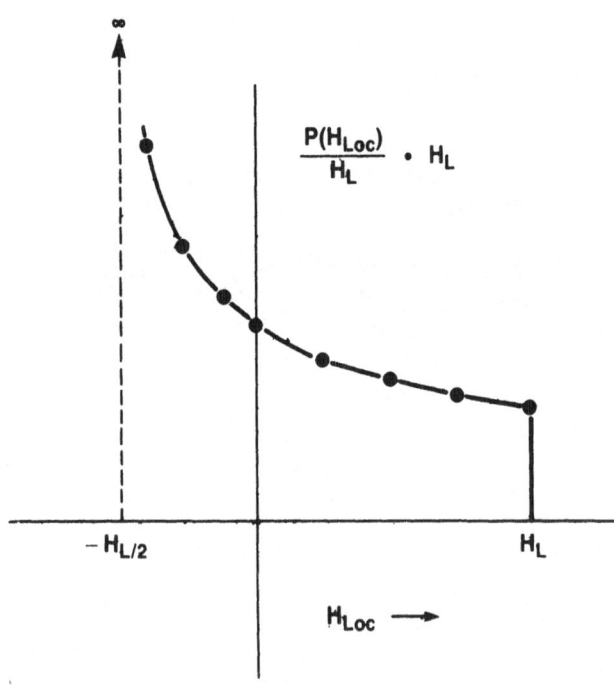

FIGURE B3: Distribution of Dipolar Field Values at a Fixed Distance From a Magnetized Sphere.

for a dipolar interaction, the average value of the local field will be H_0.

Distance variations can be determined from changes in the magnitude of H_L. Since H_L is proportional to r^{-3}, the fields will fall off quite rapidly at positions removed from the particle. However, the large values of the magnetization ($M_0 = 1707G$ for iron or $480G$ for Fe_3O_4) will cause field shifts in the vicinity of the particle which will render the carbon radical ESR signals unobservable. If we assume that field shifts in excess of 15G cause a loss of signal, we can estimate an "exclusion radius" -- i.e. a distance around a given particle that will not give an ESR signal

$$H_L = \frac{8\pi}{3} \ 1750 \ G \ \left(\frac{a}{r}\right)^3 = 15 \ G \qquad\qquad (B\text{-}6)$$

which yields $r \simeq 10a$. Thus, all radicals within ten particle diameters are unobservable.

The total field profile is built-up by a superposition of these individual contributions. Polarizations at the surface are responsible for the observed shifts.

THE EFFECT OF THE COMBUSTION CHAMBER DEPOSITS ON OCTANE REQUIREMENT INCREASE AND FUEL ECONOMY

Yoshinobu Nakamura, Yoshiaki Yonekawa and
Nobukazu Okamoto
Toa Nenryo Kogyo K. K. Central Research Laboratory

175 Tsurugaoka, Ohi-machi, Iruma-gun, Saitama, Japan

SUMMARY

In order to investigate the effect of combustion chamber deposits on octane requirement increase (ORI) and fuel economy, we carried out fleet test, bench engine test and mini-engine test which was developed in our laboratory.

Cars whose initial octane requirements were small exhibited a tendency of giving rise to larger ORI values compared with higher IOR cars. Observed ORI did not correlate with the total deposits, but ORI modified with IOR correlated fairly well with the carbon content of the deposits (mg/cm^2). This would account for the low thermal conductivity and high heat capacity of the carbonaceous material in the deposits.

Fuel economy of various passenger cars with different amounts of combustion chamber deposits was measured using 10 Mode Test method. It was found that the improvement of fuel economy followed correlative trend with ORI. Combustion chamber deposits increase ONR of an engine, which could result in an improved engine efficiency under conditions employed. This effect was evidenced with mini-engines whose cylinder head surface were coated with PTFE film. Most of this fuel benefit can primarily be ascribed to the thermal effect of the deposits accumulated in the combustion chambers.

INTRODUCTION

It is generally accepted that octane requirement increase (ORI) is caused almost entirely by combustion chamber deposits. With leaded fuels, many papers[1-8] have been reported for the composition and the properties of deposits, mechanisms of deposits formation, and modes of deposit action for ORI. It was reported that the volume effect of leaded deposits on ORI was in the range of from 20 to 40 % of ORI[3], while the thermal effect of deposits in the range of 50 - 60 % of ORI[1,8]. On the other hand, the volume effect of unleaded deposits was about 10 % of ORI and most of ORI was caused by the thermal effect in unleaded case[1,9]. The relation between the thermal properties of unleaded deposits and ORI is, however, still obscure.

Although ORI problem has long been thought of as an impediment to be overcome, one paper[10] has recently mentioned on the relation between engine deposits and fuel economy. This chapter discusses these problems from the view point of compositions and thermal properties of combustion chamber deposits.

EXPERIMENTAL

A. Field Test

Cars, Mileage Accumulation, Octane Ratings and Measurement of Fuel Consumption —— Eleven popular Japanese cars listed in Table 1 were employed in field tests[11]. Car 5 - 11 were equipped with exhaust emission controllers of the respective model year but Cars 1 - 4 of 1974 model were not.

Each of 11 cars (9 cars traveled 1 5,000 km, while Cars 8 and 11 accumulated 10,000 km) had a similar driving pattern in the course of testing; ca. 30 % urban, 50 % suburban and 20 % highway. Engines of all test cars were cleaned before mileage accumulation. Octane requirements (at borderline knock) were measured at every 2,500 km accumulation using primary reference fuels (PRF) on a chassis dynamometer. None of the data for octane requirements was

Table 1. Cars Employed in the Field Test

Car*	Model year	Engine displ. (CC)	Cylinder No.	Compression ratio	IOR** PRF
1	1974	1166	4	9.0	90.0
2	1974	1166	4	9.0	89.5
3	1974	1166	4	9.0	89.5
4	1974	1166	4	9.0	89.5
5	1976	1166	4	9.0	82.0
6	1976	1588	4	9.0	81.5
7	1975	1808	4	8.5	87.5
8***	1975	1808	4	8.5	79.5
9	1975	1808	4	8.5	84.5
10	1975	1968	4	8.5	86.5
11***	1975	1988	6	8.6	80.5

*Car 1-4 without hardware for exhaust emmision control.
**Initial octane requirement.
***Automatic transmission.

Table 2. Physical and Chemical Properties of Crankcase
Oils Employed in Field Test

	Oil A*	Oil B	Oil C
Specific gravity, 15/4°C	0.896	0.875	0.874
Viscosity, cst. 37.8°C	109.17	72.87	67.60
98.9°C	11.48	10.89	10.34
Viscosity index	100	139	151
Total acid No., KOH mg/g	2.51	2.22	1.01
Total base No., KOH mg/g	9.72	7.09	4.17
Sulfated ash, wt%	1.34	0.76	0.57
Elemental anal.,wt%			
Ca	0.373	--	0.120
Mg	--	0.122	--.
Zn	0.086	0.089	0.065
P	0.074	0.078	0.052
N	0.024	0.046	0.043
S	1.02	0.64	0.60

*Contained ca. 20% of bright stock by volume.

corrected for seasonal changes in an ambient temperature, humidity
or pressure. Measurements of fuel consumptions were made at every
2,500 km accumulation with 10 Mode Test on a chassis dynamometer
and fuel consumptions were calculated from the carbon balance in
the exhaust gas.

Crankcase Oils and Fuel Gasolines —— Crankcase oils used in
the field test are shown in Table 2 with their microanalytical data.
Oil A contained ca. 20 % bright stock, but oil B and oil C were all
distillate oils.

Physical and chemical properties of unleaded fuels are shown
in Table 3. Combinations of cars, crankcase oils and fuels are
shown in Table 4.

Table 3. Physical and Chemical Properties of Fuel
Gasolines Employed in the Field Test

	Fuel X	Fuel Y
Parafins, vol.%*	40.5	50.5
Olefins, " *	26.0	20.5
Aromatics, " *	33.5	29.0
RON	93.8	90.3
MON	81.7	80.1
Sulfur, wt%	0.02	0.02
ASTM Dist., °C		
IBP	37.0	31.0
10%	56.5	47.5
50%	101.0	92.5
90%	164.0	162.5
EBP	208.0	202.0

*FIA method

Table 4. Stabilized ORI Results for Combinations of
Lubricant, Fuel and Car in the Field Test

Car	Octane Requirement, PRF	Stabilized ORI, PRF	Lubricant*	Fuel**
1	94.8	4.8	A	X
2	95.5	6.0	A	X
3	94.2	4.7	B	X
4	94.0	4.5	B	X
5	89.0	7.0	C	Y
6	92.5	11.0	C	Y
7	91.0	3.5	C	Y
8	86.5	7.0	C	Y
9	90.5	6.0	C	Y
10	93.2	6.7	C	Y
11	88.0	7.5	C	Y

*Refers to Table 1.   **Refers to Table 2.

B. Laboratory Tests

Bench engine (1,800 cc, 4 cylinders, 4 cycle) and mini-engine
were employed in this study. Combinations of 12 crankcase oils and
12 fuels were used[12]. Mini-engine (one cylinder, L-head, 4 cycle,
200 cc, compression ratio 6.5) was equipped with a thermocouple, a
knock meter, a speed meter and a power generator[12]. A constant speed
operating mode (speed 2,000 rpm., brake load 750 W.; ca. 1/3 full
load) was used to accumulate deposits in the combustion chamber of
mini-engine. Teflon coated cylinder head was also employed in the
mini-engine to examine its thermal effect on fuel economy.

To accelerate deposit build-up, fuels employed were made to
contain an appropriate amount of lubricant. Octane requirements
for the two engines were measured with primary reference fuels after
the deposit accumulation.

In order to investigate the relationship between fuel consumption and octane requirements, an additional test car (1,600 cc, 4
cycle engine) was employed and made to travel 10,000 km under a
constant speed mode (60 km/hr) on a chassis dynamometer. A commercial unleaded fuel and a commercial 10W-30 oil were used in this
test. Octane requirements and fuel consumption were measured at
every 2,500 km.

C. Deposit Analysis

SEM and EPMA used to investigate the morphology of deposits
were Shimazu ASM-ST and EMX-SM respectively. Differential scanning
calorimeter (Perkin Elmer 1B) was used to measure the heat capacity
of deposits, and the thermal conductivity of deposits was measured
with Shibayama SS TC-18 of Tokyo National Chem. Lab. for Industry.

RESULTS AND DISCUSSION

Combustion Chamber Deposits and ORI

Octane requirements for Cars 1 - 11 are shown in Fig. 1
(Tabulated data refer to Table 4 in which "stabilized" octane requirements and ORI derived by the method utilized in the CRC paper[13]
are given). Several cars appeared to have attained "stabilized"
octane requirements at around 6,000 - 8,000 km accumulation, but some
cars, paticularly, Cars 2, 4 and 6, did not approach equilibrium
values even after 15,000 km accumulation.

The total amount of combustion chamber deposits collected from
Cars 1 - 11 ranged from 3.45 to 8.26 g., and ca. 1/3 - 2/3 of the
total deposits were found at the piston top except Car 7. No direct
relation was found in the field test between the total amount of
deposits and ORI, although Keller et al.[14] reported a correlation
between the cylinder-head deposits and ORI. As long as one employs
a single engine, a linear correlation was found between the total
deposits and ORI (Fig. 2).

"Stabilized" ORI in Table 4 are plotted against IOR (initial
octane requirements) in Fig. 3, and the following correlation could
be obtained in the present study;

"Stabilized" ORI = 43.44 - 0.44 (IOR)   1)

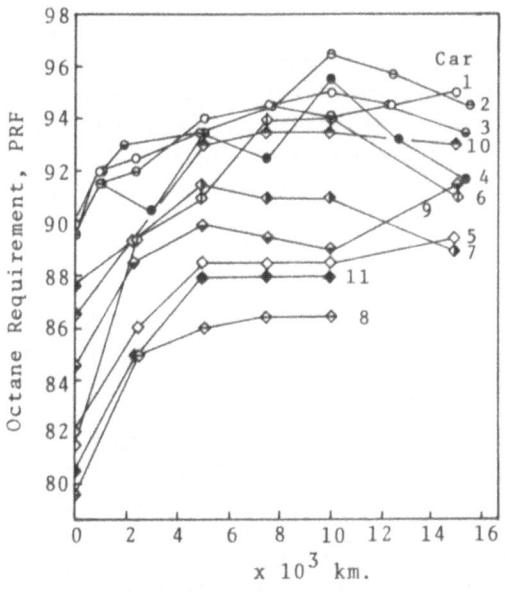

Fig. 1. Octane Requirement vs. Travel Distance

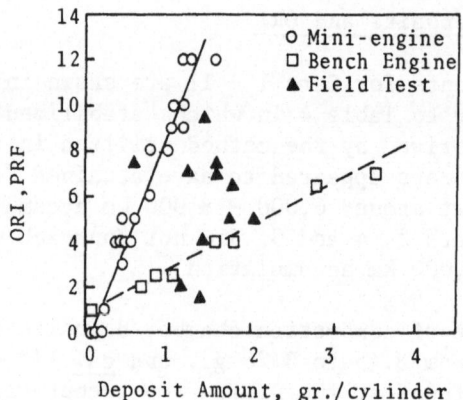

Fig. 2.   Relationship between ORI and
Deposits in Three ORI Tests

Similar observations were reported by Kunc[15], Sailent et al.[16] and
Alguist et al.[17].   Reasons for the negative relation between ORI and
IOR are not entirely clear, but it is possible that engines of higher
IOR (higher compression ratio) tend to afford more "inorganic" deposits due to higher combustion chamber temperatures than engines of
lower IOR (lower compression ratio) as noted by Kunc[15].   The deposit
removal process would, thus, occur earlier through light knock as
pointed out by Dumont[3] in the former engines than in the latter to
result in smaller ORI.

Microanalytical data of deposits from Car 1 and Car 3 are shown,
as representatives, in Table 5 together with typical data of deposits
from leaded and low-leaded fuels for comparison.   Carbon contents in
the deposits of Car 1 - 11 are shown in Table 6.   It is clear that
there are considerable differences in carbon content among them.

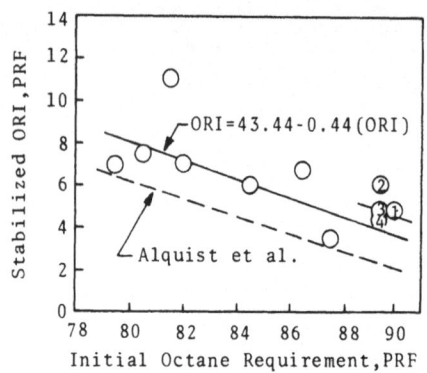

Fig. 3.   ORI vs. IOR. Car 1-4 employed
different fuel from other cars

Table 5. Microanalytical Data of Combustion Chamber
Deposits from Car 1 and Car 3

		%C	%H	C/H	%Ca	%Zn	%Mg	%Pb	%Fe	%P	%S
Car 1,	Piston top	45.85	3.11	14.7	9.4	2.1	-	-	1.0	0.1	4.08
	Cavity	43.62	3.28	13.3	9.6	2.1	0.1>	0.1>	1.8	0.1	4.26
	End gas region	54.58	4.64	11.7	5.3	1.1	0.1>	0.1>	0.8	0.6	3.09
Car 3,	Piston top	51.14	4.24	12.1	0.1>	3.3	3.9	0.1>	0.2	1.5	2.39
	Cavity	51.36	4.40	11.7	0.1>	1.3	2.3	0.1>	0.1	0.7	2.14
	End gas region	53.50	4.68	11.4	0.1>	1.1	3.4	0.1>	0.2	1.1	2.16
Leaded*,	Piston top	25.07	1.64	15.3	2.3	1.3	0.3	35.2	0.1	0.9	3.12
Low-leaded**,	Piston top	47.62	3.80	12.5	2.4	1.9	0.1>	9.8	0.7	0.2	1.89

*1.8ml TEL/USG, **0.4 TEL/USG. (Blended fuel).

A reasonable relation could be obtained between the carbon content per unit surface area of combustion chamber (mg/cm^2) and "modified" ORI which was calculated from the observed ORI using the following equation;

$$\text{"Modified" ORI} = \text{ORI} - 0.44 \left[\left(\overset{n}{\Sigma} \text{IOR} \right)/n - \text{IOR} \right] \qquad 2)$$

where $\left(\overset{n}{\Sigma} \text{IOR} \right)/n$ is an average of IOR, n, number of cars, and -0.44, noted earlier in Eq. 1) as $\Delta\text{ORI}/\Delta\text{IOR}$. This is shown in Fig. 4.

Fuel Consumption and ORI

Fuel consumption for Cars 1 - 11 were measured at every 2,500 km with 10 Mode Test on a chassis dynamometer. Fuel consumption improvements were calculated by dividing actual fuel consumption with the base value which were determined before accumulation.

Table 6. Carbon and Hydrogen Contents of Combustion
Chamber Deposits

	Piston top		Cavity		End gas region		Weighted average
Car	%C	%H	%C	%H	%C	%H	%C
1	45.85	3.11	43.62	3.28	54.58	4.68	46.1
2	47.22	3.78	41.37	3.28	52.82	4.49	45.6
3	51.14	4.24	51.36	4.40	53.50	4.68	51.6
4	55.24	5.26	48.56	4.67	56.81	5.02	53.0
5	61.46	4.38	56.48	3.92	62.07	4.46	59.7
6	56.01	3.61	61.04	4.48	64.39	5.72	58.1
7	63.52	4.50	63.13	3.89	64.42	4.40	63.9
8	62.53	4.12	60.25	3.60	61.71	4.09	61.7
9	61.70	4.59	61.70	4.47	62.94	4.75	62.0
10	63.43	4.80	63.89	4.21	64.23	4.61	64.7
11*	63.42	4.31	--	-	--	-	

*Not sufficient amount of sample available.

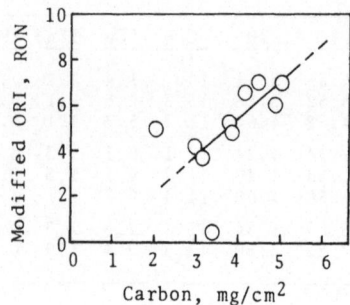

Fig. 4. Modified ORI vs. Carbon Content

Fuel consumption improvements are plotted against ORI in Fig. 5. This shows that fuel benefit tends to increase with ORI. Some showed more than 10 % fuel benefit. This tendency was also evidenced with a separate test car (1,600 cc engine) experiment which traveled 10,000 km under a constant speed of 60 km/hr on a chassis dynamometer as mentioned before, and this is shown in Fig. 6.

When deposits were completely removed from the combustion chamber, fuel benefits were only 2 % or less and octane requirements of engines were normally resumed the initial levels or gave at most less than 2 RON as ORI in the field test. This high fuel benefit is certainly ascribed to the deposits accumulated in the combustion chamber. This prompted us to investigate how the properties of the deposits, paticularly their thermal properties, influence on fuel consumption as well as ORI.

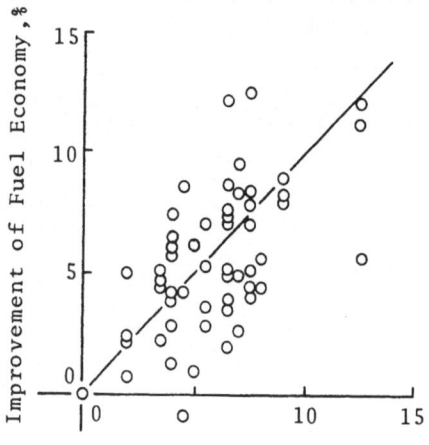

Fig. 5. Relation between ORI and Fuel
Consumption in the Field Test

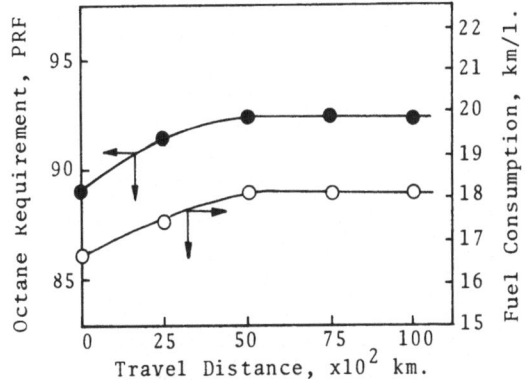

Fig. 6. Change of Octane Requirement and
Fuel Consumption during 10,000 km
Accumulation on Chassis Dynamometer

Thermal Effect of Combustion Chamber Deposits

It is believed that a considerable portion of ORI is caused with the thermal insulation effect of the combustion chamber deposits as Mikita et al.[2] noted.

Embedding a thermocouple in the combustion chamber wall, deposit accumulation test runs were carried out with the mini-engine to measure temperature changes on metal wall surface. An example of the results obtained in the present study is shown in Table 7. When the deposit on top of the thermocouple tip alone was removed, the observed temperature was found to be 13 °C higher than that of without deposits (refers to Table 7; 1 and 3). This temperature of condition 3 in Table 7 does not indicate the real temperature of the very surface of the deposits, but this temperature difference (13 °C) at

Table 7. Effect of Deposit Accumulation on Wall Temperature
with Mini-engine

Condition	Wall Temperature (°C)		Cylinder Head Outside Temp.(°C)	Room Temp.(°C)
	End Gas Region	Cavity		
1. Clean (at start)	225	240	123	15
2. Deposit Accumulated (Test end)	190	218	120	12
3. Removed Deposit from only Thermo. Couple Tip	238*	220	120	13
4. Removed Deposit from All over Comb. Chamber	220	240	120	13

*This temperature does not indicate the real temperature(see above paragraph).

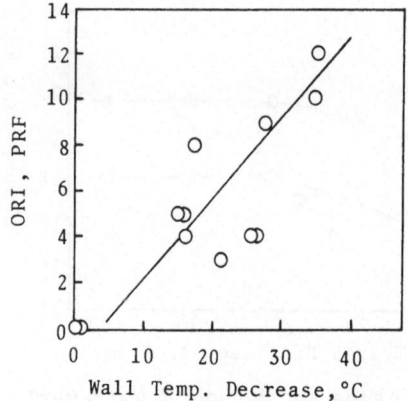

Fig. 7. Relation between ORI and
Wall Temperature Decrease at
End gas region with Mini-engine

least suggests that the incoming fuel-air charge experiences higher
temperatures under deposited conditions. As deposits accumulated,
observed temperature decreased as shown in the same Table (condition
2) due to the thermal insulation effect of deposits. Fig. 7 shows
the relation between ORI and temperature decrease at metal wall surface around the end gas region. A linear correlation was obtained
(Fig. 7).

Thermal conductivity of combustion chamber deposits was measured
by the ASTM (C 177 – 45) method. The results obtained are shown in
Table 8. Both deposits from Car 1 and Car 3 gave almost the same
thermal conductivity, 0.16 – 0.17 W/m·°C. It is interesting to note
that thermal conductivity of leaded deposits (obtained from a car
after 50,000 km accumulation) was found to be 0.15 W/m·°C. This
value was less than a half of the values reported by Mikita et al.[2],
and a synthetic deposit (PbO, $PbSO_4$/Char-coal, picth binder = 90/10

Table 8. Thermal Conductivity of Combustion Chamber Deposits

	Thermal Conductivity, W/m·°C	
Deposit Type	This Work***	Mikita [1,2]
Unleaded	0.16 – 0.17	0.24
Leaded	0.15*	0.38
Synthetic Leaded**	0.20	– –
Teflon	– –	0.26

*Obtained for a car after 50,000 km accumulation.
**PbO, $PbSO_4$/Charcoal, pitch binder=90/10 by weight.
***Measured at 68.4°C.

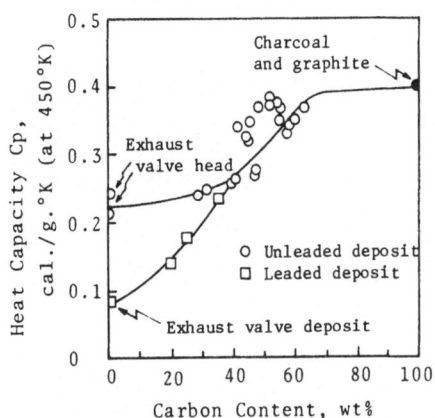

Fig. 8. Heat Capacity vs. Carbon
Content of Deposits

by weight) pelleted under <u>ca.</u> 8,900 psi. gave thermal conductivity
of 0.2 W/m·°C at 68.4 °C. This difference in thermal conductivity
between the reported value and the present result for leaded deposits
is largely attributable to the difference in the sample preparation
in which thermal conductivity was effected by the load employed in
making the sample pellets in the ASTM method. In any case, thermal
conductivity did not differ much between unleaded and leaded deposits
in the present study.

Heat capacities of combustion chamber deposits are summarized
in Fig. 8 as a function of the carbon content in the deposits. Heat
capacity of deposits depended, to some extent, on the location where
they deposited in the combustion chamber, and they ranged from 0.23
- 0.37 cal/g·°C (at 450 °K) measured with differential scanning
calorimeter. The heat capacity of leaded deposit found in the
present work was only 0.15 cal/g·°C (less than a half of unleaded
deposit) at 450 °K. The heat capacity of exhaust valve head deposit
(leaded) was 0.07 cal/g·°C, which was close to those of pure PbO and
$PbSO_4$, whereas the corresponding unleaded deposit exhibited 0.22
cal/g·°C which was close to pure $MgSO_4$ (0.22 cal/g·°C at 100 °C),
$CaSO_4$ (0.19) and $ZnSO_4$ (0.17). It should be noted here that the heat
capacity of exhaust valve head deposit (unleaded) was found to be
<u>ca.</u> 3 times greater than that of the corresponding leaded deposits.

Fig. 9 shows a secondary electron images of cross section of
unleaded deposits. Sharp concentration gradients of elements were
seen from Fig. 9. This observation was similar to that described by
Mikita et al.[1,2] for leaded deposits. They described that lead compounds were distributed primarily at the layer close to the deposits
surface. In unleaded case, inorganic stuff instead of leaded compounds was concentrated towards deposit surface found in this study.

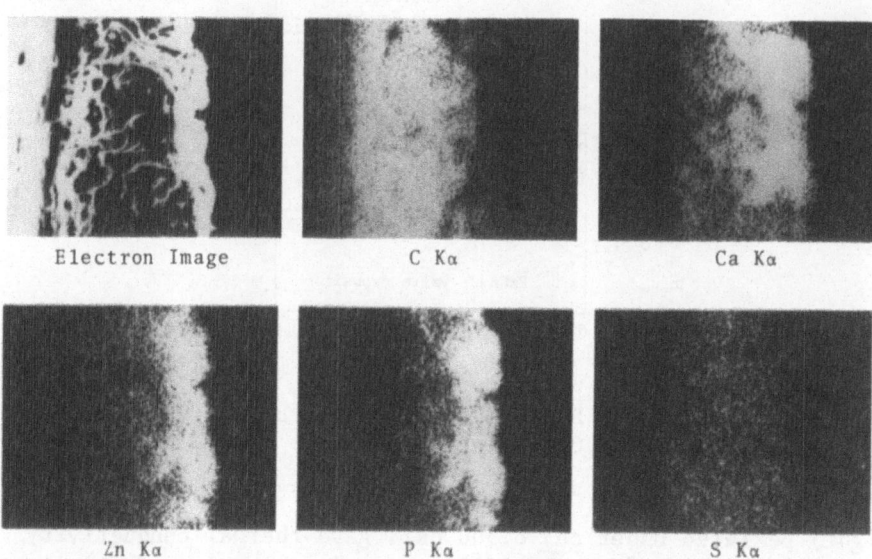

Fig. 9. Secondary Electron Image and Distributions of Five
Elements at the Cross-section of Typical Unleaded
Deposit. Combustion Chamber Wall is the Extreme Left Side

To investigate the thermal effect of deposit on octane requirements and fuel economy, the cylinder head surface of the mini-engine
was coated with Teflon (thermal conductivity 0.26 W/m·°C, heat capacity 0.25 cal/g·°C). Two cylinder heads were prepared, which were
coated with 0.05 mm. and 0.08 mm. thick, respectively. A vanishingly
small volume contraction of combustion chamber occurred by these
coatings. Octane requirements were measured at the full throttle.
Measurements of fuel economy were carried out under constant conditions (1,400 rpm., 350 W) by reading time required to consume 20 ml.
of isooctane fuel. Results obtained are shown in Table 9. These

Table 9. Effect of Teflon Coating Heads on Octane Requirement
and Fuel Economy with Mini-engine

Cylinder Head	Octane Requirement, PRF[*]	Fuel Economy[**] Improvement, %
1. Standerd Head	65	Base
2. Teflon coating (0.05 mm)	68	5.8
3. Teflon coating (0.08 mm)	74	15.4

*Measured by accelerating with full thrttle after briging the engine to
 steady operating conditions (2,000 rpm, 750 W)

**Measured in steady operating conditions (1,400 rpm, 350 W) with isooctane.

results strongly suggest that combustion chamber deposits play on important role in controlling octane requirement as well as fuel consumption of an engine.

CONCLUSION

We investigated the effect of combustion chamber deposits on ORI and fuel economy using field test, bench engine test and mini-engine test which was developed in our laboratory. Thermal properties of combustion chamber deposits were discussed to explain their actions to ORI and fuel economy.

(a) ORI was found to correlate well with the amount of combustion chamber deposits in bench engine test and mini-engine test. On the other hand, in the field test, observed ORI did not correlate with the total deposits, but ORI modified with IOR correlated fairly well with the carbon content of the deposits per unit surface area of combustion chamber (mg/cm^2).

(b) The effects of combustion chamber deposits on ORI and fuel economy were discussed. The present study provided evidence of combustion chamber deposits affecting positively on fuel benefit.

REFERENCES

1. J. J. Mikita, W. E. Bettoney, The 41st Annual Meeting Western Peter. Ref. Assoc., San Antonio, Texas, Mar. (1953).
2. J. J. Mikita, B. M. Sturgis, Proc. 4th World Petrol. Congr., VI 357, (1955).
3. L. F. Dumont, SAE Quarterly Trans., 5, 4, 565 (1951).
4. J. L. Lauer, P. J. Friel, Combustion Flame, 4, 107 (1960).
5. J. P. Bartlesson, E. C. Hughes, Ind. Eng. Chem., 45, 1501 (1953).
6. W. E. Newby, L. F. Dumont, ibid., 45, 6, 1336 (1953).
7. J. W. Orelup, O. I. Lee, ibid., 17, 7, 731 (1925).
8. J. Warren, SAE Trans., 62, 582 (1954).
9. J. D. Benson, SAE paper 750933.
10. L. B. Graiff, ibid., 790938.
11. Y. Nakamura, Y. Yonekawa, N. Okamoto, J. of Japan Petrol. Inst., 22, 2, 105 (1979).
12. Y. Yonekawa, Y. Nakamura, N. Okamoto, ibid., 24, 2, 85 (1981).
13. CRC Report, No. 451, June (1972).
14. B. D. Keller, G. H. Meguerian, C. B. Tracy, J. B. Smith, SAE paper 760195 (1976), ibid, 770195 (1977).
15. J. F. Kunc, Jr., SAE Quart. Trans., 5, 4, 584 (1951).
16. R. B. Sailant, F. J. Pedrys, H. E. Kidder, SAE paper 760196 (1976).
17. H. E. Alguist, C. E. Holman, D. B. Wimmer, ibid., 750932 (1975).

MODELLING THE EFFECT OF ENGINE DEPOSIT

ON OCTANE REQUIREMENT

Anthony DeGregoria

Exxon Research & Engineering Co.

Linden, N.J.

INTRODUCTION

We will describe work developing a model which determines the effect of engine deposit on octane requirement. Since experimental and observational sources (Mikita and Bettoney, 1953; Warren, 1954; Benson, 1975) indicate that only thermal and volumetric effects of deposit cause ORI (Octane Requirement Increase), the problem is tractable. The model quantitatively investigates the hypothesis that the increase in unburned gas temperature and pressure due to deposit, decreases the autoignition (knock) delay time sufficiently to account for ORI. The model also provides instantaneous deposit and gas temperatures which are necessary in any attempt to determine the chemistry of deposit formation, even though the model itself contains no chemistry.

THE MODEL

The model (DeGregoria, 1982) is an expansion of existing quasi-dimensional thermodynamic engine models to include the thermal and physical (volumetric) effects of deposit as well as a calculation of engine octane requirement.

A typical quasi-dimensional engine model contains the following assumptions and effects:

1. The gas is a single thermodynamic unit for all parts of the engine cycle except combustion.

2. During combustion, the gas is two thermodynamic units, burned and unburned gas, separated by an infinitely thin flame front.

3. Mass burning, during combustion, is treated empirically. In our case, we have chosen a simple expression which depends only on the crank angle at which ignition begins and the burn duration in degrees of crank angle (Blumberg and Kummer, 1971).

4. Instantaneous heat transfer between the gas and wall surfaces is treated through an empirical heat transfer coefficient for the gas boundary layer (Woschni, 1968).

5. Instantaneous flow through valves is treated.

We have added the effects of deposit to the quasi-dimensional model in the following way:

In Figure 1, we illustrate the combustion chamber, during combustion, with the piston at its uppermost position. Deposit forms on the piston top and head. Models with no deposit assume constant surface temperatures over the engine cycle which are input parameters. With deposit present, we have to specify deposit surface temperatures as a function of time over the engine cycle. These deposit surface temperatures are determined by one dimensional time dependent heat transfer calculations. Inputs to these calculations are the gas temperature and heat transfer coefficient of the gas boundary layer as functions of time. Consequently, we must perform a self-consistent calculation; gas temperatures effect deposit surface temperatures and vise versa.

For practical reasons, we only take a two point deposit approximation, performing the one dimensional time dependent heat transfer calculations at the two points shown in Figure 1. We determine the instantaneous deposit surface temperature for an arbitrary point on the surface by linear interpolation between these two points.

In Figure 2, we illustrate the one dimensional time dependent heat transfer calculations performed at the two points shown in Figure 1. We assume we have certain thicknesses of deposit and steel. On one side we have the gas;on the other side, we have water (coolant). On

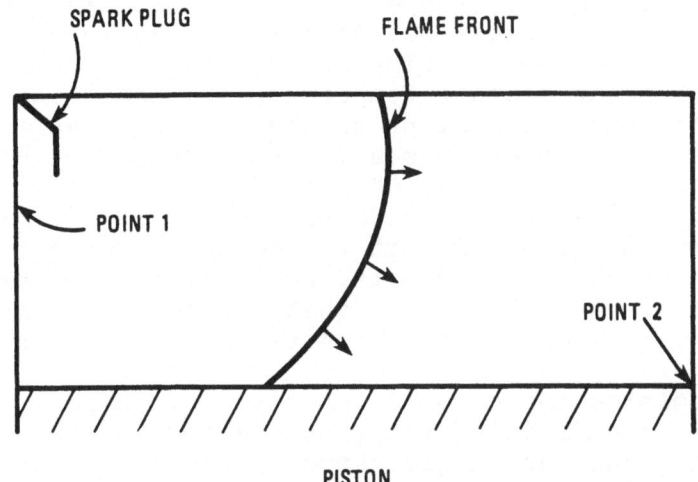

SPARK PLUG FLAME FRONT

POINT 1

POINT 2

PISTON

Fig. 1. Schematic of combustion chamber, during
combustion, showing the location of points 1 and
2 in the two point approximation to the heat
transfer between gas and wall. The piston is
assumed to be at top dead center.

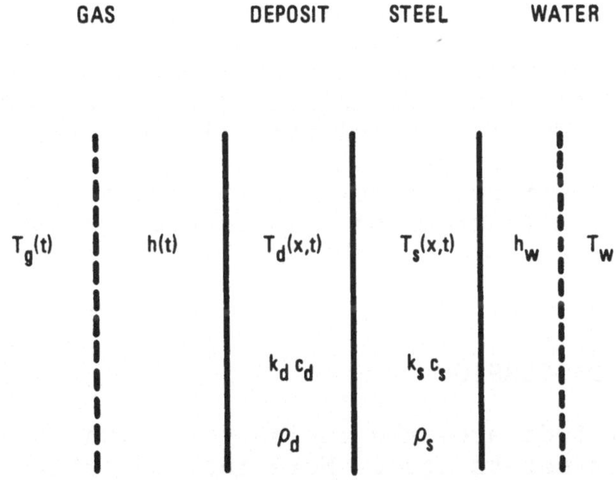

GAS   DEPOSIT   STEEL   WATER

$T_g(t)$ $h(t)$ $T_d(x,t)$ $T_s(x,t)$ h_w T_w

$k_d\, c_d$ $k_s\, c_s$

ρ_d ρ_s

Fig. 2. Schematic for the one dimensional time dependent heat transfer calculation performed at
points 1 and 2.

215

the gas side, the gas temperature and heat transfer
coefficient for the gas boundary layer, as functions of
time, come from the gas simulation. On the coolant side,
we have constant coolant temperaure and heat transfer
coefficient for the coolant boundary layer. We assume
the deposit is homogeneous and, therefore, its thermal and
physical properties are constants. We solve the one
dimensional time dependent Fourier heat transfer equation
in deposit and steel subject to the usual boundary
conditions at the gas-deposit, water-steel, and deposit-
steel interfaces. We assume that there is no contact
thermal resistance at the deposit-steel interface.

In solving the coupled deposit-gas problem, a simple
relaxation scheme is effective since deposit only produces small changes in gas temperatures and pressures.
We start by assuming constant deposit surface temperatures over the cycle and solve the gas simulation. The
resulting gas temperatures and heat transfer coefficients
for the gas boundary layer are used in the one dimensional heat transfer calculations through deposit and steel
at the two points described above. The resulting time
dependent deposit surface temperatures are used as the
new boundary conditions and the gas simulation is
repeated. Iterating this procedure, we rapidly converge
to the self-consistent solution.

We have now accomplished the development of a model
which determines the effect of deposit on gas temperature and pressure. To determine the deposit's effect on
octane requirement, we have to convert gas temperature
and pressure histories to time of knock. This is accomplished using the simple integral expression formulated by
Livengood and Wu (1955), relating time of knock of the
unburned gas to the autoignition delay time, along with
an accurate empirical expression for the autoignition
delay time as a function of temperature, pressure, and
research octane number of the fuel (Douaud and Eyzat,
1978).

RESULTS AND DISCUSSION

Table 1 indicates the engine specifications and
operating parameters used. Note that an intake manifold
pressure of 1 atmosphere corresponds to wide open
throttle (WOT). The burn duration of 40 degrees of crank
angle is on the short side, corresponding to a high
turbulence engine.

Table 1. Engine Specifications and Operating Parameters

Intake Manifold	- temperature	100 deg F
	- pressure	1 atm
Exhaust Manifold	- temperature	1500 deg F
	- pressure	1 atm
Engine Speed		1500 RPM
Compression Ratio		8 to 1
Start of Combustion		10 deg BTDC
Burn Duration		40 deg

Table 2 shows the assumed deposit thermal and physical properties. These properties are representative of a solid carbonaceous material. Two conductivities were tried in an attempt to measure the sensitivity of results to the uncertainties in deposit properties. The value of 1.4 is a value for unleaded deposit quoted in the literature (Mikita and Bettoney, 1953). The value of 1.0 is an approximate theoretical lower limit to the conductivity of a solid (Cohen, 1980).

Table 2. Assumed Deposit Thermal and Physical Properties.

Conductivity	1.4(1.0) Btu/ft°Fhr
Heat Capacity	.5 Btu/lb °F
Density	50 Lb/ft^3

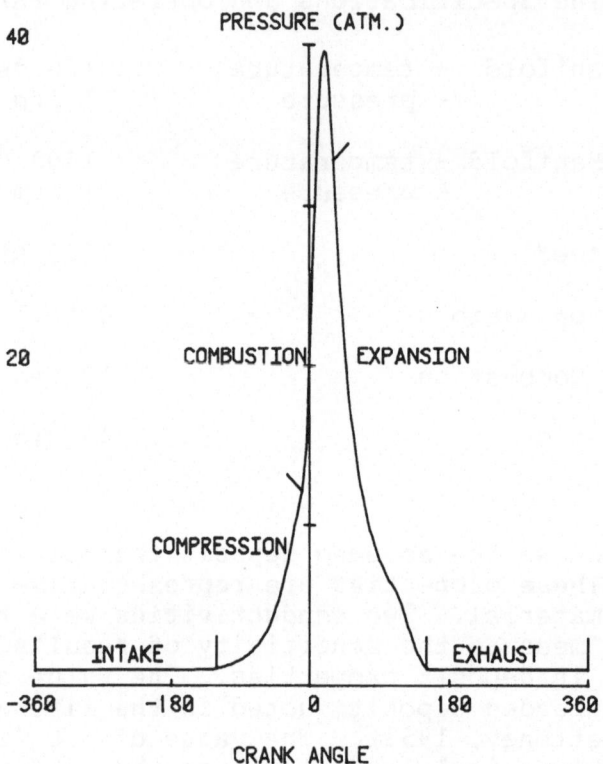

PRESSURE (ATM.)

40

20

COMBUSTION EXPANSION

COMPRESSION

INTAKE   EXHAUST

-360   -180   0   180   360

CRANK ANGLE

Fig. 3. Gas pressure vs. crank angle for the engine
model with no deposit.

     Figure 3 shows the gas pressure as a function of
crank angle over the engine cycle for the clean engine.
The parts of the engine cycle are indicated along the
curve. Figure 4 shows the corresponding gas temperature
for the clean engine over the engine cycle. Note that the
curves splits during combustion when there is both burned
and unburned gas separated by the flame front. Figure 5
shows the heat transfer coefficient for the gas boundary
layer over the engine cycle. Again, the curve splits
during combustion. Note that the heat transfer coefficient is higher on the unburned side of the flame front.
This is due to the higher density of the unburned gas
relative to the burned gas.

     Figure 6 shows surface temperature vs. crank angle
over the engine cycle at the two points adjacent to the
spark plug and furthest from the spark plug. We show
surface temperatures when there is no deposit and when

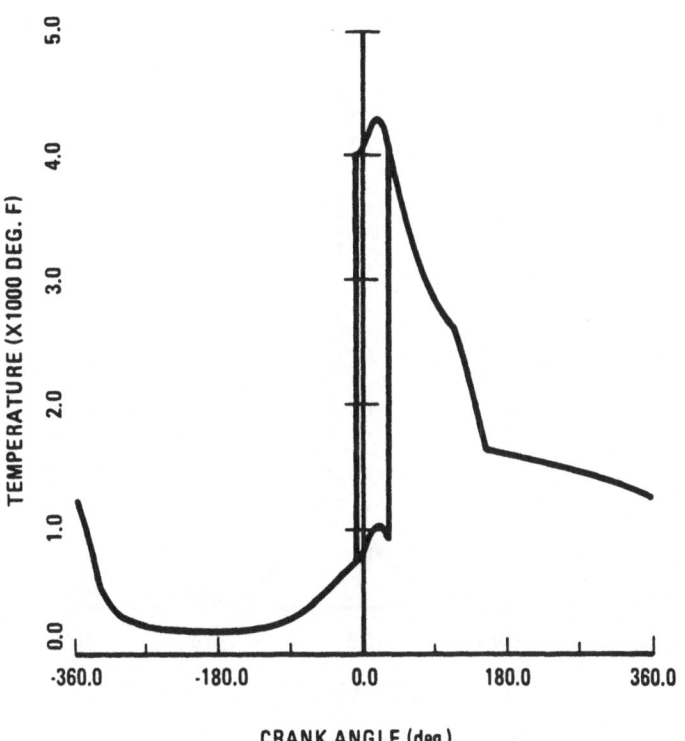

Fig. 4. Gas temperature vs. crank angle for the engine
         model with no deposit. The curve splits during
         combustion, the upper (lower) curve representing
         the burned (unburned) gas.

there is 3 grams of deposit. Note that there is a
significant difference in clean steel wall surface
temperature across the chamber ($\sim$ 100F). This is due to
the fact that the surface adjacent to the spark plug sees
hot burned gas during combustion while the end gas
surface sees only unburned gas.

   With deposit, the deposit thicknesses at the two
points were adjusted so as to produce approximately
equivalent peak surface temperatures (25 microns at point
1, 475 microns at point 2). The assumption here is that
peak deposit surface temperature limits deposit growth.
It is interesting that the peak deposit surface temperatures of approximately 900F for this heavily deposited
cylinder (3g) are not much below exhaust valve surface
temperatures of approximately 1000F or more where

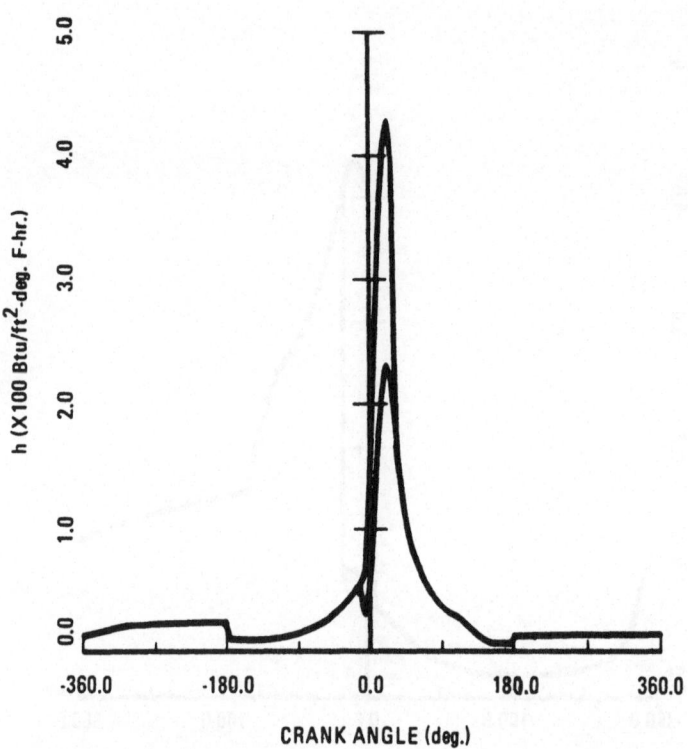

Fig. 5. Heat transfer coefficient vs. crank angle. The
curve splits during combustion with the lower
(upper) curve representing the burned (unburned)
gas side of the flame front.

carbonaceous deposit does not form. We have assumed no
contact thermal resistance between deposit and steel
wall. Such a resistance would increase deposit surface
temperatures. Since the engine is running under the
severe condition of wide open throttle, we would expect,
on average, deposit surface temperatures to run much less
than those obtained here in a road vehicle. This
suggests that there is a significant thermal contact
resistance between deposit and steel.

Figure 7 shows the effect of deposit on the temperature-pressure history of the unburned gas ahead of the
flame front (end gas track). The effect of 3 grams of
deposit is to increase peak unburned gas temperature and
pressure by approximately 40F and 1 atm., respectively.
In calculating octane requirements, we assume that 10% of

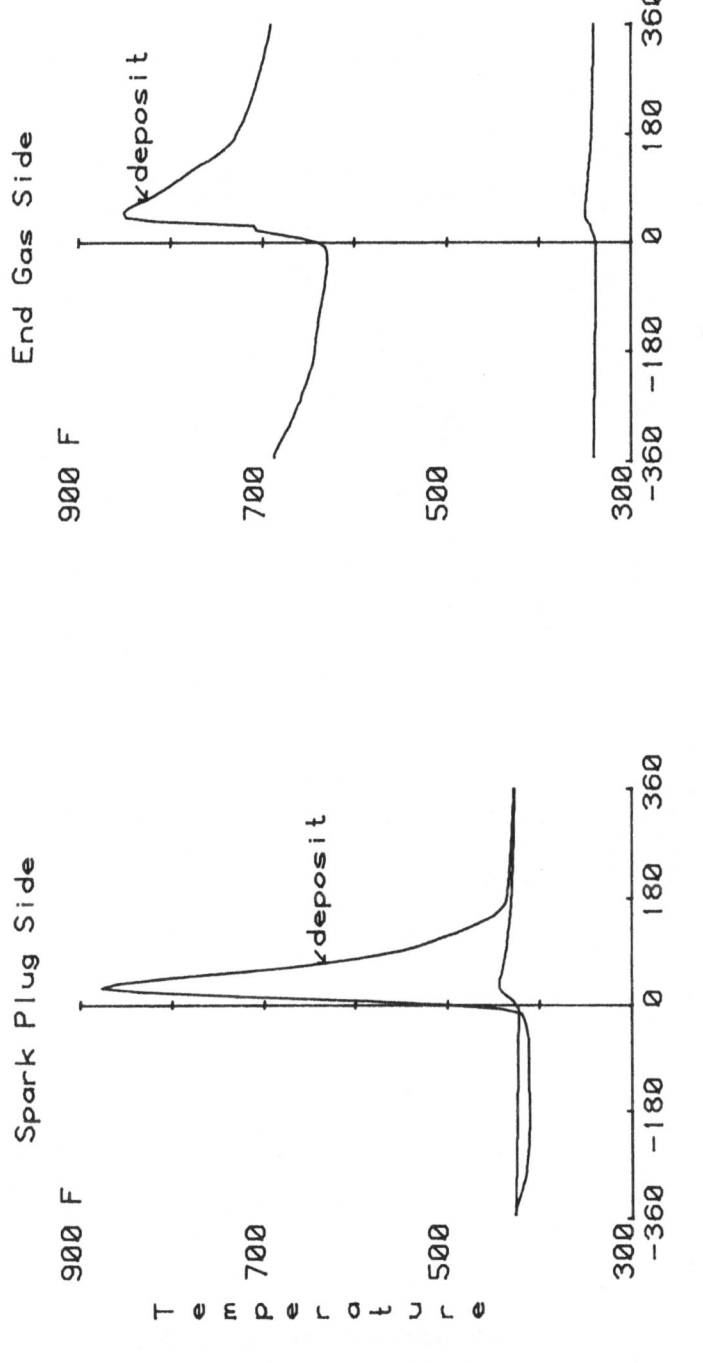

Fig. 6. Surface temperature vs. crank angle at the point adjacent to the spark plug and furthest from the spark plug for a clean engine and for an engine with 3.2g of deposit. Deposit thicknesses are 25 and 475 microns at the spark plug and end gas sides, respectively. Deposit conductivity is .10.

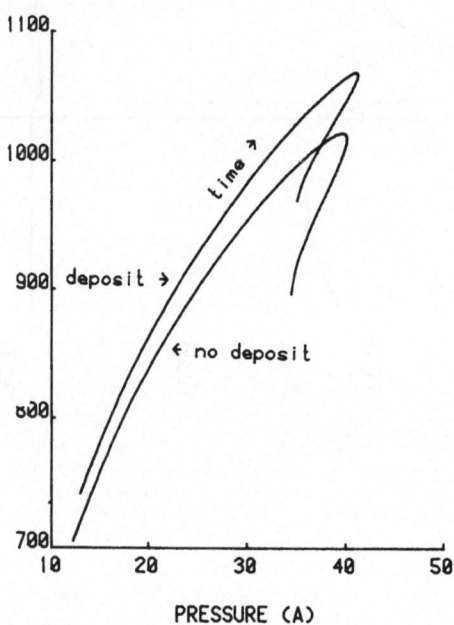

TEMPERATURE (F)

Fig. 7.   Temperature-pressure history of the unburned
         gas ahead of the flame front (end gas track) for
         the clean engine and for the deposit configura-
         tion described in Figure 6.

the original charge must autoignite to produce percep-
tible knock. This corresponds to a time of knock somewhat
after the temperature-pressure peak on the end gas track.

Table 3. Octane Requirements for Various Deposits.

Deposit Thickness (μ) Point 1	Point 2	Deposit Weight(g)	Deposit Conductivity	Octane Requirement
0	0	0	--	85.5
100	100	1.2	.10	87.4
100	200	1.9	.10	88.6
100	400	3.2	.10	90.8 88.2*
100	400	3.2	.14	90.1
25	475	3.2	.10	90.8

*Without the compression ratio increase (from 8.0 to 8.4) due to deposit volume.

Table 3 shows the engine octane requirement for various deposits including no deposit. We show the effect of changing only the deposit distribution across the cylinder and changing only the deposit conductivity. We also separate out the volumetric contribution of ORI when there is 3.2 grams of deposit. Changing the deposit distribution somewhat produced no change in ORI to 2 significant figures. Increasing the conductivity by 40% decreases total ORI by 15%, thermal ORI by 26%. With 3.2 grams of deposit, corresponding to a heavily deposited cylinder, ORI runs around 5. The thermal ORI is approximately 50% of the total. Experiment and observation indicates that thermal ORI is 70-90% of the total, while peak ORI can be as high as 12. Since the model's volumetric contribution to ORI is reasonable, we conclude that the model's thermal ORI is down by a factor of 2 or 3 from observation.

This significant discrepancy in thermal ORI between model and observation may well be accounted for by the existence of a thermal contact resistance between deposit and steel wall. The model assumes zero contact resistance. Since deposit is known to peal from the wall, there are times, at least, when this contact resistance can be quite large. There is poor correlation between ORI and deposit weight. The stochastic nature of deposit pealing may well account for this. That is, there should be times when deposit is in good thermal contact with the wall, resulting in relatively low ORI. There should also be times when deposit is significantly separated from the walls, resulting in high ORI. The relatively low ORI predicted by the model when there is good contact between deposit and steel supports this hypothesis.

CONCLUSION

We have developed a self-consistent model for ORI which gives semiquantitative results. That is, thermal ORI is low by a factor of 2 or 3. The thermal ORI form the model actually represents a lower bound since thermal contact resistance between deposit and steel is assumed to be zero. Since deposit is known to peal from the wall, this contact resistance should be quite variable, achieving large values at times. This can resolve the discrepancy between model and observation as well as explain the poor observed correlation between ORI and deposit weight.

ACKNOWLEDGEMENTS

I would like to thank the following people from Exxon Research and Engineering Co. for many interesting and helpful discussions of the ORI problem: R. S. Lunt, J. M. Pecoraro, D. J. Martella, B. S. Berkowitz, L. B. Ebert, W. G. May, B. G. Silbernagel.

REFERENCES

Benson, J. D., 1975, Some factors which affect octane requirement increase, SAE Transactions 750933.

Blumberg, P. N., and Kummer, J. T., 1971, Prediction of NO formation in spark-ignited engines - an analysis of methods of control, Combustion Science and Technology, 4, pp. 73-95.

Cohen, R. W., 1980, Exxon Research and Engineering Company, private communication.

DeGregoria, A. J., 1982, A theoretical study of engine deposit and its effect on octane requirement using an engine simulation, SAE Technical paper 820072.

Douaud, A. M., and Eyzat, P., 1978, Four octane number method for predicting the anti-knock behavior of fuels and engines, SAE Paper 780080.

Livengood, J. C., and Wu, P. C., 1955, Correlation of autoignition phenomena in internal combustion engines and rapid compression machines, Fifth (International) Symposium on Combustion, Reinhold Publishing Corp., New York, p. 347.

Mikita, J. J., and Bettoney, W. E., 1953, How combustion-chamber deposits affect engine-fuel requirements, The Oil and Gas Journal, Vol. 52, pp. 84-86.

Warren, J., 1954, Combustion chamber deposits and octane number requirement, SAE Transactions, Vol. 62, pp. 582-594.

Woschni, G., 1968, A universally applicable equation for the instantaneous heat transfer coefficient in the internal combustion engine, Paper 670931 of SAE Transactions, 76.

DEPOSIT FORMATION BY DIFFUSION

OF FLAME INTERMEDIATES TO A COLD SURFACE

J.D. Bittner, S.M. Faist, J.B. Howard and J.P. Longwell

Department of Chemical Engineering
Massachusetts Institute of Technology
Cambridge, Massachusetts 02139

INTRODUCTION

The problem of deposit formation on cold surfaces is important in a number of combustion systems and is of special interest
in the Otto-cycle engine, where octane requirement is significantly increased above that for a clean engine by the formation of
deposits on combustion chamber walls. Past work on this problem
supplies information on the effects of fuel, lubricant, and engine
variables on the formation of deposits. Elemental analysis and
other limited information on deposit composition are also available; however, very little information is available on the chemical species that reach the surface from the flame to cause these
deposits.

Techniques for direct sampling from a low-pressure flame into
a mass spectrometer are providing a wealth of new information on
the chemical species found in the flame zone. In the course of
flame zone studies, it was observed that some stoichiometric
flames formed deposits on the surface of the flat flame burner
used. The physical nature and elemental composition of these deposits were similar to those found on the cold walls of a spark
ignition gasoline engine. Flame temperatures in these studies
were comparable to those in engines, but the pressure, 2.67 kPa,
is much lower.

Scaling considerations (Fristrom and Westenberg, 1965) indicate that for second-order reactions and flames of the same
unburned gas velocity V_0, distance z should scale inversely with
pressure P. Therefore, similar mole fraction and temperature profiles are expected when plotted as z/P. Also, since the diffusion

coefficient $D_{i, mix}$ is inversely proportional to pressure, the mass flux fraction G_i of species i as defined in Eq. 1 for one dimensional flat flames of different pressures should be independent of pressure at constant z/P.

$$G_i \equiv \frac{PX_i M_i}{RT\rho_o V_o} \left[V - \frac{D_{i,mix}}{X_i} \frac{dX_i}{dz} \right] \qquad \text{(Eq. 1)}$$

In Eq. 1, X_i is the mole fraction of species i, M_i is the molecular weight of species i, T is temperature, R is the ideal gas constant, ρ is mass density, and V is the bulk gas velocity defined in Eq. 2 with V_o being the inlet value.

$$V = \frac{\rho_o V_o RT}{P \sum_i X_i M_i} \qquad \text{(Eq. 2)}$$

Since the absolute convective and diffusional fluxes are expected to be proportional to pressure, the mass deposition rate is also expected to be proportional to pressure if the process is diffusion controlled. However, as a fraction of the total mass flux from the burner, the mass deposition is expected to be independent of pressure. It is therefore believed that the results discussed in this paper are relevant to the chemistry of deposit formation in practical equipment.

EXPERIMENTAL

The molecular beam mass spectrometer apparatus (Figure 1) used for collection of deposits and analysis of the flame structure has been described in detail elsewhere (Bittner and Howard, 1981; Bitner, 1981). The burner was constructed from a 1.25 cm thick by 7.1 cm diameter copper plate, drilled with 1 mm diameter holes in a close-packed hexagonal array with a 14.5% void area. Since the sampling probe orifice diameter was 0.7 mm the mole fraction profiles within several millimeters of the surface may be distorted and the gradients measured may not be those present in the absence of the probe. The solid deposit was removed from the burner surface after one to three hours of flame operation by careful scraping with a scapel and stored in glass and teflon.

Deposits were collected and detailed gas composition profiles were measured for two benzene flames, (1) a stoichiometric (fuel equivalence ratio, $\phi = 1.0$) flame (8.2 mol % C_6H_6 - 61.8 mol % O_2 - 30.0 mol % Ar) and (2) a nearly sooting $\phi = 1.8$ flame (13.5 mol % C_6H_6 - 56.5 mol % O_2 - 30.0 mol % Ar). For both flames the

228

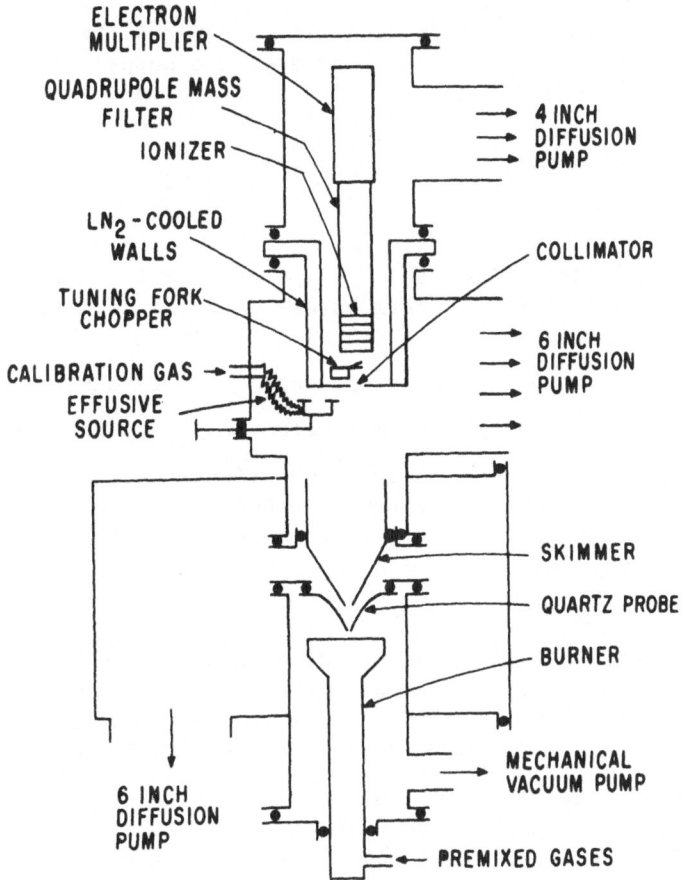

ELECTRON MULTIPLIER

QUADRUPOLE MASS FILTER

IONIZER

LN_2-COOLED WALLS

TUNING FORK CHOPPER

CALIBRATION GAS

EFFUSIVE SOURCE

4 INCH DIFFUSION PUMP

COLLIMATOR

6 INCH DIFFUSION PUMP

SKIMMER

QUARTZ PROBE

BURNER

MECHANICAL VACUUM PUMP

6 INCH DIFFUSION PUMP

PREMIXED GASES

Figure 1. Molecular Beam Mass Spectrometer System (Source: Bittner, 1981; Bittner and Howard, 1981a, 1981b).

unburned gas velocity was 0.5 m s^{-1} and the burner chamber pressure was 2.67 kPa. The maximum flame temperatures were 1800K for the ϕ = 1.0 flame and 1900K for the ϕ = 1.8 flame. Argon was used as a diluent to prevent interference of N_2 with CO at mass 28.

DEPOSIT CHARACTERIZATION

The deposits collected from both benzene flames after several hours of operation had a velvety black appearance prior to scraping that suggests a finely divided porous structure, perhaps similar to the agglomerated spheres observed in engine deposits (Lauer and Friel, 1960). After scraping, the solid material was reddish brown in color. When exposed to room air the deposit became increasingly gummy as it took up water. From qualitative observations made during the scraping procedure, the deposit from the ϕ = 1.0 flame was more hygroscopic than that from the ϕ = 1.0 flame. The deposition rate from the ϕ = 1.0 flame was 7 x 10^{-7} g cm^{-2}s^{-1}. This is only about 0.02% of the carbon flux from the burner (3.2 x 10^{-4} g cm^{-2}s^{-1}).

Both the deposits from the ϕ = 1.0 and the ϕ = 1.8 benzene flames were quite different from burner deposits collected from very rich and sooting (2.5 < ϕ < 3.0) C_2H_2 flames. (Stoichiometric C_2H_2 flames formed no deposits under similar conditions). The deposits from the rich C_2H_2 flames were very tarry and gummy, with a large fraction of methylene chloride soluble material. Very little material was extracted from the benzene flame deposits by methylene chloride. However, a large amount of material was extracted by oxygenated solvents such as water, ethylacetate, and acetone.

The presence of large amounts of oxygen in the deposits, as suggested by their qualitative behavior in the solvents and their hygroscopic nature, was confirmed by elemental analysis after drying over P_2O_5 under vacuum at 383K. The results from two deposits from the stoichiometric flame and one from the ϕ = 1.0 flame are shown in Table 1. As might be expected, the O/C ratio of the deposit decreases with increasing ϕ. The more hygroscopic nature of the ϕ = 1.0 deposits is verified by the weight loss on drying and correlates with higher O/C ratios.

Table 1. Composition of Burner Deposits for Different Equivalence Ratios ϕ.

ϕ	Weight Loss on Drying %	Weight % C	H	O	Molar Ratios H/C	O/C
1.0 (a)	16.0	52.85	3.11	41.19	0.71	0.58
1.0 (b)	21.6	50.24	3.83	43.12	0.95	0.64
1.8	7.41	60.05	4.92	34.02	0.98	0.43

The high oxygen content of these deposits is similar to that found in engine deposits from lead-free fuels. White (1955) found engine deposits from benzene, toluene, xylene, unleaded gasoline, and a pure paraffinic fuel (C_7-C_8) to contain about 20 wt % oxygen, 65 to 70 wt % carbon, and 3 to 5 wt % hydrogen. Fuel composition did not have an appreciable effect on the elemental analysis. Similarly, Lauer and Friel (1960) found engine deposits from toluene, isooctane, and a commercial unleaded fuel using both paraffinic and naphthenic lube oils, to have 20 to 36 wt % oxygen and 57 to 72 wt % carbon, with the higher oxygen content corresponding to the use of naphthenic lube oil. Although toluene formed much more deposit, the deposits from isooctane had higher O/C ratios.

In addition to elemental analysis, several other techniques were used to characterize the burner deposits that gave results similar to those found for engine deposits. The infrared (IR) absorption spectrum obtained by the KBr pellet method for the dried deposit from the $\phi = 1.0$ flame had the broad band absorption characteristic of high molecular weight material. However, strong bands at about 3400 cm^{-1} and 1750 cm^{-1} indicated the presence of hydroxyl and carbonyl groups as found in engine deposits by Dimitroff et al. (1969) and Lauer and Friel (1960). An absorption at about 1200 cm^{-1} indicated the presence of single C-O bonds. Strong broad band absorption between 675 and 1000 cm^{-1} indicated the presence of probably both alkene and aromatic C-H bonds. Reaction of the deposits with 2,4-dinitrophenylhydrazine gave a reddish-orange precipitate indicating the presence of aldehyde and/or ketone groups (Morrison and Boyd, 1975).

FLAME STRUCTURE

To provide information about the flame gases from which the deposit is formed, the molecular beam spectrometer apparatus was used to measure profiles of mole fraction versus distance from the burner for many intermediate and product species formed in the $\phi = 1.0$ and $\phi = 1.8$ flames. The mole fraction profiles of the major stable and radical species observed in the $\phi = 1.0$ flame are shown in Figures 2 and 3, respectively. The end of the primary reaction zone occurs at about 6 mm from the burner surface where benzene has been consumed and the mole fractions of the major radicals H, OH, and O have almost reached their maxima. In the region between the burner and the end of the primary reaction zone many intermediate species are observed (Figures 4 to 6). The mole fraction profiles of those with masses between 26 and 30 amu are shown in Figure 4. Relatively large mole fractions of formaldehyde (H_2CO) are present near the burner surface. The profile labelled mass 29 may have contributions from both formyl radical (HCO) and ethyl radical (C_2H_5). The absolute values and the gradients of mole fractions of acetylene (C_2H_2) and ethylene (C_2H_4) indicate

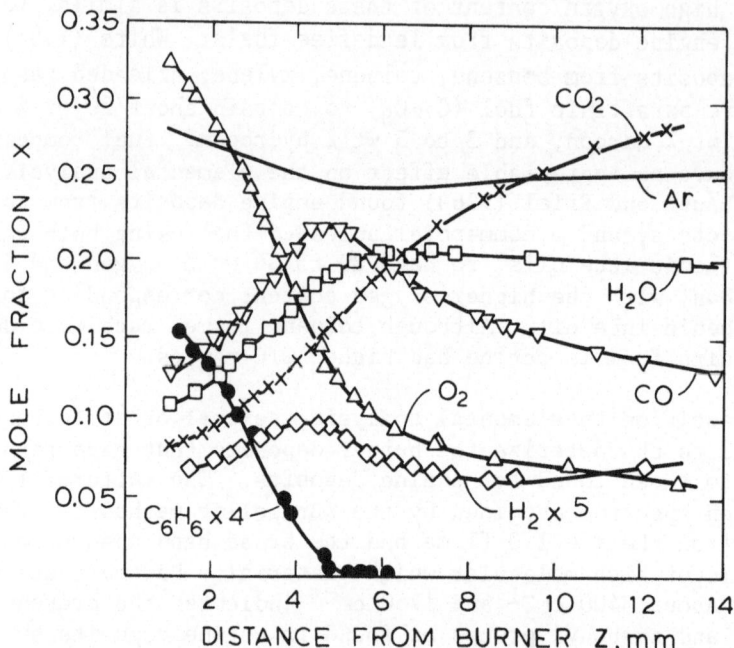

Figure 2. Mole Fractions of Major Stable Species versus Distance from Burner in a Stoichiometric (ϕ = 1.0) Benzene (8.2 mol %) - Oxygen (61.8 mol %) - Argon (30.0 mol %) Flame. Unburned gas velocity, 50 cm s^{-1}; pressure, 2.67 kPa (source: Bittner, 1981).

large amounts of polymerizable unsaturated hydrocarbons near the burner surface. Several unsaturated C_4 hydrocarbons, butadiyne (C_4H_2), 1-buten-3-yne (C_4H_4), and the 1-buten-3-ynyl radical (C_4H_3) are also observed in the primary reaction zone (Figure 5).

The mole fraction profiles of intermediate species with masses between 65 and 94 amu are shown in Figure 6. The early forming oxygenated species, C_6H_6O, may stabilize as phenol in stable species sampling probes. The second most abundant intermediate in this mass range is observed at mass 66 (C_5H_6). Although another possible source of the mass spectrometer signal at mass 66 is a $C_5H_6^+$ fragment ion from phenol (C_6H_6O) or an energy-rich adduct formed by reaction of benzene with O or OH, the ionization efficiency curve indicated

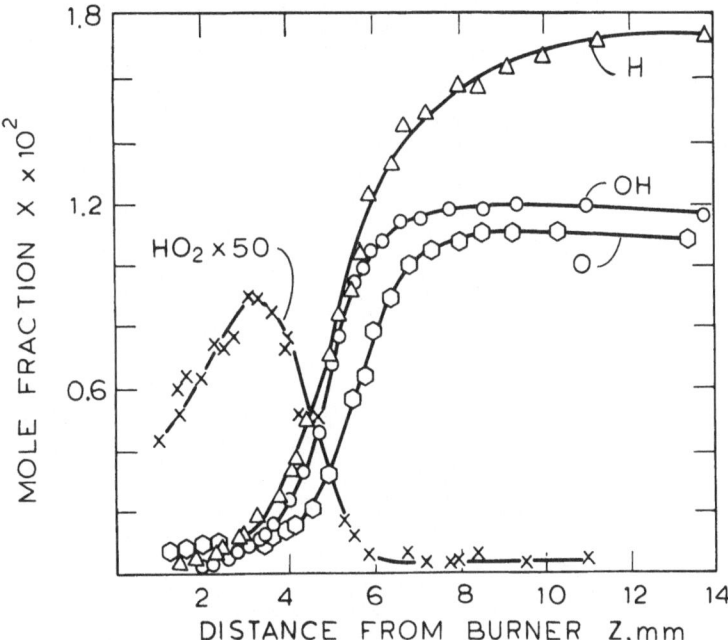

Figure 3. Mole Fractions of H, OH, O, and HO_2 versus Distance from Burner in a Stoichiometric (ϕ = 1.0) Benzene (8.2 mol %) - Oxygen (61.8 mol %) - Argon (30.0 mol %) Flame. Unburned gas velocity, 50 cm s^{-1}; pressure, 2.67 kPa (source: Bittner, 1981).

contributions from only one source. Two other observations support the argument that the mass 66 signal is from a neutral species generated in the flame. Firstly, the presence of two C_5H_6 hydrocarbons was confirmed by the GC-MS analysis of samples withdrawn from this flame with a conventional quartz microprobe (Faist, 1979). The most abundant isomer had an open chain structure, either 2-methyl-1-buten-3-yne or 3-penten-1-yne. The other C_5H_6 isomer was probably cyclopentadiene. Since the sample collection system and chromatographic column were not suited for analysis of oxygenated species, a contribution to the mass 66 signal from a compound containing oxygen cannot be excluded. Secondly, a natural flame ion unique to benzene flames has been observed at mass 67 (Olson and Calcote, 1981). This ion is presumably formed by protonation of

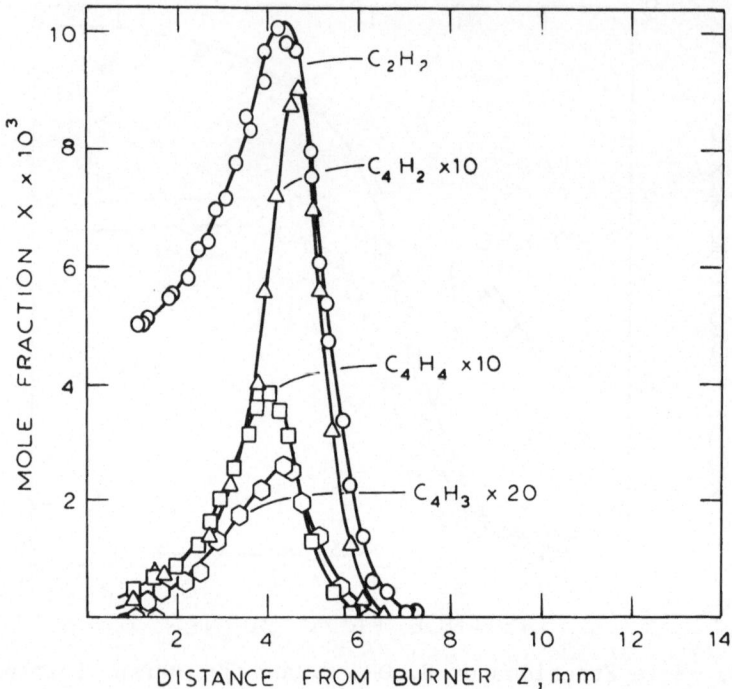

Figure 4. Mole Fractions of Intermediate Species with Masses between 26 and 30 amu versus Distance from Burner in a Stoichiometric (ϕ = 1.0) Benzene (8.2 mol %) – Oxygen (61.8 mol %) – Argon (30.0 mol %) Flame. Unburned gas velocity, 50 cm s^{-1}; pressure 2.67 kPa (source: Bittner, 1981).

an unsaturated neutral species with a mass of 66 amu.

The presence of toluene (C_7H_8) and five (C_6H_8) isomers that included cyclohexadienes, hexatrienes, and methylcyclopentadienes was also confirmed by GC-MS analysis of microproble samples (Faist, 1979). The mole fraction profile of phenyl radical formed by hydrogen abstraction (C_6H_5, Figure 6) maximizes further from the burner than the stable species C_6H_6O, C_6H_8, and C_7H_8, which are formed by addition to the benzene ring. The C_6H_5 mole fraction

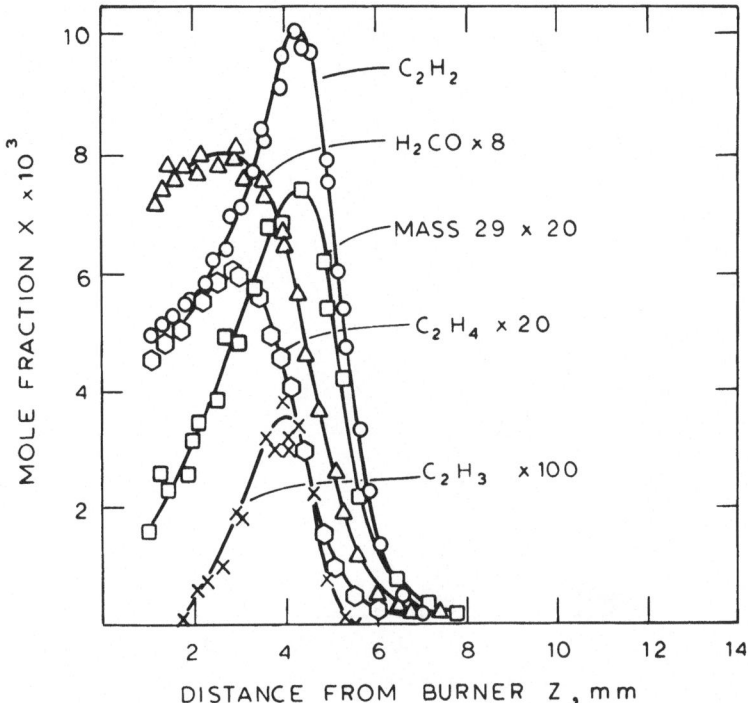

Figure 5. Mole Fractions of C_2H_2, C_4H_2, C_4H_3, and C_4H_4 versus Distance from the Burner in a Stoichiometric ($\phi = 1.0$) Benzene (8.2 mol %) – Oxygen (61.8 mol %) – Argon (30.0 mol %) Flame. Unburned gas velocity, 50 cm s^{-1}; pressure 2.67 kPa (source: Bittner, 1981).

extrapolates to zero while the stable species have non-zero mole fractions near the burner surface.

The mole fraction profiles of these same species and many others were measured in the $\phi = 1.8$ flame and have been reported elsewhere (Bittner and Howard, 1981). Profiles of several species of possible importance to the deposit formation that were observed in both flames, but for which profiles were measured only in the $\phi = 1.8$ flame, are shown in Figure 7. In addition to C_6H_6O, at mass 94, signals were observed at masses 108 (C_7H_8O) and 110 ($C_6H_6O_2$)

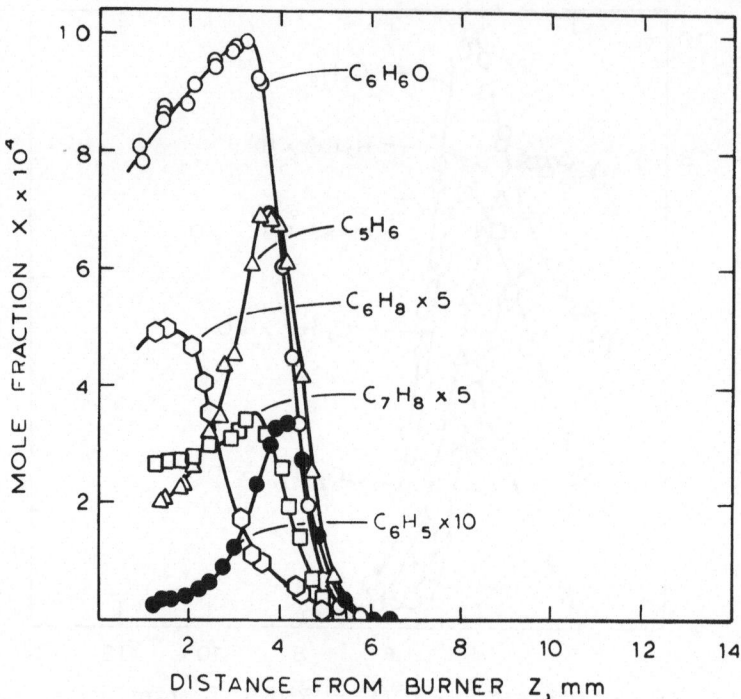

Figure 6. Mole Fractions of C_5 and C_6 Intermediates versus Distance from Burner in a Stoichiometric ($\phi = 1.0$) Benzene (8.2 mol %) - Oxygen (61.8 mol %) - Argon (30.0 mol %) Flame. Unburned gas velocity, 50 cm s^{-1}; pressure, 2.67 kPa (source: Bittner, 1981).

that are probably due to the cresols (hydroxytoluenes) and dihydroxybenzenes, respectively. The mole fractions of these species maximize even closer to the burner than C_6H_6O.

Polycyclic aromatic hydrocarbons were measured in both flames near the maximum of the phenyl radical mole fraction. In the ϕ = 1.8 flame (Figure 7), the maximum mole fractions of PAH ranged from 10^{-4} for naphthalene ($C_{10}H_8$) to about 3×10^{-6} for pyrene ($C_{16}H_{10}$). As shown by the profile labelled $I_{m>700}$, species with masses higher than 700 amu were detectable in the ϕ = 1.8 flame. This signal was

236

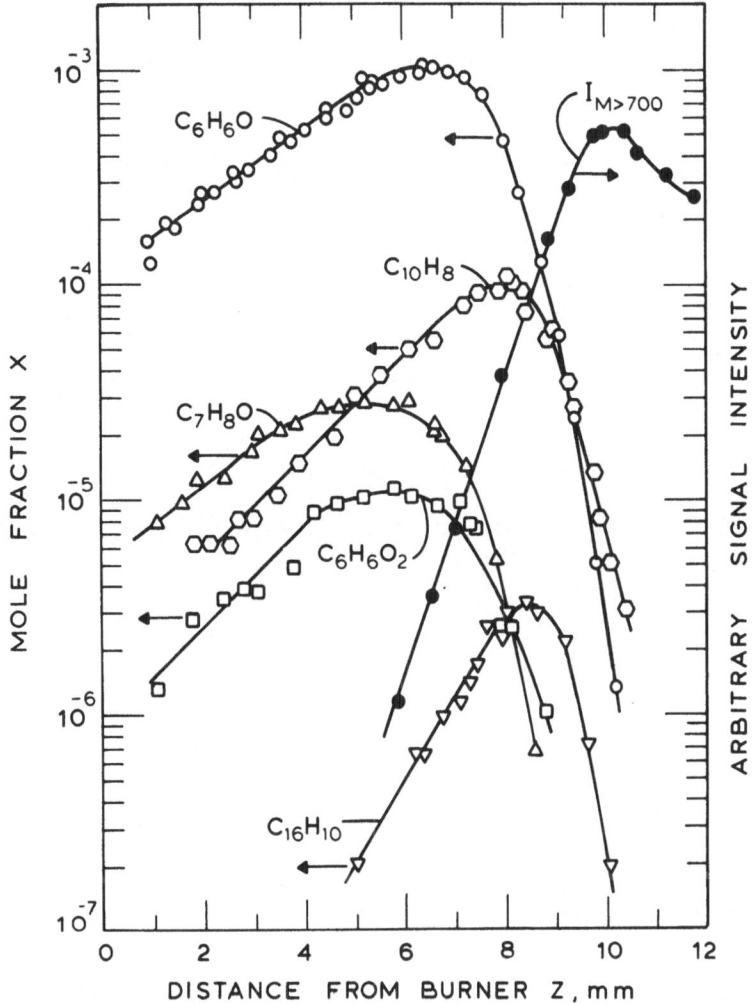

Figure 7. Mole Fractions of Several Intermediate Species and the
Signal from Species with Masses greater than 700 amu,
$I_{M>700}$, versus Distance from the Burner in a Near Sooting (ϕ = 1.8) Benzene (13.5 mol %) – Oxygen (56.5 mol %)
– Argon (30.0 %) Flame.   Unburned gas velocity, 50 cm
s^{-1}; pressure, 2.67 kPa (source: Bittner, 1981).

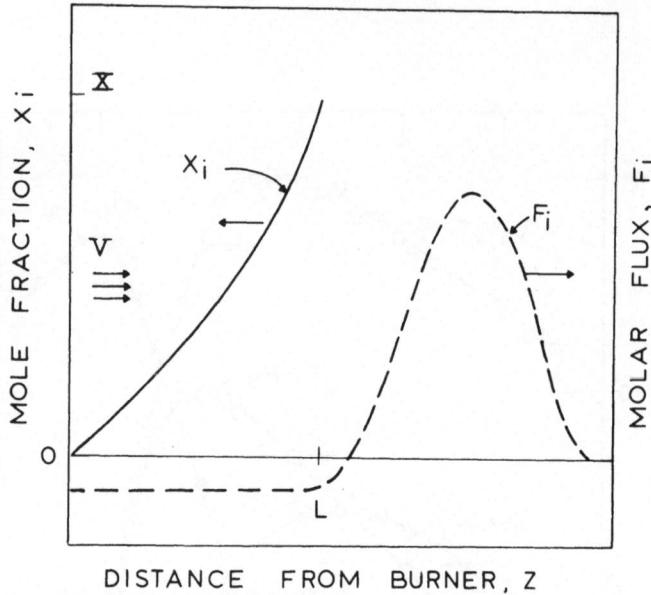

Figure 8.   Schematic Representation of Model of Species Diffusion
through the Preheat Zone.

measured by operating the mass spectrometer as a high pass filter
in which all ions of a specified mass and larger were transmitted.
It was not possible to estimate the mole fraction of this high
molecular weight material.

POSSIBLE DEPOSIT FORMATION MECHANISMS

The observation that deposits form on the surface of the burner
from ϕ = 1.0 benzene flames but not ϕ = 1.0 acetylene flames is consistent with the finding of Dimitroff et al. (1969) that aromatic
compounds are the main precursors to engine deposits. The major
compositional differences between the stoichiometric and rich (ϕ =
1.8) benzene flames reported here and rich (ϕ = 2.4 and 3.0) acetylene flames reported by Homann and colleagues (Bonne et al., 1965)
and Bittner and Howard (1981a), are the presence of large mole
fractions of C_6H_6O and C_5H_6, and larger mole fractions of PAH in the
benzene flames.

238

The mechanism of deposit formation must include an oxidation step as well as polymerization steps to account for the large amount of oxygen present in the dried solid. Either one or both of these steps may occur either homogeneously in the gas phase or heterogeneously on the burner surface. For example, partial oxidation of the aromatics forms oxygenated species such as phenol (C_6H_6O) and formaldehyde (H_2CO). These species may form high molecular weight oxygenated polymers. As discussed in the previous section, an oxygenated compound at mass 66, unique to benzene, may also be present and could participate in polymerization.

Deposition of the polymer could be either a physical condensation process or may be aided by further cross-linking at the surface. It is also possible that as the high molecular weight polymers diffuse into the cooler region of the flame, an aerosol is formed either by a physical condensation process or by coagulation and growth by reactions with lower molecular weight material. In fact, Homann et al. (1973) have visually observed a pale mist in the preheat zone in low pressure benzene-oxygen flames similar to those of the present study. Homann et al. measured profiles of light absorption at 518.4 nm versus distance from the burner and observed a maximum at the beginning of the oxidation zone in benzene flames but not in acetylene flames. The maximum absorption occurred near the point where the mole fraction of the C_6H_6O intermediate formed in a similar benzene flame (Homann et al., 1963). Light scattering measurements may help determine whether a particulate phase is formed from the oxygenated aromatics in the preheat zone.

Other mechanisms for deposit formation can be envisioned in which the oxygen is incorporated into the deposit by oxidation at the burner surface. Oxygen atoms and OH radicals may diffuse from the primary reaction zone to the surface where they oxidize polymeric deposits. Hydrogen atoms diffusing toward the burner react with O_2 to form hydroperoxyl radicals (HO_2, Figure 3) that may participate in a low temperature oxidation mechanism at the surface.

Species that may be important in a mechanism in which the depositing polymeric material is composed primarily of hydrogen and carbon are the unsaturated C_2's and C_4's shown in Figures 4 and 5, and the polycyclic aromatic hydrocarbons (PAH) represented by $C_{10}H_8$ and $C_{16}H_{10}$ in Figure 7. Since the concentrations of the C_2 and C_4 species are similar to those found in C_2H_2 flames (Bittner and Howard, 1981a), it is unlikely that they are important deposit precursors. On the other hand, the maximum PAH mole fractions in the oxidation zone of the benzene flames are several order of magnitude higher than in even the much richer C_2H_2 flames. The $I_{M>700}$ signal of Figure 7 indicates that species with molecular masses larger than 700 amu are formed, presumably from PAH, and diffuse toward the burner surface.

SIMPLE DIFFUSION MODEL

Deposition of solid material on the burner surface requires the diffusion of one or more flame intermediates from the primary reaction zone where it is formed, through the preheat zone, to the surface. A simple model of species diffusion counter to the flow of gas from the burner was used to determine which species are present in high enough concentrations to account for the measured rate of deposition from the $\phi = 1.0$ flame ($\dot{m} = 7 \times 10^{-7}$ g cm^{-2} s^{-1}) if the process were limited by the rate of diffusion to the surface. The model was based on the assumptions that the concentration of the depositing species is zero at the surface and that no reactions occur in the preheat zone. The zero surface concentration approximation would hold for the case of a surface reaction which is very fast relative to the arrival rate of the precursors and for the case of deposition by physical condensation of a species that has a very low vapor pressure at the temperature of the surface.

The primary reaction zone, depicted in Figure 8 by the region of increasing molar flux F_i is assumed to be a source of deposit precursors i that maintains a mole fraction X at a distance L from the burner. Between the burner surface ($z = 0$) and L, the species conservation equation for i is

$$\frac{d}{dz}\left[C\left(VX_i - D_{i,mix}\frac{dX_i}{dz}\right)\right] = 0 \qquad \text{(Eq. 3)}$$

where V is the bulk gas velocity, C is the molar concentration, $D_{i,mix}$ is the diffusion coefficient of species i in the mixture, and z is the distance from the burner. For this order of magnitude calculation it is assumed that V is constant and Eq. 3 can be integrated, using the boundary conditions $X_i = X$ at $z = L$ and $x_i = 0$ at $z = 0$, to give Eq. 4.

$$X_i = X\left[\frac{e^{zV/D_{i,mix}}-1}{e^{LV/D_{i,mix}}-1}\right] \qquad \text{(Eq. 4)}$$

The mass deposit rate $\dot{m}_i$ at the surface ($z = 0$) can be expressed as

240

$$m_i = M_i CD_{i,mix} \left[\frac{dX_i}{dz} \right]_{z=0} = \frac{M_i CXV}{e^{LV/D_{i,mix}} - 1} \qquad \text{(Eq. 5)}$$

The measured mass deposition rate was used in Eq. 5 to calculate the mole fraction X at distances of L = 2 and 4 mm required for several species observed in the flame to account for the measured deposition rate. Mean temperatures T_m of 500K and 1000K at L = 2 and 4 mm, respectively, were used to evaluate V from Eq. 6.

$$V = 50 \frac{cm}{s} \left[\frac{T_m}{300K} \right] \qquad \text{(Eq. 6)}$$

The diffusion coefficients $D_{i,mix}$ were assumed to be equal to D_{i,N_2} since the mean molecular weight of the flame gases was about 30g/mole. The Chapman-Enskog kinetic theory (Bird et al., 1960) was used to calculated D_{i,N_2} at T_m.

This analysis indicated that only species with masses less than 100 amu are present in the ϕ = 1.0 flame in concentrations within an order of magnitude of those sufficient to account for the observed deposition rate. This model probably underestimates the total flux since the 1 mm holes in the burner head would give additional convective mixing. Thermal diffusion, due to the temperature gradient, will also add to the flux. Although this analysis must be considered preliminary, it suggests that relatively low molecular weight species are diffusing to the surface and forming polymeric material at the surface, rather than polymerizing in the gas phase far from the surface.

It should also be recognized that diffusive transport to the surface in a steady burner-stabilized flame may be quite different from the unsteady case of flame propagation to a wall. A recent numerical study by Westbrook et al. (1981) has shown that stable intermediate species such as formaldehyde and methane with mole fraction profile maxima in the primary reaction zone of freely propagating unquenched methanol flames have maximum mole fractions at the wall at the time of quench (when the flame no longer propagates toward the cold wall) that are close to the maximum mole fractions calculated for the freely propagating case. Thus, intermediate species observed in a burner stabilized flame, that may be limited by diffusion counter to the flowing stream, may be able to reach the wall through the nearly stagnant gas as the flame approaches the wall and slowly moves away during the diffusion-limited fuel oxidation phase that follows the quench.

SUMMARY

Deposits have been collected from low pressure laminar premixed benzene flames that are similar to engine deposits. Like engine deposits they have a high oxygen content and IR adsorption bands that indicate the presence of carbonyl groups, OH groups, and C-O linkages. Deposits at stoichiometric fuel-air mixtures were not observed in similar C_2H_2 flames.

Analysis of the flame gases near the burner surface by molecular beam mass spectrometry has revealed the presence of many polymerizable intermediate species that may form the deposit. Compounds at mass 94 (C_6H_6O) and mass 66 (C_5H_6 or C_4H_2O) that are found in high concentrations in benzene flames but not acetylene flames may be important deposit precursors.

A simple model of the diffusion of intermediate species from the reaction zone to the surface suggests that only species with masses less than 100 amu are present in high enough concentrations to account for the measured deposition rate under conditions of mass diffusion control; however, convection and thermal diffusion could greatly increase the access of high molecular weight species to the surface.

Continuing studies will examine the effect of fuel structure, temperature level, and higher pressures.

ACKNOWLEDGEMENTS

Grateful acknowledgement is given to Exxon Research and Engineering Company for support of the application of our molecular beam sampling and analysis technique to this problem. The thesis research of J.D.B. was initiated under Grant R-803242 from the Environmental Protection Agency and carried out with in-house support, both of which sources are gratefully acknowledged.

REFERENCES

Bird, R.B., Stewart, W.E., and Lightfoot, E.N., Transport Phenomena, John Wiley and Sons, New York, 1960, p. 511.

Bittner, J.D., and Howard, J.B., "Pre-Particle Chemistry in Soot Formation," in Particulate Carbon-Formation during Combustion, General Motors Research Symposium, Warren, Michigan (to be published), 1981a.

Bittner, J.D., and Howard, J.B., "Composition Profiles and Reaction Mechanisms in a New Sooting Premixed Benzene-Oxygen-Argon Flame," Eighteenth Symposium (International) on Combustion, The Combustion Institute, Pittsburgh, Pennsylvania, 1981b, p. 1105.

Bittner, J.D., "A Molecular Beam Mass Spectrometer Study of Fuel-

Rich and Sooting Benzene-Oxygen Flames," Sc.D. thesis, Department of Chemical Engineering, Massachusetts Institute of Technology, Cambridge, Massachusetts, 1981.

Bonne, U., Homann, K.H., and Wagner, H.Gg., "Carbon Formation in Pre-mixed Flames," Tenth Symposium (International) on Combustion, The Combustion Institute, Pittsburgh, Pennsylvania, 1965, p. 503.

Dimitroff, E., Moffitt, J.V., and Quillian, R.D., Jr., "Aromatic Compounds Identified as Main Precursors of Engine Varnish," SAE Journal 77, 52 (1969).

Faist, S.M., "Analysis of Stable Species in a Benzene-Oxygen-Argon Laminar Premixed Flame by Chemical and Spectroscopic Techniques: Applications to Soot Formation and Combustion Chamber Deposits," M.S. thesis, Department of Chemical Engineering, Massachusetts Institute of Technology, Cambridge, Massachusetts, 1979.

Fristrom, R.M. and Westenberg, A.A., Flame Structure, McGraw-Hill, New York, 1965, p. 336.

Homann, K.H., Morgeneyer, W., and H.Gg. Wagner, "Optical Measurements on Carbon Forming Benzene-Oxygen Flames," Combustion Institute European Symposium, Academic Press, New York, 1973, p. 394.

Lauer, J.L. and Friel, P.J., "Some Properties of Carbonaceous Deposits Accumulated in Internal Combustion Engines," Combustion and Flame 4,107 (1960).

Morrison, R.T. and Boyd, R.N., Organic Chemistry, Allyn and Bacon, Boston, Massachusetts, 1965, Third Edition, p. 645.

Westbrook, C.K., Adamczyk, A.A., and Lavoie, G.A., "A Numerical Study of Laminar Flame Wall Quenching," Combustion and Flame 40,81 (1981).

White, P.C., discussion following paper by J.J. Mikita and B.M. Sturgis, Proc. Fourth World Petrol. Cong., Sect. VI, (1955), p. 376.

JET AIRCRAFT FUEL SYSTEM DEPOSITS

Robert N. Hazlett and James M. Hall

Chemistry Division, Naval Research Laboratory

Washington, D. C. 20375

INTRODUCTION

Fuels for use in turbine powered aircraft require many limitations on properties to obtain desirable engine behavior and reliable fuel handling. One of the most critical of the jet fuel properties is that of thermal oxidation stability. This requirement became evident in the middle fifties.

Degradation of the fuel due to poor stability produces solid materials, both fuel system varnishes and particulate matter. These materials deposit on surfaces of heat exchangers in which the fuel cools vital aircraft fluids or parts. For instance, the lubricant for the turbine engine bearings is cooled in an oil/fuel heat exchanger. The insoluble material can also exert a detrimental effect on spray patterns by depositing in the combustor nozzles. Thermal oxidation stability is influenced by three major factors – fuel composition, dissolved oxygen, and temperature. Early studies on fuel stability chemistry have been reviewed by Nixon (1) and Schwartz and Eccleston (2).

REVIEW OF DEPOSIT CHEMISTRY

The composition of deposits affords clues to the molecular species involved in deposit formation and the mechanism of formation. The composition of jet engine fuel nozzle deposits has been discussed by Nixon (1) and Rogers (3). In both cases, hetero atoms – oxygen, nitrogen and sulfur – and ash comprised over 40 percent of the deposit. Oxygen was second only to carbon in concentration. In fact, Nixon's data (1) indicated the

O/C atom ratio for the engine deposit was 0.35. The hydrogen contents reported by Nixon (1) and Rogers (3) were relatively high, 1.5 and 1.9 H/C ratio, respectively. The high oxygen and hydrogen contents may be due to absorbed water. Rogers reported the ash contained large amounts of lead and iron but much lower amounts of copper.

Considerable other analytical information has been developed for fuel simulators and smaller scale devices. For flow devices (1,4,5), the hetero-atoms are major components of the thermally formed deposits. Values reported have ranged between 0.5 and 12% for nitrogen, 12 and 36% for oxygen, 0.3 and 9% for sulfur, and 4 and 8% for hydrogen. Remarkable O/C atom ratios have been observed, in excess of 0.5 (1,5). Hydrogen/carbon atom ratios in the deposits are usually less than 1.5 and many analyses find ratios between 0.5 and 1.0 (1). These low results suggest that aromatic compounds contribute significantly to deposit formation.

For deposits formed in bombs at elevated temperatures (160–200°C) similar compositional data was observed – high oxygen, sulfur and nitrogen contents (1,6,7).

Definitive work on the chemical functionalities present has not been reported. Nixon (1) detected the presence of carbonyl and aldehyde groups and suggested that strong acids are also present. The lack of knowledge in this type of analysis is due to low solubility of the deposits in any solvent that has been tried.

SOURCES OF DEPOSIT SAMPLES

The Naval Research Laboratory has recently examined fuel deposits formed in several different types of fuel system hardware. Two of the deposit samples were taken from engines during maintenance operations. The first sample, from a commercial airliner operated on Jet A, was deposited in the valve cavity of a fuel nozzle from a CF6-50A engine. The second sample, from a military aircraft operated on JP-5, was taken from the fuel manifold adjacent to the combustor nozzles of an engine.

Additional deposits were obtained from fuel test devices operated in the laboratories of the Naval Air Propulsion Center (NAPC) and the Air Force Aero Propulsion Laboratory (AFAPL).

The NAPC device used a single heat exchange tube from the TF30 turbine engine (F-14 Tomcat) (Fig. 1). Fuel flowed on the inside of the tube, countercurrent to deaerated hot lubricating oil on the exterior. Dimples on the inside and semicircular fins

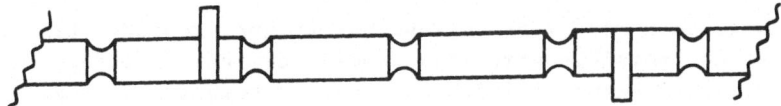

O.D.	2.5mm (0.10 in)	NO. SEGMENTS	33
I.D.	2.0mm (0.08 in)	NO. FINS	15
FIN O.D.	6.1mm (0.24 in)	TOTAL LENGTH	35cm (13.8 in)
SEGMENT LENGTH	9.7mm (0.38 in)		

Fig. 1 Section of TF30 Engine Heat Exchange Tube Showing Four Segments and Fins

or baffles on the exterior of the tube encourage turbulent flow of the two fluids. After several hundred hours of test with accompanying heat exchange measurements (8), the stainless steel tube was cut into 33 one-cm segments. There was a black coating on the inside but no deposit on the exterior. Any residue of lube oil and fuel was removed by soaking in benzene and then drying in an oven at 100°C. Selected specimens were subject to C-H-N analysis, to oxygen analysis, or to sulfur analysis.

The AFAPL deposits were laid down in a 5/16 inch (7.9mm) O.D., 1/4 inch (6.5mm) I.D., 10ft (3.05m) long straight stainless steel fuel manifold which was incorporated into the AF Advanced Aircraft Fuel System Simulator (9). At the conclusion of 40 to 100-hour heat exchange tests using electrical heating, 10-inch (25 cm) long sections at 64(163), 74(188) and 101(257) inches (cm) from the manifold inlet were removed and sliced in half lengthwise. The deposit could then be removed for analysis by scraping the inner wall. It was a thin layer of brown to black dry powdery material. Three to twelve mg of deposit were obtained from a 9 cm length of the various specimens (1/2 of tube only). The tubes were determined to be free of fuel and were analyzed without drying or heating. Separate portions of the solid fuel deposit were used for C-H-N, oxygen and sulfur analyses.

ANALYTICAL PROCEDURES

Carbon, hydrogen, nitrogen and oxygen in the deposits were determined with a Perkin-Elmer Elemental Analyzer, Model 240B. C,H and N were determined simultaneously while oxygen was measured on a separate sample after modifying the instrument.

247

Since each NAPC tube segment was unique, reproducibility data was not possible with this deposit device. The deposit pattern with segment position was usually consistent, however, giving us confidence in the analyses. The reproducibility for the AFAPL deposits was good. Duplicate carbon analyses agreed to within one percent for several samples. Those which gave larger differences (± 4 to 6%) always had a high but variable ash content. This is thought to be due to inhomogeneities in the sample, probably metal particles from the cut tube.

Sulfur in the deposits was determined on a separate sample by modifying ASTM method D3120-75. A Dohrmann microcoulometer was mated to a Single Boat Inlet and furnace in order to oxidize the sample. Duplicate sulfur analyses on the AFAPL specimens differed by 0.7% or less absolute.

The range of the Elemental Analyzer was not sensitive enough to determine nitrogen content of fuels. Such analyses of the Air Force fuels were carried out at NRL using an Antek Nitrogen Detector (chemiluminescent) with a Dohrmann furnace and Single Boat Inlet (10).

RESULTS

Engine Deposits

Table I lists data for the deposits removed from engines. Duplicate data is included where available. The sum of the C,H,N,O and ash totals 95% for the CF6 engine and 93% for the TF-30 engine. Sulfur, which was not determined for these materials, undoubtedly accounts for a portion of the missing 5 and 7%.

Table I. Composition of Jet Engine System Deposits

Element	CF-6		TF-30	
Carbon	50.3%	55.9%	56.4%	55.9%
Hydrogen	2.4	3.0	3.1	2.9
Nitrogen	0.6	0.8	1.4	1.4
Oxygen	20.5	–	26.5	26.5
Ash	20.9	–	5.6	6.7
TOTAL	94.7		93.0	93.4
Atom Ratio				
H/C	0.57	0.64	0.66	0.62
N/C	0.01	0.01	0.02	0.02
O/C	0.31	–	0.35	0.36

The amount of oxygen is quite high, over 20 percent for both engine samples. On an atom basis, we find about one oxygen atom for three carbon atoms. Nitrogen is much lower than oxygen but still significant. The hydrogen to carbon ratio is distinctly lower than the 2.0 which is typical of jet fuels.

Deposits From NAPC Device

The NAPC single tube heat exchange device has been run on seven fuels (Table II). The heat exchange results for the first four tests have been reported by Delfosse (8). The same base fuel was used for tests B,C,D. The stability of fuels used in Tests A-D was degraded by the addition of copper in various chemical forms. The maximum fuel temperature (at outlet), which should be a significant factor in deposit formation, was maintained within narrow limits. Results are illustrated in Figs. 2 to 7. They reveal the relative deposition tendency for the several fuels and the pattern of deposition with rising fuel temperature along the heat exchanger tube. The results should be interpreted in terms of fuel stability, fuel test temperature and test duration.

Fig. 2 shows carbon per segment for selected segments for each tube. In most cases the amount of carbon and each other element, and hence the amount of deposit, increased with temperature as fuel flowed toward the exit, as would be expected. The relationship is approximately linear. Tests A and D gave atypical patterns. Test A showed a sharp rise at segment 28, giving a V-shaped curve overall, especially obvious with the elements other than carbon. Test D exhibited significantly greater amounts of deposit in the cooler part of the tube, segments 2-5. This suggests that breakdown began before entering the heat exchanger tube, apparently due to preheating. Fuel D was a particularly poor one and resulted in excessive deposits in the preheater ahead of the heat exchange tube (8). The shapes of corresponding curves for hydrogen (Fig. 3), nitrogen (Fig. 4), and oxygen (Fig. 5) were generally similar to those shown for carbon. Data for sulfur were limited. A sulfur plot for Test C is shown in Fig. 6. The drop in deposition at segments #31-33 is attributed to reduced temperature due to proximity to the tube end fitting.

The data shown are for segments without fins after solvent cleaning and oven drying. Cleaning removed any traces of lube oil and unreacted fuel. Generally segments with exterior fins gave slightly higher results. Tube C showed the maximum effect for fins (Fig. 7). Oxygen was affected similarly. This effect is ascribed to two causes. Fins on segments would produce hot spots arising from greater heat transfer from the lube oil through the fin. Secondly, some finned segments even after

Table II Fuels in NAPC Heat Exchanger Tests (a)

Test No.	Fuels	JFTOT (b) Breakpoint °F	Fuel Temp °F In/Out	Test Duration hr	Deposit Rate (c) ppb C x10^2/hr	Total Deposit mg C	Hrs. to 1% Heat Loss Xchgr
A	JP-5R + 50 ppm Cu (e)	460	286/366	280	6.5	7.4	31
B	JP-5(d) + 50 ppm Cu(e)	520	250/330	600	1.4	7.8	107
C	JP-5(d) + 250 ppm Cu(e)	470	260/340	717	2.7	22	73
D	JP-5(d) + 250 ppm Cu(f)	416	260/340	360	6.7	15	29
E	JP-5 + 10% DFM (g)	470	176/345	382	–	49	37
F	JP-5	525	176/345	395	–	16	158
G	Shale II JP-5	545	176/345	400	–	0.3	1000(est)

(a) Data from Ref. 8 except next to last column
(b) High temperature flow test for jet fuel stability (Jet Fuel Thermal Oxidation Tester, ASTM Test No. D3241)
(c) ppb C or mg C refers to carbon
(d) Same JP-5 used in these fuels
(e) As bis(1-phenyl 1,3-butanediono) Cu II
(f) From exposure to Cu metal
(g) Navy diesel fuel marine added to same JP-5 used in Test F

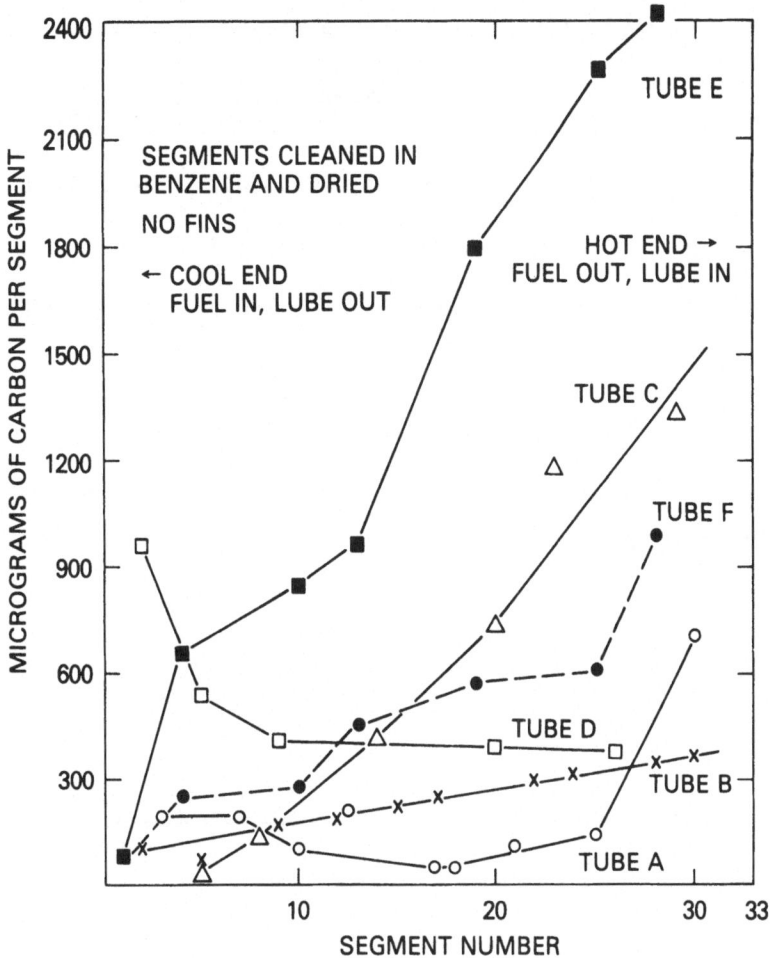

Fig. 2   Carbon in Fuel Deposits on NAPC Heat Exchanger Tube Segments

cleaning showed a slight coating on the fin. Any such residue
from the lube oil would add to the amounts of C, H, N and O found.

The atom ratios for six NAPC tests are presented in Table
III. The seventh test, using highly refined JP–5 from shale oil
(11), gave such small amounts of hydrogen, nitrogen and oxygen
that ratios were meaningless.* The H/C ratios in Table III vary
widely, both between tests and between different parts of a
specific tube. The higher H/C ratios for tube B are unrealistic
and are probably due to absorption of water. The N/C atom ratios
were quite high for tests A, B and C. The ratios for tests D,
E and F were similar among themselves but much lower than the

---

*Maxima per segment: carbon 14 µg, hydrogen 2 µg, nitrogen 1 µg,
 oxygen 20 µg. These are 15–60 times less than tube F data.

Table III. NAPC Heat Exchanger Deposits

| Tube | Atom Ratio | | | |
	H/C	N/C	O/C	S/C*
A	0.4–1.7	0.08–0.13	0.3–0.7	0.02–0.10
B	1.4–2.4	0.08–0.11	0.7–1.5	---
C	1.1–1.6	0.11–0.18	0.2–0.5	0.04–0.08
D	0.5–1.5	0.03–0.04	0.2–0.4	---
E	0.7–1.4	0.03–0.04	0.3–0.6	---
F	0.7–0.8	0.03–0.04	0.2–0.5	---

*Limited data for S/C

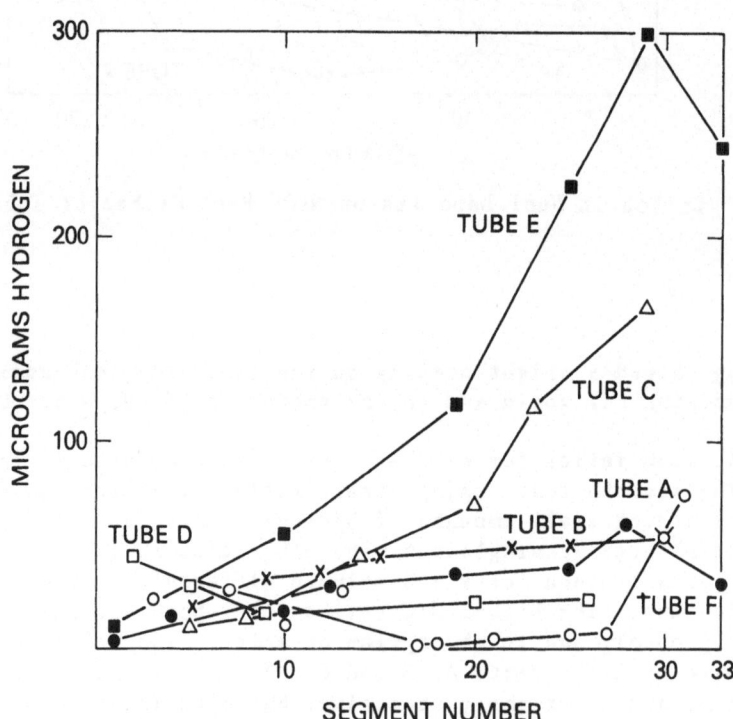

Fig. 3   Hydrogen in Fuel Deposits on NAPC Heat Exchanger
Tube Segments

other three. The oxygen-to-carbon ratios also varied somewhat
and were generally quite high. The O/C ratios for tube B were
abnormally high, reinforcing the suggestion that this deposit
readily absorbed moisture. The limited data available for sulfur
show that the compounds of this element are also involved in jet
fuel deposit formation.

The test conditions for tests A to G were not uniform.
However, tests E, F and G were carried out at identical stress
conditions of 176°F fuel-in temperature and 345°F fuel-out temperature. Thus a valid comparison of data for these tests can be
made. The total amount of deposit (sum of all segments) decreased from 49 to 16 to 0.3 mg of carbon for tubes E, F and G. The
hours to one percent loss in the heat exchange coefficient gave a
pattern consistent with this data. Test E required only 37 hours
to reach this loss, whereas tests F and G required 158 and 1000
hours, respectively. A fuel specification test (JFTOT) for the
three fuels gave corroborating data. The Jet Fuel Thermal

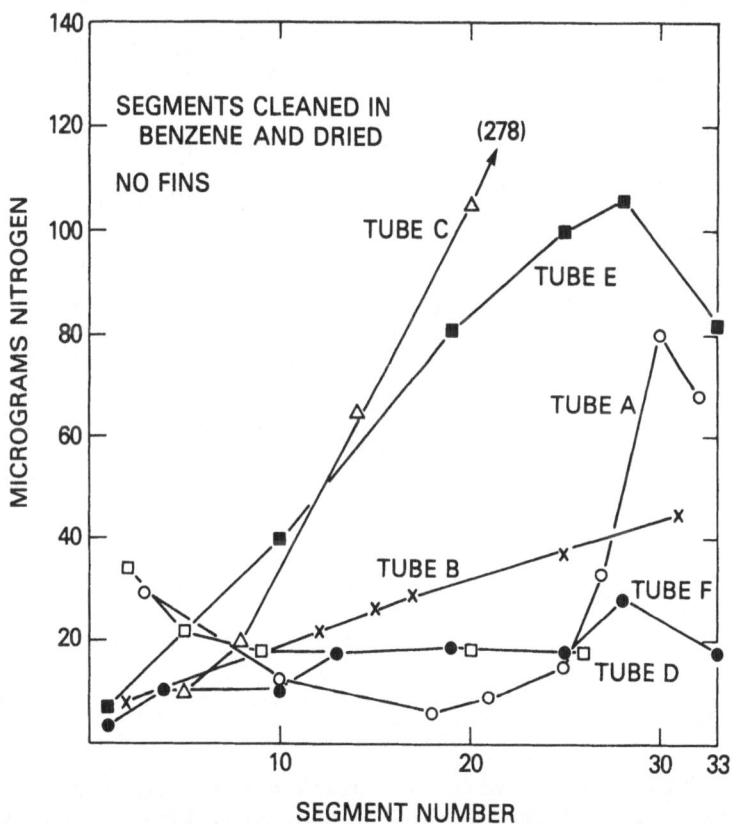

Fig. 4 Nitrogen in Fuel Deposits on NAPC Heat Exchanger
Tube Segments

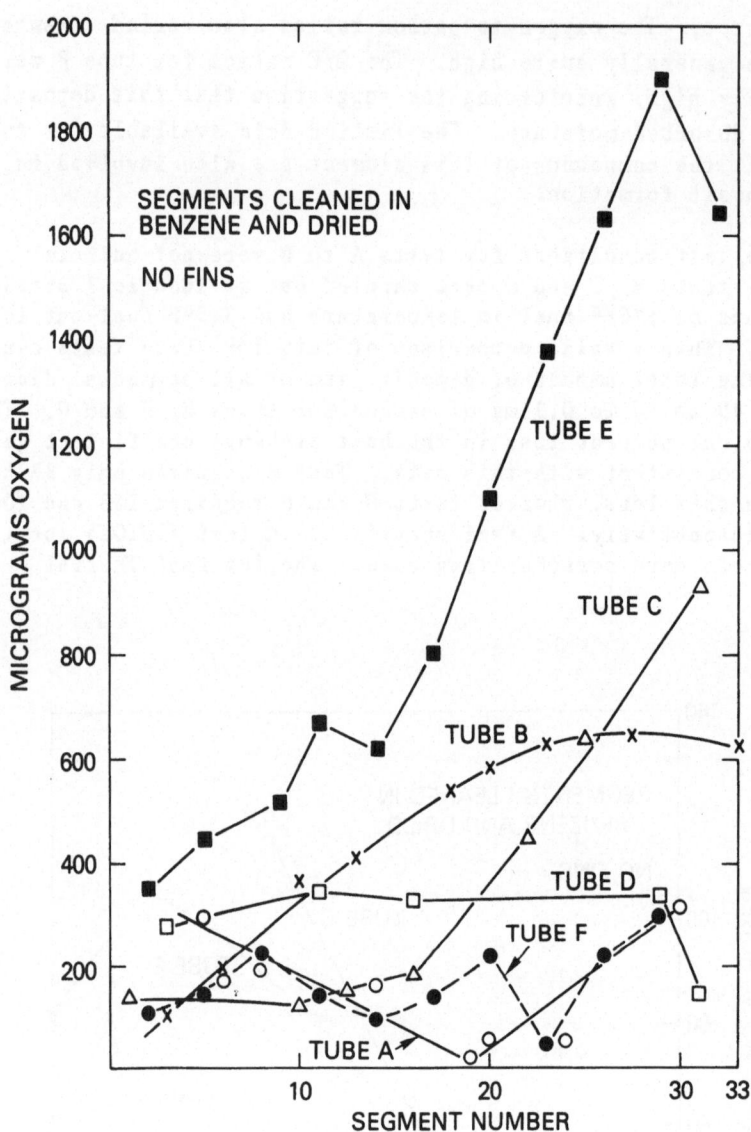

Fig. 5   Oxygen in Fuel Deposits on NAPC Heat Exchanger
Tube Segments

Oxidation Test (JFTOT) results confirmed that fuel E had the
worst stability and fuel G had the best stability. Thus, the
deposit amount, the heat exchange loss and the JFTOT data, rate
fuels E, F and G in the same order.

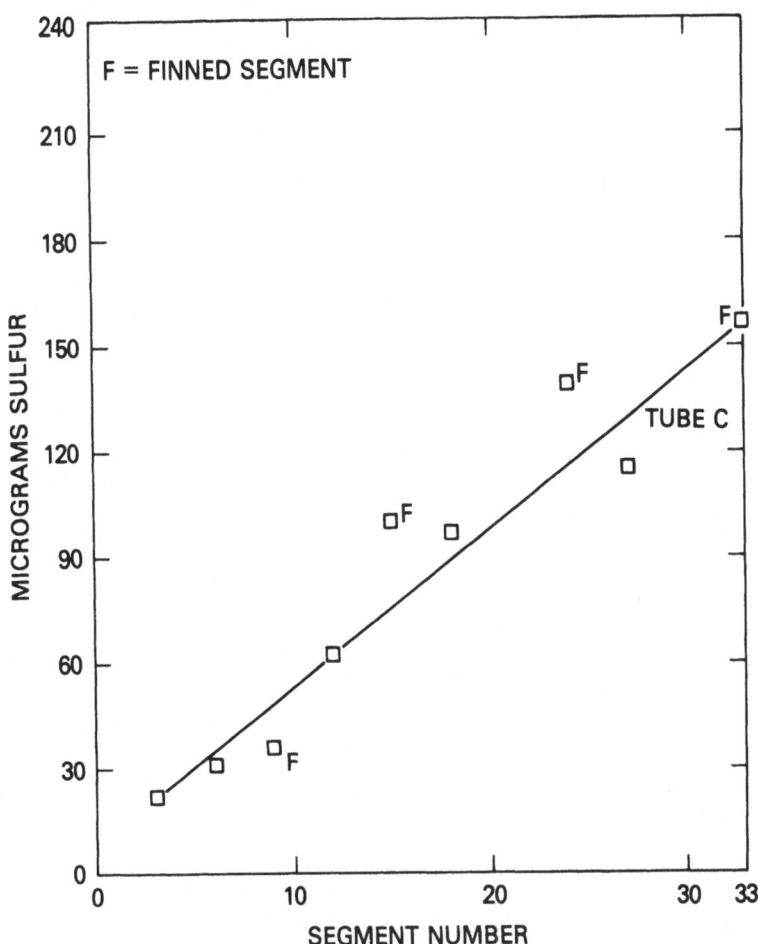

Fig. 6   Sulfur Content of Fuel Deposits on Segments of NAPC Heat
Exchanger Tube C

Deposits from AFAPL Simulator

The AF simulator has been run on a variety of jet fuels-
JP-4, JP-5, JP-7, JP-8 and Jet A (9, 12, 13). This allowed the
formation of deposits at a range of temperatures, 512-688°F tube
temperature. One fuel, AFFB-14, a Jet A, gave so little deposit
(0.04 mg per 9 cm of split tube vs 3-12 mg for the others) that
analyses were precluded. However, sulfur content of this fuel
(0.020%) was not unusually low. The tube temperatures and
corresponding percentage compositions and elemental ratios of the
deposits are given in Table IV. The sum of C,H,N,O,S and ash
gave a good material balance (94-97%) except in three cases.
"Ash" was due to metal particles from the tube wall. Duplicate
analyses gave elemental totals within 3% of each other.

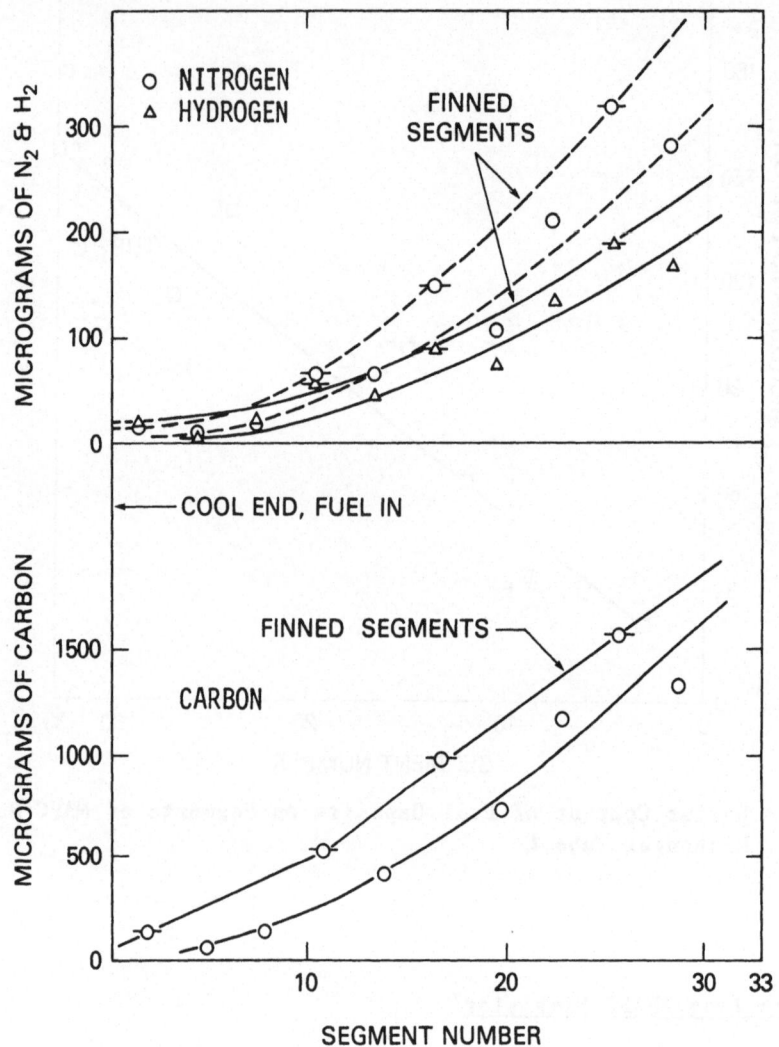

Fig. 7   C–H–N Analyses of NAPC Heat Exchanger Tube C Segments

Table IV. Air Force Heat Exchanger Deposits

AFFB Fuel	Fuel Type	Film Temp. (°F-Calc.)	H/C	N/C	O/C	S/C
			Atom Ratio			
10	aged JP-7	512	0.60	0.053	0.28	0.01
9	JP-5	544	0.43	0.019	0.22	0.06
8	Jet A-1	550	0.52	0.027	0.25	0.01
16	JP-4	554	0.53	0.017	0.23	0.01
13	Jet A	564	0.45	0.006	0.22	-
11	JP-7	665	0.57	0.005	0.09	0.001
12	JP-7	688	0.62	0.011	0.10	0.003

	C	H	N	O	S	Ash	Total
	% Found						
10	57.6	2.8	3.3	17.6	1.5	0.0	82.8
9	61.0	2.4	1.3	15.9	8.3	0.1	89.0
8	65.6	2.9	2.1	21.4	1.6	2.8	96.4
16	68.1	3.1	1.3	20.9	1.0	0.3	94.7
13	69.3	2.6	0.5	20.4	-	0.3	93.8
11	73.6	3.5	0.4	8.4	0.3	1.7	87.9
12	79.9	4.2	1.0	10.8	0.7	0.2	96.8

These results are generally similar to the engine deposit results in Table I allowing for their higher ash content. Carbon is higher in four cases, oxygen is lower and nitrogen is both higher and lower. The percentages of nitrogen and sulfur in the AFFB deposits parallel each other reasonably well (except for AFFB-9) but oxygen does not correlate with either.

The H/C ratios are uniformly low, 0.43-0.62. This may be interpreted as indicating a highly aromatic deposit. The O/C ratios, which vary over a wider range, 0.09 to 0.28 with an average of 0.20, would indicate a significant involvement by oxygen compounds in deposit formation. The nitrogen compound involvement is less since the N/C ratio is much lower, ranging between 0.005 and 0.053. The S/C ratios are still lower, varying from 0.001 to 0.01 except for AFFB-9. All four elemental ratios are generally lower than those found in the NAPC device but are similar to the engine deposit values. AFFB-11 and AFFB-12 gave especially low percentages and ratios for O and S. The nitrogen ratios also were low. The low values may be related to the quality of these fuels. These two materials were JP-7 jet fuels which were produced to have improved thermal oxidation stability. Consequently, they were tested at much higher temperatures than the other fuels. AFFB-10, an aged JP7, gave O/C and N/C ratios that were higher than any others although this deposit came from an area stressed at a low temperature.

Two explanations can be offered for the low oxygen, nitrogen and sulfur content of the JP-7 deposits. The first is that these fuels contained only small amounts of polar compounds, the precursors to deposits containing oxygen, nitrogen, and sulfur. An alternate explanation is that the deposits formed at higher temperatures are derived via hydrocarbon pyrolysis rather than polar compound involvement.

The hetero elements O, N and S become enormously concentrated in fuel deposits. Oxygen content of deposits varied from 16 to 22% (except for AFFB-11 and AFFB-12 where it was 8 and 11% respectively). The amount of dissolved oxygen in a fuel (at saturation) is about 60 ppm. Thus the concentration factor may be 3300 or more. However, oxygenated compounds may also be present in the fuel. The nitrogen and sulfur content of the AFFB deposits is compared with the nitrogen and sulfur concentration in the fuels in Table V. The enhancement factor for nitrogen is tremendous, greater than 10,000/1 for all seven fuels. The sulfur enhancement factor is much lower but still substantial. It varied from 54 to 120 except for AFFB-11 and AFFB-12 where it was much higher although the actual percentages were low. This fact suggests, with support from the nitrogen data, that the percent nitrogen or sulfur in deposits is relatively insensitive to wide variations in the fuel. Note that in one case deposit

Table V. Nitrogen and Sulfur Concentrations in Jet Fuel Deposits

Fuel Sample	Nitrogen			Sulfur		
	In Fuel*	In Deposit*	Concen. Factor	In Fuel*	In Deposit*	Concen. Factor
AFFB- 8	0.000063	2.1	33,000	0.019	1.6	82
AFFB- 9	0.000089	1.2	13,000	0.066	8.0	120
AFFB-10	<0.00001	3.0	>300,000	0.014	1.2	88
AFFB-11	<0.00001	0.4	>40,000	0.0003	0.24	970
AFFB-12	0.000042	0.95	23,000	0.0003	0.74	2500
AFFB-13	<0.00001	0.5	>50,000	0.014	-	-
AFFB-16	0.00014	1.3	10,000	0.019	1.0	54

*Percent by wt.

sulfur reached the high value of 8%. There was no correlation between fuel mercaptan sulfur (less than 0.001%) and deposit sulfur.

IMPLICATIONS FROM DEPOSIT COMPOSITION

The high amount of oxygen, nitrogen and sulfur found in the deposits is noteworthy. The data reported above generally corroborates the information from the literature. All of the deposits from the engines and the test devices point to the importance of compounds containing hetero atoms.

The importance of oxidation in triggering solids formation has been reviewed (14). It would appear that the trace primary oxidation reactions occurring in a fuel system, those due only to involvement with dissolved oxygen (ca 60 ppm), would be insufficient to give the high concentration of oxygen in the deposit. However, if the compounds undergoing oxidation were oxygen-containing compounds rather than hydrocarbons, the high oxygen concentrations in the deposit would be more reasonable. Of course, many oxygenated compounds oxidize more readily than hydrocarbons.

Some nitrogen and sulfur compounds in fuel increase deposit formation (15,16). Thus, it is not surprising to find N and S in the deposits. The high enhancement factors, however, are quite remarkable. Thus, we must conclude that some nitrogen and sulfur compounds found in fuels are very susceptible to oxidation and subsequent deposit formation.

The high concentrations of hetero atoms implies that the deposits have highly polar characteristics. Since such material would have little attration for the non-polar fuel, the insolubility of deposits may be due primarily to polarity differences rather than high molecular weight.

ACKNOWLEDGEMENTS

The authors thank Mr. C. J. Nowack, Mr. R. J. Delfosse, and Mr. Larry Maggitti of the Naval Air propulsion Center for furnishing the heat exchanger test tubes and the TF-30 deposits. Mr. Royce Bradley of the Air Force Aero Propulsion Laboratory supplied the AF simulator specimens and Mr. Maury Shayeson of the General Electric Co. donated the CF-6 deposits.

REFERENCES

1. Alan C. Nixon, "Autoxidation and Antioxidants of Petroleum," Ch. 17 in Autoxidation and Antioxidants, W. O. Lundberg, Editor, John Wiley, New York, 1962.
2. F. G. Schwartz and B. H. Eccleston, "Survey of Research on Thermal Stability of Petroleum Jet Fuels," BuMines Information Circular 8140, 1962.
3. J. D. Rogers, Jr., "Turbine Fuel Thermal Stability. CFR Coker and Flight Evaluations," Society of Automotive Engineers Meeting, Preprint 103T, Los Angeles, CA, October 5-9, 1959.
4. H. B. Minor, A. C. Nixon and R. E. Thorpe, "Stability of Jet Turbine Fuels," Wright Air Development Center, Tech. Report 53-63, Pt. 3, August 1955.
5. B. A. Englin et al, "Service Properties of Jet Fuels Obtained by Vacuum Gas Oil," Khim. i Tekhnol. Topliv i Masel, p. 16 (May 1975), Engl. transl. p. 352.
6. Z. A. Sablina and A. A. Gureev, "The Problem of Methods of Investigating the High Temperature Stability of Aviation Fuels," Khim. i Tekhnol. Topliv SSSR p. 63, No. 9 (1957).
7. Y. R. Tereshchenko and M. Y. Yararyshkin, "Investigation of the Thermal Stability of Sulfurous Fuels at Temperatures Above 100°C," Chemistry of Sulfur Organic Compounds Contained in Petroleum and Petroleum Products, 5, 149-182 (1963). Air Force Systems Command, Foreign Technology Division transl., FTD-MT-64-215.
8. R. J. Delfosse, "Performance of Hot Fuel in a Single Tube Heat Exchanger Test Rig," Rpt. No. NAPC-PE-11, August 1978.
9. H. Goodman and R. Bradley, "High Temperature Hydrocarbon Fuels Research in an Advanced Aircraft Fuel System Simulator," AFAPL-TR-70-3, March 1970.

10. H. V. Drushel, "The Determination of Nitrogen in Petroleum Fractions by Combustion with Chemiluminescent Detection of NO," Anal. Chem. 49, 932 (1977).

11. J. Solash, R. N. Hazlett, J. C. Burnett, E. Beal and J. M. Hall, "Relation Between Fuel Properties and Chemical Composition. Chemical Characterization of U. S. Navy Shale-II Fuels," ACS Symposium Series No. 163, Am. Chem. Society, 237 (1981).

12. CAPT L. Lampman and CAPT J. C. Ford, "High Temperature Hydrocarbon Fuels Research in an Advanced Aircraft Fuel System Simulator on Fuel AFFB-13-69," Air Force Aero Propulsion Laboratory, AFAPL-TR-70-70, March 1971.

13. R. Bradley, R. Bankhead and W. Bucher, "High Temperature Hydrocarbon Fuels Research in an Advanced Aircraft Fuel System Simulator on Fuel AFFB-14-70," Air Force Aero Propulsion Laboratory, AFAPL-TR-73-95, April 1974.

14. Coordinating Research Council Inc., "CRC Literature Survey on the Thermal Oxidation Stability of Jet Fuel," Rpt. No. 509, Chap. IV, Atlanta, GA, April 1979.

15. W. F. Taylor and T. J. Wallace, "Kinetics of Deposit Formation from Hydrocarbons. Effects of Trace Sulfur Compounds," Ind. & Eng. Chem. Prod. R&D, 7, 198 (1968).

16. W. F. Taylor, "Mechanisms of Deposit Formation in Wing Tanks," Society of Automotive Engineers Trans., 76, 2811 (1968).

SOOT REDUCTION IN DIESEL ENGINES BY CATALYTIC EFFECTS

Richard S. Sapienza
Thomas Butcher
C. R. Krishna and
Jeffrey Gaffney

Department of Energy and Environment
Brookhaven National Laboratory
Upton, New York 11973

INTRODUCTION

The diesel engine's proven fuel economy and lower emissions of unburned hydrocarbons and carbon monoxide makes it a viable alternative to a gasoline automotive plant, but the inherent production of particulate matter (soot) threatens the expanded use of the diesel engine.

A variety of methods have been proposed over the years for reducing soot emissions. These methods include engine modifications such as fuel injection optimization, combustion chamber shapes and turbocharging; fuel modifications such as the use of water injection, fuel additives, and fuel fumigation; and more recent suggestions which stress exhaust treatments including "tray oxidizers" similar to catalytic converters and various types of replaceable filters. Fuel blending such as methanol-diesel emulsions, as well as separate introduction of methanol, have also been tried.

In the diesel engine the air charge is compressed to a high temperature and pressure, preceeding fuel injection and combustion begins spontaneously shortly after. The rate of burning and heat release is controlled by the rate of fuel air mixing. This mixing-limited operation leads to not only many of the diesel's desirable characteristics but also to hot, fuel rich zones despite the air charge being in excess of the stoichiometric requirements. Cracking occurs, leading to soot formation in these fuel

263

rich zones. This can be perceived as an autothermal cracking process in which heat, from combustion of the fuel, serves to crack the remainder endothermally (similar to the manufacture of ethylene and acetylene via the pyrolysis of hydrocarbons), but a precise mechanism of soot formation during combustion in a diesel engine in not completely defined.

In this work, a tentative approach was formulated relating the partial combustion of liquid hydrocarbons to the initiation of soot formation in the diesel combustion process coupled with the ample evidence that soot formed in combustion processes consists of mixture of surface and gas formed carbons.[1] Therefore, the influence of both homogeneous and wall reactions in the combustion chamber were considered in order to define the fuel and engine properties which contribute to the formation of particulate matter. If soot is to be avoided complete combustion must be attained. The introduction of either oxygen carriers or catalytic surfaces would increase the probability of contact between fuel and oxygen. The addition of hydroxyl (OH) carriers to the fuel was felt to be a direct way to verify or illustrate the proposed chemistry. The possible catalytic influence of wall and valve materials to promote the surface ignition of carbon and soot precursors was also investigated.

EXPERIMENTAL

A standard Waukeska CFR engine coupled to a D.C. dynomometer was used in the tests. Cetane was used instead of diesel oil as fuel for obvious chemical reasons and n-butanol was used as the fuel additive. The engine was operated at the ASTM Cetane number test conditions making a concerted effort to keep the engine operating variables constant. Typical engine operating conditions were: speed, 900 rpm, fuel flow, 13 ml/min., compression ratio, 19:1.

Sampling of the engine's particulate effluent was done by drawing 15.6 l.p.m. through a 47mm Millipore FA filter directly from the exhaust flow. This procedure minimized distortion of the sample and maintained the integrity of the structurally-weak filter with no particulate break-off. The filters were equilibrated at constant humidity and weighed to determine the soot by difference. Duplicate samples were reproducible to better than 10% and at least two samples were taken at each data point.

After the cetane/n-butanol mixtures were studied, the engine was dismantled and platinum was deposited onto the piston crown and valve faces by vacuum sputtering to a thickness of approximately 6,000 Å with an MRC model SEM-8620 RF Bias sputter-etch

etch module. Prior to insertion in the chamber the components were cleaned mechanically in a glassblaster and washed with acetone and alcohol as a final step. Sputter etching was not possible.

RESULTS AND DISCUSSION

Use of Alcohols

The technoogy of alcohol fuels utilization in diesel engines is not nearly as advanced as that for spark ignition engines. Since our tentative approach to homogeneous soot formation may be conveniently formulataed in terms of the free radical theory of the cracking of hydrocarbons, alcohol could suppress the formation of soot by providing species which remove or retard radical fuel pyrolysis intermediates.

Alcohols Action

1. Radical Retardation

   a. Transfer agent

   $$P\bullet + RCH_2OH \longrightarrow PH + R\bullet CHOH$$

   $$P\bullet + R\bullet CHOH \longrightarrow PH + RCHO$$

   $$P\bullet + RCHO \longrightarrow RCH + PH$$

   b. Copolymerization agent

   $$P\bullet + R\bullet CO \longrightarrow P-\overset{\overset{\textstyle O}{\|}}{C}R$$

   $$P\bullet + R\bullet CHOH \longrightarrow P-\overset{\overset{\textstyle OH}{|}}{\underset{\underset{\textstyle H}{|}}{C}}-R$$

2. Precursor Trap

   a. reaction with acetylene

   $$HC=CH + RCHO \longrightarrow H_2C=C-\overset{\overset{\textstyle OH}{|}}{\underset{\underset{\textstyle H}{|}}{C}}-R$$

    b. inhibits diene cyclization reactions

    c. reaction with aromatic radicals

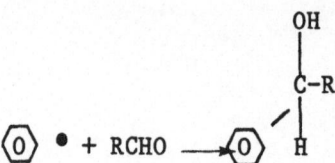

    The observed data as presented in Figure 1, reveal a rapid and then gradual decrease in soot with n-butanol addition with an optimum 30-40% reduction at ∿7% n-butanol addition.

    Although recent tests using methanol and ethanol/diesel emulsions have shown particulate reduction[2],[3] to our knowledge this represents the first systematic investigation of specific alcohol addition versus reduction of soot particulates in a diesel engine. These results seem to support our concept of controlling the gas-phase mechanism of soot formation utilizing the radical retarding effect of alcohols.

Use of Metals

    Catalytic combustion can promote the oxidation of long chain hydrocarbons and soot precursors minimizing polymerization reactions which result in soot formation.

    The mechanism by which metal containing fuel additives reduce smoke is obscure[4], but British Petroleum tests suggest that successful soot suppressants lower the soot ignition temperature while other ineffective additives yield little change in this temperature.[5]

    Alkaline-earth metal fuel additives such as barium salts affect homogeneous soot formation in a manner similar to that proposed for alchols. These metals which reduce soot in all oxygen-fuel ratios, act by the gas-phase catalysis of the decomposition of hydrogen or water vapor to yield hydroxyl radicals.[6] Ignition delay is often the result of the radical retardation caused.

Alkaline Earth Catalysis

M-OH + P• ⟶ M-O• + PH
M-O• + P• ⟶ M-OP

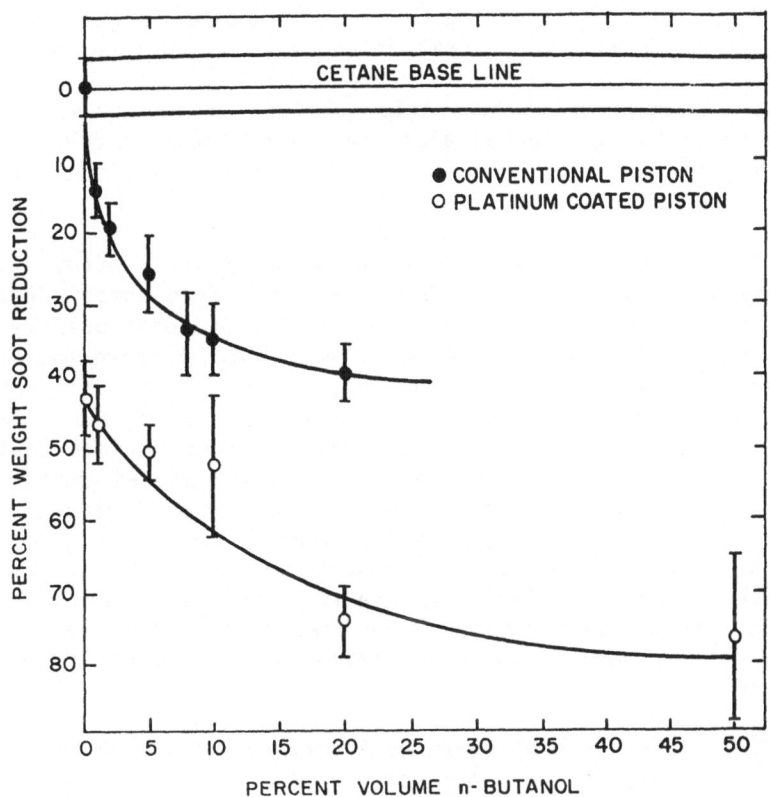

Figure 1.   Effect of Alcohol and Catalyst (Platinum)

Other effective metal fuel additives such as molybedemum, tungsten and chromium perform only at high oxygen-fuel ratios, (>.9) suggesting a different mechanism for their action.[6] It is our contention that these metals act as surface catalysts in the combustion process.

Considering that the nature of the reactor wall has a significant effect on the rate of carbon formation and, the composition of the exit gas in a cracking reaction,[7] the oxidation properties of the combustion chamber surface could determine the quantity and quality of the particulate emissions in a diesel engine, that is the combustion chamber surface functions as an oxidation catalyst. Therefore, the coating of the combustion chamber surfaces with platinum, a well known combustion catalyst, was tried.

The platinum coated combustion surface produced $\sim$ 40% reduction in soot with no alcohol addition. Addition of alcohol to the platinized reactor was found to further reduce soot with reduction up to $\sim$80% being measured (see Figure 1).

A cyclic oxygen transfer mechanism appears to offer reasonable explanations of these results since the free energy decrease on reduction of iron oxides with carbon (or hydrocarbons) is less than that for the reduction of platinum oxide. Because platinum will markedly lower the ignition temperature of carbon[9] and promote ultraclean combustion of hydrocarbons,[8] soot and unburned fuel reaching the platinized combustion chamber surface could be partially or completely oxidized leading to reduced particulate emissions. Heterogeneous reaction of this soot will be limited by the amount of carbon reaching the surface by molecular and turbulent transport but may produce products which also affect the homogeneous chemistry. That is, reaction intermediates initially formed at the surface, break-off to take part in homogenous reactions. This chemistry would be similar to the action of alcohol described earlier.

After eight hours of continuous operation the platinum coated surface was lost, and the observed soot emission found to be identical with the original iron surfaced piston and valves. Although the precise mechanism by which the platinum was removed is unknown, we belive that surface impurities such as, iron carbide and oxide may have caused problems in the platinum adhesion since no stringent cleaning of the piston crown was performed prior to coating.

A simple mechanism generated from various slow combustion studies can explain the surface catalyst behavior.

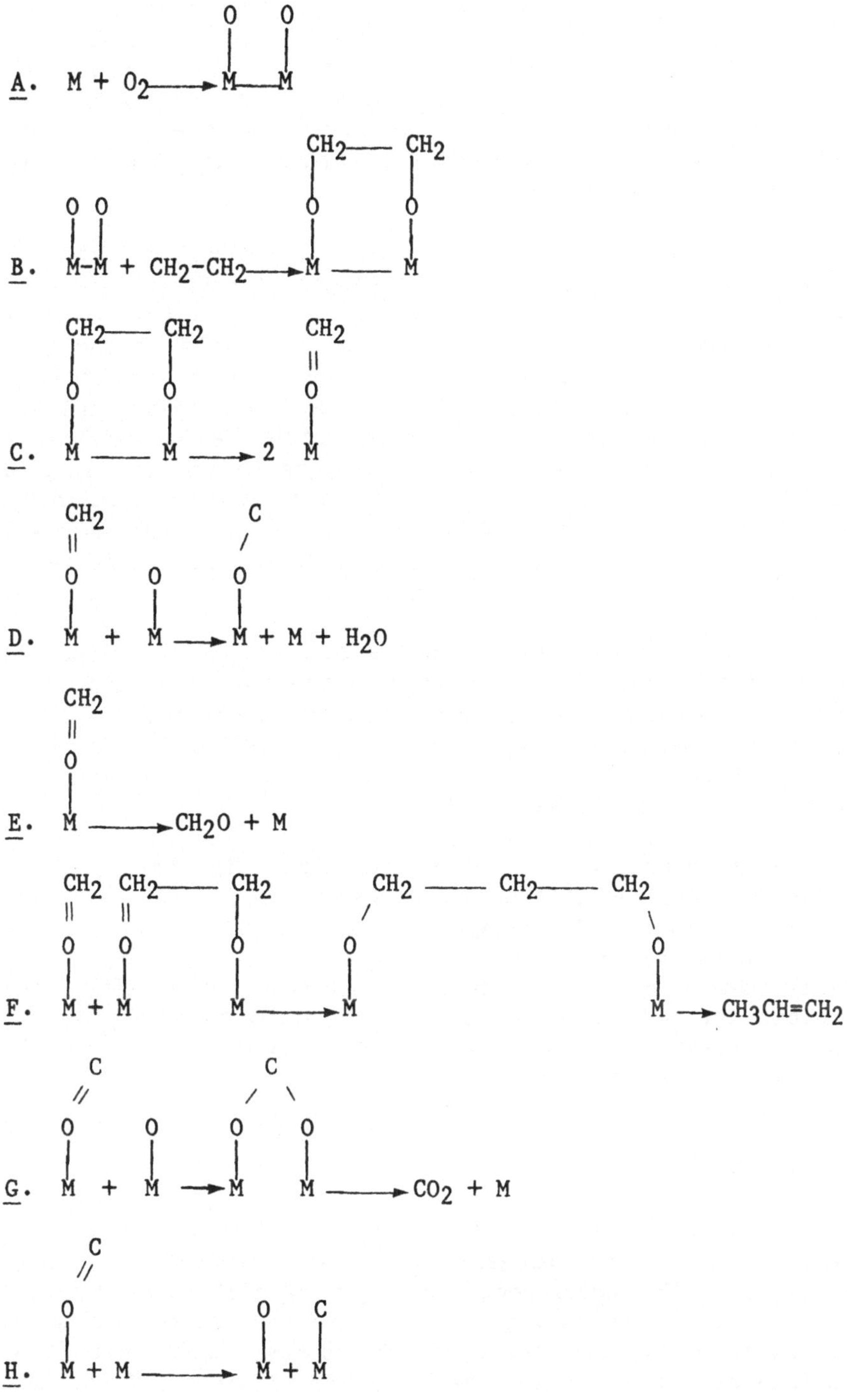

$\underline{A}$. $M + O_2 \longrightarrow$

$$
\begin{array}{cc}
O & O \\
\| & \| \\
M & M
\end{array}
$$

$\underline{B}$.

$$
\begin{array}{cc}
O & O \\
\| & \| \\
M\!-\!M
\end{array}
+ CH_2\!-\!CH_2 \longrightarrow
\begin{array}{cc}
CH_2 & \!\!\!\!\! CH_2 \\
| & | \\
O & O \\
| & | \\
M & \!\!\!\!\! M
\end{array}
$$

$\underline{C}$.

$$
\begin{array}{cc}
CH_2 & \!\!\!\!\! CH_2 \\
| & | \\
O & O \\
| & | \\
M & \!\!\!\!\! M
\end{array}
\longrightarrow 2
\begin{array}{c}
CH_2 \\
\| \\
O \\
| \\
M
\end{array}
$$

$\underline{D}$.

$$
\begin{array}{c}
CH_2 \\
\| \\
O \\
| \\
M
\end{array}
+
\begin{array}{c}
 \\
 \\
O \\
| \\
M
\end{array}
\longrightarrow
\begin{array}{c}
C \\
\diagup \\
O \\
| \\
M
\end{array}
+ M + H_2O
$$

$\underline{E}$.

$$
\begin{array}{c}
CH_2 \\
\| \\
O \\
| \\
M
\end{array}
\longrightarrow CH_2O + M
$$

$\underline{F}$.

$$
\begin{array}{cc}
CH_2 & CH_2\!\!-\!\!CH_2 \\
\| & \| \\
O & O \\
| & | \\
M & + M
\end{array}
\quad
\begin{array}{c}
CH_2 \\
| \\
O \\
| \\
M
\end{array}
\longrightarrow
\begin{array}{c}
CH_2 \\
\diagup \\
O \\
| \\
M
\end{array}
\quad
\begin{array}{ccc}
CH_2\!\!-\!\!CH_2\!\!-\!\!CH_2 \\
 \\
 \\
\diagdown \\
O \\
| \\
M
\end{array}
\longrightarrow CH_3CH\!=\!CH_2
$$

$\underline{G}$.

$$
\begin{array}{cc}
C & \\
\| & \\
O & O \\
| & | \\
M & + M
\end{array}
\longrightarrow
\begin{array}{cc}
C & \\
\diagup\diagdown & \\
O & O \\
| & | \\
M & M
\end{array}
\longrightarrow CO_2 + M
$$

$\underline{H}$.

$$
\begin{array}{cc}
C & \\
\| & \\
O & \\
| & | \\
M & + M
\end{array}
\longrightarrow
\begin{array}{cc}
 & C \\
 & \| \\
O & O \\
| & | \\
M & + M
\end{array}
$$

This proposal starts with the assumption that oxygen must become activated upon a surface, M (step A). This may be the metal in catalyzed combustion or even the walls of the reaction vessel. It has been found that reaction initiation in the absence of walls may be severe[10] and most radical mechanisms use the walls to convert intermediates to products.

The carbonaceous material (e.g., ethylene in diagram) reacts with the surface oxygen forming a saturated intermediate (B) which undergoes carbon-carbon bond cleavage to give a surface formaldehyde species, (C). The possible reaction paths of the formaldehyde intermediate result in the various products observed in the slow combustion of ethylene.[11] This would be associated with the induction period of combustion, and would be an exothermicreaction. Similarly, saturated hydrocarbons could form acetaldehyde as an intermediate. In fact, the addition of formaldehyde or acetaldehyde to ethane combustion reactions eliminates the induction period.[12] This species could explain chain branching (F) in the combustion of many simple fuels.[11]

Carbon dioxide arises from the reaction of the oxygen coordinated carbon monoxide with surface oxide (G) in a step consistent with the mechanism of carbon monoxide oxidation.[13] Carbide or soot formation occurs (H) when the adjacent surface oxide is not present (e.g., thru reduction). If the oxide site were available soot formation would be reduced.

An interesting feature of this scheme is that it is closely related to the recently proposed oxide Fischer-Tropsch (F-T) mechanism.[14] Since both proposals assume similar reaction intermediates correlations of metal catalysts should be noted. Indeed, metals that form strong oxides are not good F-T catalysts and will inhibit the oxidation of carbon. This relationshiop is particularly attractive in the examination of controlled combustion for chemical conversion.

CONCLUSION

This work towards reducing soot emission from diesel engines suggests that both homogeneous and heterogeneous approaches are feasible. The homogeneous approach consisted in mixing small quantities of alcohol with the fuel and the heterogeneous approach consisted of coating parts of the combustion chamber (bounding) surface with an oxidation catalyst like platinum. Both approaches produced significant reductions in soot emissions. These tests were conducted over a small range of operation in a single cylinder engine (a CFR type), with most of the testing done using pure cetane as fuel at constant speed and load. Although the engine

engine and test conditions are not representative of an automotive diesel engine, the results may be useful to assess the feasibility of soot reduction in automotive and truck diesel engines, using these methods.

The major features of the reaction mechanisms proposed for both fuel additive and surface catalyst effectiveness could be applied to various combustion problems.

NOTE: In a paper published since the present work was completed,[5a] the results obtained with a catalyst mesh in a Ricardo Comet engine have been compared with that from a conventional comet diesel engine. The catalytic comet gives lower Bosch smoke numbers than the comet diesel over a BMEP range of 3-4 Bar. The BMEP in the present work also ranged from 3 to 4.4 Bar and showed significant soot reductions with a platinum coated piston.

REFERENCES

1. H. B. Palmer, _J. Chem. Phys._,(1969).
2. H. A. Havemann, M. R. Rao, A. Natarajan, and T. L. Narasimhan, Alcohol and normal diesel fuels, in: "Gas and Oil Power," 50:15-19, (1955).
3. A. Lawson and A. J. Last, Development of an on-board mechanical fuel emulsifier for utilization of diesel/ methanol and methanol/gasoline fuel emulsions in transportation, Proceedings of the Third International Symposium in Alcohol Fuels Technology, Asilomar, California (1979).
4. J. Lahaye and G. Prado, Mechanisms of carbon black formation, in: "Chemistry and Physics of Carbon," P. L. Walker, Jr. and P. A. Thrower, eds., Marcel Dekker, Inc., New York (1978).
5. G. McConnell and H. E. Howells, "Diesel Fuel Properties and Exhaust Gas - Distant Relations," SAE 670091 (1967).
5a. R.H. Thring, The catalytic engine, Platinum Metals Review, 24:126-133 (1980).
6. D. H. Cotton, N. J. Friswell, and D. R. Jenkins, The suppression of soot emission from flames by metal additives, Comb. Flame, 17:87-98 (1971).
7. D. L. Trimm, The formation and removal of coke from nickel catalyst, Cat. Rev. Sc. Eng., 16:155 (1977).
8. W. C. Pfefferle, The catalytic combustor: an approach to cleaner combustion, J. Energy, 2:142-146 (1978).
9. A. J. Robell, E. V. Ballou, and M. Boudart, Surface diffusion of hydrogen on carbon, J. Phys. Chem., 68:2748-2753 (1964).

10. M. Boudart, Chemical kinetics and combustion, Proceedings of the Eight Symposium (International) on Combustion, The Wilkams and Wilkins Co., Baltimore, Maryland (1962).
11. G. J. Minkoff and C. R. H. Tipper, "Chemistry of Combustion Reactions," Butterworths, London (1962).
12. B. Lewis and G. VonElbe, "Combustion, Flames, and Explosions in Gases," Academic Press, New York (1951).
13. J. Happel, S. Kiang, J. L. Spencer, S. Oki, and M. A. Knatow, Transient rate tracer studies in heterogeneous catalysis: oxidation of carbon monoxide, _J. Catalysis_, 50:429-440 (1979).
14. R. S. Sapienza, M. J. Sansone, L. D. Spaulding, and J. F. Lynch, Novel interpretation of carbon oxide reductions, _Fundamental Research in Heterogeneous Catalysis_, 3:179 (1979).

Work performed under the auspices of the United States Department of Energy under Contract No. DE-AC02-76CH00016.

ENGINE DEPOSITS AND THE DETERMINATION OF THEIR

ORIGIN BY ATOMIC EMISSION SPECTROMETRY

E. Guinat

"Panel Laboratories" - I.C.T. Ltd
Tel Aviv 61392 - POB 39330, Israel

INTRODUCTION

It is sometimes cumbersome,to determine the origin of engine deposits.A simple and direct spectrometric method was devised which enables us to identify engine deposits as to their origin,without necessity to establish elemental composition,molecular weight or functional groups.The results obtained are very clearcut,but we are fully aware of the fact that this might not always be so,and more deposits,obtained with different fuels and other oils, would have to be examined.Other types of deposits might be found,as mentioned by Spilners et al.(1) but it seems to us,that by following the proposed spectrometric procedure,the origin of the deposits could be established in all cases where fuel as well as oil are involved.

EXPERIMENTAL

The engine

A Volvo passenger-car engine was found to give trouble with deposit formation from the very beginning.Several times inlet lines had to be opened and cleaned,to permit proper operation.

The lubricating oil was consumed in regular quantities but its contribution to varnish or deposit formation could be significant. However,Spilners et al.(2) had already noted,that incomplete combustion of fuel,contributed more deposits than decomposition of the lubricant.

The fuel

The fuel used was a commercial 94-Octane gasoline,meeting Israëli standard IS 90. Physical and chemical requirements are given in Table 1.

Table 1. Physical and Chemical Requirements - IS 90.

Property	Requirement Octane No.94	Test method ASTM
Distillation		
a. 10%(by volume) shall be distilled up to a temperature of (degrees C) max.	70	
b. 50%(by volume) shall be distilled up to atemperature of (degrees C) max.	125	
c. 90%(by volume) shall be distilled up to a temperature of (degrees C) max.	180	D 86-67
d. Final boiling point(degrees C) max.	195	
e. Distillation residue(percent by weight)	1.5	
f. Distillation loss (percent by weight)	1.5	
Vapour pressure according to Reid (1b) max.	8.5	D 323-72
Total sulphur content (percent by mass) max.	0.20	D 1266-70
Corrosion test-copper strip- 3 hours at 50°C	shall pass the test	D 130-75
Existent gum(mg per 100cc) max	7	D 381-70
Oxidation stability:		
-Induction period (minutes) min.	240	D 525-74
-Potential residue in 3 hours (mg per 100 cc)	shall be reported	D 873-74
Octane No.(Research Method) min.	94	D 2699-68
Lead content (g per litre) max.	0.42	D 526-70 or D 3341-74

The oil

The lube-oil used was API-grade SE,15W/40

The problem of aromatics in the fuel

As mentioned in (3),the concentration of aromatic hydrocarbons in the gasoline could possibly play an important role in the formation of the deposits.It is stated there (as abstracted in CA) that "the concentration of aromatic hydrocarbons in AI 93 gasoline must be $\leqslant$ 40%; higher concentrations cause depositions of Carbon in the engines".

It seems clear to us, that the total aromatics content should not be considered as the sole criteria for deposit formation , as the nature of the individual aromatic compounds should be taken into consideration.

We made a quick determination of aromatic hydrocarbons according to (4)- Note 7.,and found our sample of gasoline to contain about 41% of total aromatics.

The origin of our engine deposits

Deposits from gasoline engine exhaust inlet and pistons were taken for analysis.They showed the characteristic property as defined by CRC engine rating (5):they could not be removed by wiping with a soft cloth.

It was the aim of this study to trace the origin of the deposits, i.e. wether they were due to a resin-forming tendency of the fuel, or to residues due to incomplete combustion or decomposition of the lube oil.

The samples of residues were finely ground and suspended in Conostan Base oil,which is used for spectrometric analyses.They were tested by direct reading emission spectrometric analysis,with a BAIRD-instrument (Spectromet 1000).The results are given in Table 2.

Table 2. Spectrometric determination of metals in
engine deposits,and in used oil (in ppm)

	Fe	Cu	Cr	Al	Si	Pb	Mg	Ni	Sn	Ca	Zn	P
Engine depo-	150	35	0	21	50	950	14	0	2	50	50	23
sits.	160	3	1	24	21	1000	50	0	2	0	70	38
Used oil	50	6	6	12	8	1000	45	4	19	1750	1100	900

Interpretation of results

The results in Table 2 show clearly,that the analysed residues are mainly composed of gasoline components although containing also a very small proportion of additive elements,Zn and P,contributed by the oil.A comparison is made between the element distribution of Pb,Zn and P found in the deposits and those present in the lube oil from the same engine,found by the same spectrometric procedure.If the deposits were due to the oil,they would be expected to contain Zn and P in amounts practically equivalent to Pb.

CONCLUSION

1. By submitting engine deposits to a spectrometric determination of elements,it was possible to establish their origin.The evaluation is based on a comparison of the concentration ratio of typical elements such as Pb,Zn and P,in the deposits and in the used oil sample. The determination has to be made with the same instrument and mode of excitation. Other elements could be used alternatively as indicators.

2. In the present case it appeared,that the deposits were mainly due to decomposition or incomplete burning of the gasoline. The contribution of the oil to the deposits was minimal.

REFERENCES

1. Ilgvars I.Spilners,John F.Hedenburg and Carl R.Spohn. "Evaluation of engine deposits in a modified Simple-cylinder engine test" ACS Div.Petrol.Chem.Atlanta Meeting Marsh 29,1981
2. Ilgvars I.Spilners,John F.Hedenburg."Effect of fuel and lubricant composition on engine deposit formation"AmChem.Soc.Div. Pet.Chem.1981,26(2)632-8.
3. Kuznetsova,L.M;Mikheeva,E.G.;Ioffe B.V. "Refractometric determination of aromatic hydrocarbon content in AI-93 gasoline". Khim.Tekhnol.Topl.Masel 1982 (4) 35-7.see CA 96,1982,220202.
4. ASTM D 936-Note 7.
5. CRC Handbook,Coordinating Research Council,New-York 1946

EVALUATION OF ENGINE DEPOSITS IN A MODIFIED SINGLE-CYLINDER ENGINE TEST

Ilgvars J. Spilners, John F. Hedenburg and Carl R. Spohn

Gulf Research and Development Company
P. O. Drawer 2038
Pittsburgh, PA 15230

INTRODUCTION

A new gasoline engine test was developed using a single-clyinder CLR oil test engine modified to promote rapid accumulation of deposits and degradation of the crankcase lubricant. Precise evaluation of the degree of engine cleanliness was achieved using standard CRC engine rating techniques and determining deposit weights.   Analyzing deposits for their elemental compositions, molecular weights, and functional groups provided information on which to base the deposit evaluation.

EXPERIMENTAL

Engine Test Development

The influence on engine deposits of nitrogen oxides and fuel hydrocarbons, introduced to the crankcase atmosphere, by blow-by gases, has been well documented.[1,2] Our purpose was to increase their concentration by both mechanical and chemical means in a controllable and repeatable manner. This was intended to cause accumulation of significant deposits in a relatively short test time and allow differentiation among various fuels and lubricants.

Test Engine

The engine used was a single-cylinder CLR oil test engine. It was reconditioned, basically, using CRC Low Temperature Deposit Engine Specification, 348.2.[3] In this setup, the cylinder   liner can be distorted with a jackscrew to control the flow of blow-by gases to the crankcase.

Modifications to the engine included: a) the installation of sampling tubes for analyzing exhaust gases and the crankcase atmosphere, b) the provision of a means to add fuel samples and nitrogen oxides to the oil sump without shutting the engine down, and c) the installation of removable metal specimens in critical areas to monitor deposit formations as the test progressed.

Engine Test Conditions

In the early phases of development, many trial runs were made in an effort to select conditions which would maximize the presence of nitrogen oxides and hydrocarbons in the crankcase atmosphere. Starting with the Low Temperature Deposit CLR engine test, the following operating conditions were developed and used in this study:

Speed, rpm	2000 ± 25
Load, lb-ft	30 ± 1
Spark Advance, deg.	20 ± 1
A/F Ratio	15.75 ± 0.25:1
Oil Gallery Temp., °F	260 ± 2
Coolant Out Temp., °F	190 ± 2
Blow-by Rate, ft^3/hr	36 ± 2
Total Test Hours	48
NO_x Injection Rate, cc/min	700
Fuel Additions, cc	100 every 12 hrs

These conditions were arrived at by varying parameters one at a time and monitoring the effect on the composition of the crankcase atmosphere. As might be expected, increased spark advance, lean air/fuel ratio, and increased blow-by caused the major increases in concentrations of nitrogen oxides and hydrocarbons in the crankcase. To reduce the test time to 48 hours, additional NO_x was injected and periodic additions of test fuel were made into the crankcase. The increase of nitrogen oxides caused heavier deposit formation but crankcase gas analyses indicated that the additional fuel was rapidly lost by evaporation. These conditions generated a crankcase atmosphere containing approximately 1,000 ppm of nitrogen oxides (NO_x) and 82,000 to 84,000 ppm of hydrocarbons, which corresponds to the passage of 2,136 to 2,304 g in 48 hours. Fuel additions to the crankcase increased hydrocarbons to 100,000 ppm, but as stated the increase was short-lived. In the case of the reference fuel, Fluid Catalytically Cracked or FCC, over 500 g of olefins entered the crankcase vapor space during a 48-hour test from the 45 gallons of fuel consumed. Seven fuels with totally different compositions were used (Table I).

Table I

COMPOSITION OF FUELS USED IN ENGINE TESTS

HYDROCARBON GROUP TYPES, VOLUME PERCENT OF
TOTAL SAMPLE BY CHROMATOGRAPHIC ANALYSIS

	Light FCC Fuel	Total Platformate	Heavy Platformate	Alkylate	Isooctane	Toluene	Heptane
Hydrocarbon Type Volume %							
Saturates	57.7	44.1	26.2	99.6	100	0.01	100
C_4	0.3	3.0		5.3			
C_5	13.3	3.9		4.4			
C_6	16.5	6.4		4.7			
C_7	15.8	11.2	1.2	5.6	0.6		99.7
C_8	8.7	12.2	9.2	76.7	99.4		0.3
C_9	3.1	5.0	9.8	.7			
C_{10}		2.5	6.0	1.3			
C_{11}				0.9			
Olefins	22.2	0.04	–	–	–	–	–
C_4	0.9						
C_5	13.3	0.01					
C_6	8.0	0.04					
Aromatics	20.1	55.9	73.8	–	–	99.99	–
C_6	1.6	1.2					
C_7	10.2	10.2	4.1			99.99	
C_8	5.1	18.5	24.6				
C_9	2.7	16.1	30.2				
C_{10}	0.4	6.2	10.8				
C_{11}	0.1	3.6	4.1				

The single-cylinder engine test used a total of 3.2 lbs. (1,450 g) of lubricant. The loss of lubricant in a typical 48-hour test was between 0.3 and 0.9 lbs. (140 and 410 g) or 9 to 28%. Some of this loss was due to sampling and leakage, but most of it resulted from combustion. If combustion is incomplete, formation of varnish precursors may be expected. Although the amount of oil consumed was small as compared to the hydrocarbons in the blow-by (over 2,000 g), direct material contribution of oil hydrocarbons to varnish may be significant if the incomplete combustion of oil yields reactive and nonvolatile precursors.

Test repeatability was assured by periodic check runs with the same reference fuel and lubricant. This allowed the comparison of results to the same baseline performance. Since most test parameters were precisely controlled, CRC deposit ratings for reference tests usually fell within ±0.5 numbers on the 0-10 scale.

Evaluation of Engine Deposits

Periodic inspections were made of rocker, crankcase, and pushrod covers to monitor the formation of deposits. The CRC engine rating system[4] defines varnish as a resinous deposit which cannot be removed by wiping with a soft cloth. Varnish ratings are made visually on a merit basis; a part completely free of deposits is rated 10 and a part completely covered with heavy deposit is rated 0. Intermediate ratings are based on varnish intensity (not color), by referring to CRC rating scales, and the percent of the area covered. Sludge is defined as a deposit composed of crankcase oil, varnish-like resins, water and solid contaminants which does not drain from engine parts but can be removed by wiping with a soft cloth. The sludge rating is based on sludge depth and the percent of the area of the part covered. Ratings are also on a 10 to 0 merit scale.

Deposits were also determined quantitatively. After draining of the oil and completing the merit rating, the inside of the engine was thoroughly sprayed with hexane. This removed oil film from the engine walls and any oil occluded in the sludge, but left the varnish and the resinous materials from the sludge, which were dissolved by an acetone spray. The acetone solution was filtered, concentrated, and diluted with ten times its volume of pentane. The precipitated solids were dried, weighed, and reported as varnish.

The drain oil, plus the hexane wash, was filtered and diluted with ten volumes of pentane. The precipitated brown solid was dried, weighed, and reported as pentane-insolubles.

RESULTS AND DISCUSSION

Engine Deposit Composition

Varnish and pentane-insoluble materials were collected from engine tests by the techniques described above and analyzed for their physical properties and composition (Tables II, III). Average molecular weights in the range of 500 to 2,700 showed that both types of deposits are polymeric materials. In one test, a small sample of varnish was collected which had a molecular weight of 11,700. The deposits melted over a temperature range from about 100 to 300°C or above. Generally, varnish had a higher melting range than pentane-insolubles. Engine deposits consisted of carbon, hydrogen, nitrogen, oxygen, sulfur, and some ash. The elemental composition of the deposits showed that they were composed mostly of oxidized hydrocarbon units.

Engine varnish, generally, has a slightly higher oxygen and lower carbon and hydrogen content than pentane-insolubles (Table II).

Light FCC fuel usually produced varnish and pentane-insolubles with a higher carbon and lower oxygen content than other fuels. This was the case in two widely different lubricants, 240 Neutral mineral oil (Table II) and polyalphaolefin "A" (Table III). Light FCC is distinguished from other fuels by its relatively high olefin content. Oyxgen is by far a more abundant element in both varnish and pentane-insolubles than nitrogen or sulfur. Nitrogen is derived from intake air and sulfur is found in small amounts in the fuel. Possible mechanisms of varnish formation are discussed elsewhere.[5]

As for the lubricant effect with a common fuel, deposits formed in polyalphaolefin "A" had a lower carbon content and a lower molecular weight than deposits in the 240 Neutral mineral oil. With Light FCC fuel, the average molecular weights of varnish formed in synthetic alkylbenzenes and pentaerythritol ester were very low, 380 and 770, respectively.

The infrared spectra showed that varnish and pentane-insolubles contain the same functional groups, including bonded or nonbonded carboxylic, hydroxyl, carbonyl, nitrate, and nitro groups. In some samples, ionized carboxylic groups and ether linkages were observed. The most prominent band in the spectrum of every deposit was due to the carbonyl group. The intensity of unbonded hydroxyl stretching frequency (at about 3,500 cm^{-1}) was usually associated with a parallel intensity of a doublet between 1,000 and 1,100 cm^{-1} which is typical of glycols. Since the infrared spectra of deposits showed similarities when widely

Table II

ANALYSIS OF VARNISH AND PENTANE-INSOLUBLES FORMED IN 240 NEUTRAL HIGH VISCOSITY OIL

	FUELS USED						
	Light FCC	Light FCC, 90% Methanol, 10%	Total Platformate	Toluene	Alkylate	Isooctane	Isooctane, 85% Heptane, 15%
Varnish							
Av. Mol. Wt.	2,700	1,800	1,625	570	1,100	1,475	1,500
Melting Range, °C	220–245	140–150					
Composition, % Wt.							
C	72.1	63.8	69.4	61.0	62.3	65.7	64.5
H	7.4	6.8	6.9	7.8	7.3	7.4	6.9
N	2.7	4.2	1.6	1.6	1.0	2.9	3.3
S	0.9	0.9	0.2	0.7	0.3	0.3	0.5
O	16.1	21.0	21.2	27.0	26.6	19.1	21.8
Ash	0.8	3.3	0.7	1.9	2.5	4.6	3.0
Pentane-Insolubles							
Av. Mol. Wt.	2,500		1,100	1,075	1,250		
Composition, % Wt.							
C	75.2		72.8	71.2	73.5	72.3	
H	8.0		7.7	8.2	8.4	8.3	
N	2.2		1.3	1.1	1.3	1.3	
S	0.5		0.3	0.8	0.1	0.2	
O	13.4		17.8	16.0	16.2	17.5	
Ash	0.7		0.1	2.7	0.5	0.4	

Table III

ANALYSIS OF DEPOSITS IN 6 cs POLYALPHAOLEFIN "A"

	FUELS		
	Light FCC	Total Platformate	Toluene
Varnish			
Av. Mol. Wt.	1,775	880	510
Melting Range	over 300°C		
Composition, % by Wt.			
C	60.7	58.1	58.7
H	6.1	6.7	6.5
N	4.4	2.9	3.0
O	26.3	29.4	29.9
S	0.7	0.8	0.7
Ash	1.8	2.1	1.2
Pentane-Insolubles			
Av. Mol. Wt.	1,350	700	
Melting Range	135-155°C		
Composition, % by Wt.			
C	68.7	58.0	
H	9.8	6.6	
N	3.3	2.3	
O	10.2	26.7	
S	1.0	0.6	
Ash	7.0	5.8	

different fuels, such as toluene and Alkylate, were used, it was assumed that incomplete combustion of the lubricant contributed to deposits, and that widely different fuels gave similar types of fragments, which oxidized and recombined to give similar deposits. Differences were noted, however, in the amount of deposits when different fuels were used with the same lubricant. Using the same fuel but widely different lubricants, again the infrared spectra were generally similar. Thus, the predominant building blocks of deposits in all cases appeared to be largely the same. There was no indication of a regular repeating sequence of the polymeric building units.

Varnish deposits from 90 VI 240 Neutral, with and without 1.25% zinc dialkyldithiophosphate, and Light FCC fuel were examined by carbon-13 nuclear magnetic resonance. In both cases, the aromatic region showed a broad band which is devoid of any special information. Several configurations were possible for saturated carbons and there were more lines than expected for n-alkane. Both methylene and methyl groups were present.

Varnish was completely soluble in alkali, but it was not attacked by hydrochloric acid. Saponification of varnish from tests where 240 Neutral base oil was used with 100% isooctane or with 90% Light FCC/100% methanol fuel gave a soluble portion which contained carboxylic acids and an insoluble portion which contained alcohol and nitro groups. Esterification of sodium salts of the acids was carried out to obtain materials which would be identifiable by gas chromatography. However, the esters formed had high boiling points and no comparison with known esters was possible.

Piston varnishes have been examined by the reflectance infrared method and found to have bands due to gamma-lactone, acyclic carboxylic ester, ketone, and carboxylic acid groups. When they had been detached from the piston with sodium hydroxide, the transmission infrared spectrum showed only ketone and carboxylate anion groups.[6] Our infrared determinations did not exclude the presence of esters, but the several overlapping bands in the carbonyl region did not permit definite identification.

Comparison of Merit Ratings with Quantitative Deposit Determinations in Engine Tests

In the engine tests, varnish and sludge were rated only for the pushrod, crankcase, and rocker covers. The three values were averaged for easier comparison of ratings with other tests. The spread in varnish ratings for the three covers ranged from zero to three or more numbers. There did not appear to be any regular pattern in ratings for the three covers, either for varnish or for sludge. The rating number spread for sludge was smaller.

For quantitative determinations, on the other hand, complete deposit collection was made with the help of solvents.

Deposit weights and varnish and sludge ratings were obtained on engine tests using base oils of different weights and finishes, synthetic lubricants, blends of oils, rerefined oils, and oxidatively inhibited and noninhibited oils. In most tests, the fuel used was Light FCC. When the varnish ratings were plotted against the weight of varnish, a poor relationship was found (Figure 1).

The maximum weight of varnish obtained in a single test was 15.56 g. The quantitative weight determination in this case agreed very well with the varnish rating (3.7) for these generally heavy deposits from Light FCC fuel and 240 Neutral mineral oil. The lowest weight of varnish (1.41 g) was for a heavy duty commercial formulation which agreed, generally, with the high varnish rating of 9.8 (sludge 9.5). Also, polyalphaolefin "A" as a lubricant gave high merit ratings which agreed with the low weights of varnish deposits. Between these high and low values, which showed agreement between deposit weights and ratings, there were many which disagreed widely with the rating, usually on the high side (Figure 1). For example, an unextracted base oil with a high varnish rating of 8.3 (sludge 7.6) gave 14.2 g of varnish. Similarly, a hydrotreated base oil with a varnish rating of 6.9 (sludge 7.9) gave 14.8 g of the acetone-soluble varnish.

The most pentane-insolubles (62 g) were separated from an experimental formulation made with an unextracted 240 Neutral oil and a Bright Stock. In the case of formulated oils, it was frequently difficult to judge what portion of the deposit was due to the additives.

The lowest combined weight of varnish and pentane-insolubles (1.8 g) was found for a synthetic alkylbenzene base oil, with a varnish rating of 9.2 and a sludge rating of 9.3.

Generally, the varnish ratings ranged from 3.7 to 10 and sludge ratings from 7.0 to 9.5. The varnish weights ranged from 15.5 to 1.4 g and the pentane-insolubles from 62.64 g to 0.

CONCLUSIONS

Deposit accumulation in a single cylinder gasoline engine could be greatly accelerated by distorting the cylinder liner to increase the blow-by rate. Adjusting operating conditions to favor incomplete burning of fuel in the combustion chamber and oxidations and condensations of the blow-by, raw fuel, and

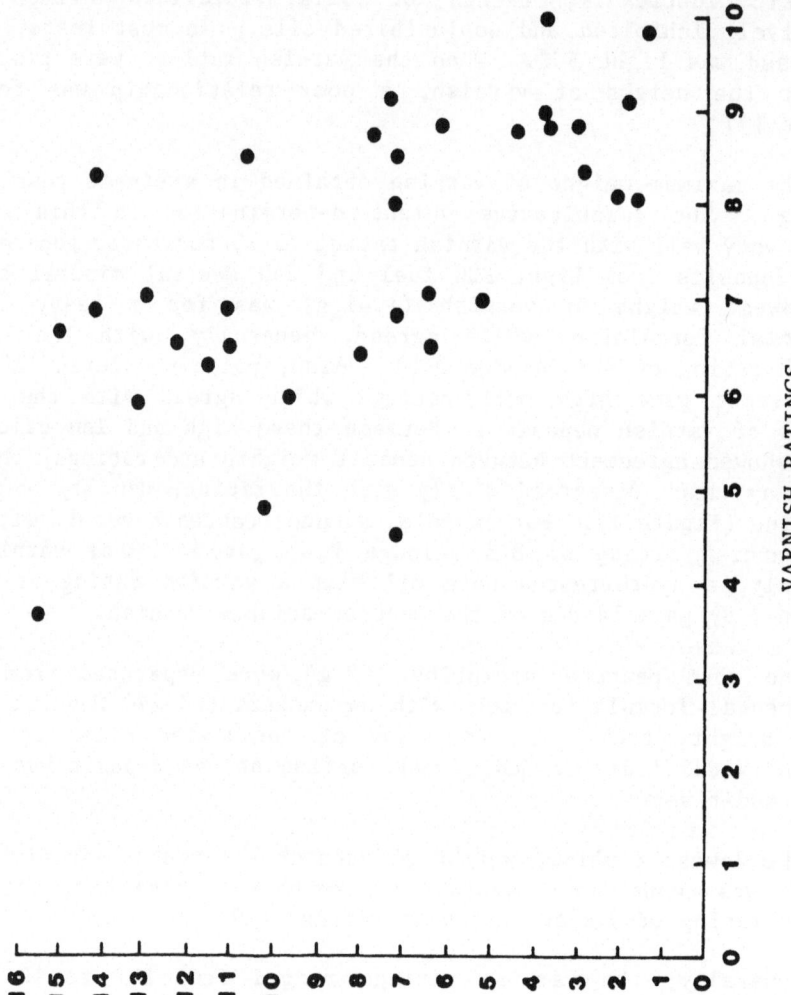

Figure 1. Engine Varnish Weights, g VS. Varnish Ratings

VARNISH RATINGS

VARNISH
WEIGHTS, g

lubricant in crankcase also contributed to deposit accumulation. The results of the 48-hour test showed that it can differentiate among fuels and oils with respect to their tendencies to form deposits.

According to their solubility characteristics, two types of deposits were collected: varnish and pentane-insolubles. These two types of materials constituted all deposits formed and their weights served as indicators of the oxidative stability of the oil and of the deposit forming tendencies of the fuel and oil combined. Both types of deposits gave similar elemental and functional group analyses. Because of the similarities, the pentane-insolubles, precipitated from the oil, may be considered as a potential varnish in deposit evaluation. The commonly used CRC varnish and sludge merit ratings allowed a quick evaluation of engine deposits, but they may not always reflect the extent of deposit formation as accurately as quantitative deposit determinations. The merit rating does not include the pentane-insolubles in the oil, although these potential deposits are strong indicators of interactions of oil and varnish precursors and may predict how much varnish may be anticipated in extended engine operation. Using the same fuel, wide variations in weights of pentane-insolubles were observed for different base and inhibited oils.

Deposits on certain engine parts may be more critical than on others. However, quantitative determinations of all deposits should serve as an indication of the oxidative stability of the oil, the deposit generating tendency of the fuel and of other engine chemistry parameters which directly influence the performance and useful life of an oil. The quantitative approach to deposit evaluation should be more reliable than the merit rating which may be influenced by localized and short term situations which are often subject to small mechanical variations.

ACKNOWLEDGEMENT

We thank Mr. C. H. Phoebe for supervising the engine tests and assuring the control and accuracy demanded by this study.

REFERENCES

1. K. L. Kruez, Lubrication, 55, 6, 53 (1969).
2. R. S. Spindt, C. L. Wolfe, and D. R. Stevens, SAE Paper No. 638, "Nitrogen Oxides, Combustion and Engine Deposits", 1955.

3. Federal Test Method Standard No. 791b, "Lubricants, Liquid
    Fuels and Related Products: Methods of Testing".
4. CRC Handbook, Coordinating Research Council, New York, 1946.
5. I. J. Spilners and J. F. Hedenburg, "Effect of Fuel and
    Lubricant Composition on Engine Deposit Formation", ACS
    Meeting, Petroleum Division, Atlanta, March 28-April 3,
    1981.
6. H. Spedding and S. F. W. Noel, Tribology 5, No. 1, 1972.

EFFECT OF FUEL AND LUBRICANT COMPOSITION

ON ENGINE DEPOSIT FORMATION

Ilgvars J. Spilners and John F. Hedenburg

Gulf Research and Development Company
P. O. Drawer 2038
Pittsburgh, PA 15230

INTRODUCTION

Unburned and incompletely burned fuel and lubricant oxidize and condense in the crankcase producing varnish and sludge. Such formation of deposits and degradation of the crankcase lubricant are primary causes of many engine performance difficulties.

Since the discovery that nitrogen oxides are formed in the combustion chamber of an internal combustion engine,[1] researchers have investigated their effect on engine operation. Emphasis has been on reactions affecting varnish formation and, in particular, on nitration of olefins by nitrogen oxides.[2-5] Other varnish intermediates and possible reactions in ignition-fired engines have not received the necessary attention. The purpose of this study was to learn more about engine chemistry from the effects of fuel and lubricant composition on engine deposit formation.

All primary chemical reactants, including fuel, base oil, lubricant additives and air, were evaluated and considered as possible sources or causes of deposits. Tendencies of deposit formation could be determined in a 48-hour test using a modified single cylinder CLR oil test engine under conditions which had been found to give a high blow-by rate and produce a crankcase atmosphere rich in nitrogen oxides and incompletely burned or raw fuel.[6]

DISCUSSION AND RESULTS

Chemical Reactants and Reactions in an Engine

During compression in the combustion chamber, burned and some unburned gases get into the crankcase, where these undergo many simultaneous and sequential reactions.[7,8] The reactions take place over a wide range of temperatures and conditions and some of them yield varnish precursors.

Fuel Hydrocarbons. In precombustion reactions, the fuel hydrocarbon molecules lose hydrogen atoms. The hydrocarbon radicals formed undergo thermal decomposition by carbon-carbon cleavage and hydrogen stripping in the presence of oxygen to yield relatively stable olefins. Further reactions with oxygen form peroxides which decompose to yield aldehydes, ketones and smaller olefins.[9-11] Thus, the olefins entering the crankcase as a part of the blow-by have been formed in incomplete combustion of fuel or may be present as components of the fuel. About 22% by volume of the Light FCC fuel, used in this study, was olefins,[6] which in the blow-by increased to 24.9%. Ethylene and propylene were the predominant newly formed olefins but an increase was noted in the higher olefins. Generally, emissions and blow-by contain low molecular weight olefins and acetylene, which form by carbon-carbon cleavages in alkanes and alkyl groups of alkylaromatics.[7,9] Benzene, toluene and xylenes, however, do not have sufficient hydrogen to yield olefins directly by carbon-carbon cleavages.

Nitrogen Oxides. An important component of blow-by is nitrogen oxide which forms from the oxygen and nitrogen in the air at the peak cycle conditions after the passage of the flame front.[12] With excess of oxygen, nitrogen oxide is converted to the dioxide, a very reactive odd-electron molecule which undergoes free radical reactions. Nitrogen dioxide associates with other free radicals, adds to olefin double bonds, and takes part in hydrogen-abstractions. Thus, it can greatly affect oxidation reactions in the blow-by and the crankcase vapor space and, depending on its concentration, change the velocity of the process.[13]

In the engine tests of this study, additional nitrogen oxide was added to the crankcase, according to the procedure described elsewhere.[6] It was recognized that some of the nitrogen oxide is readily converted to nitrogen dioxide under the operating conditions.

Oxidation of Fuel and Lubricant Hydrocarbons. In the oxidation of hydrocarbons, the initial product is a hydroperoxide. Generally, saturated hydrocarbons are much more resistant to autooxidation than olefins, but the alkyl side chains on aromatics are readily attacked.[14]

Nitrogen dioxide acts as an initiator of oxidations of hydrocarbons.[15] The vapor phase nitration in the presence of air gives both nitrated and oxidized products.[16] Secondary reactions convert hydroperoxides to alcohols and carbonyl compounds[15] which are assumed to condense, yielding polymeric engine deposits of high oxygen content.

Reactions of Olefins with Nitrogen Oxides. Nitration and oxidation of olefins take place in the gas phase in the combustion chamber, around the compression rings and in the vapor space of the crankcase.[16] The initial nitration products are unstable and undergo further reactions. Thus, vapor phase reactions of ethylene and isobutylene with nitrogen dioxide at 150°C, reportedly, produce heavy viscous oils.[17] When olefins and nitrogen oxides are picked up by the lubricant mist and the film on the cylinder walls, they may undergo liquid phase reactions which have been reported to yield nitro, nitrite and nitroso compounds and nitroolefins.[18] Earlier research has shown that definite nitration products, dinitro compounds and nitronitrites, are obtained only in low yields and then only at very low temperatures.[17] Addition of nitrogen dioxide at the terminal position of alpha-olefins occurs by C-N bond formation. The second NO_2 group is added either by C-N or C-O bond formation, giving a dinitro or a nitronitrite compound, respectively. Nitronitrites may become oxidized to nitronitrates or hydrolyzed to nitroalcohols. The latter may dehydrate to nitroolefins which undergo polymerization.

If nitration of low molecular weight olefins, from ethylene to hexene, is assumed as one of the steps by which varnish is formed, then varnish, on the average, should contain about 14% nitrogen and 32% oxygen by weight. The engine varnish in this study, however, had an elemental composition of about 2-3% nitrogen and 15% oxygen, suggesting that nitrated hydrocarbons do not serve as the principal sources of varnish.

Fuel Derived Oxygenates. A small fraction of the blow-by was found in this study to consist of organic oxygenates. Of these, the carbonyl compounds were separated by passing the blow-by through 2,4-dinitrophenylhydrazine reagent. The hydrazones formed were collected and analyzed colorimetrically for quantitative determination of the carbonyl compounds. The hydrazones of low molecular weight carbonyl compounds were identified by gas chromatography.[19] The hydrazones of acrolein, acetone and propionaldehyde have nearly the same retention times and were not easily separated.

The total carbonyl for the blow-by from several fuels were determined (Table I). Addition of nitrogen oxides and fuel to the

Table I

Fuel Derived Oxygenates in the Engine Blow-By Relationship of Carboxyl Formation to Fuel Composition

Fuel	n-Alkanes In Fuel, Volume %	PPM of Carbonyls by Colorimetric Determination as 2,4-Dinitrophenylhydrazones		
		Without Additional NO$_x$ and Fuel Into Crankcase	With Additional NO$_x$ Into Crankcase	With Additional NO$_x$ And Fuel Into Crankcase
Isooctane, 85% and Heptane, 15%	15.0	695	735	715
Light FCC, 90% and Methanol, 10%	6.2	434	455	N.D.
Light FCC	6.8	402	328	N.D.
Alkylate	5.3	375	348	360
Heavy Platformate	5.3	258	250	N.D.
Isooctane	none	226	235	235
Toluene	none	59	59	N.D.

N.D. = Not Determined

crankcase did not change the total carbonyl content significantly. Hence, it may be concluded that volatile aldehydes and ketones are formed before the blow-by enters the crankcase. The source of the carbonyl compounds in our tests appeared to be primarily n-alkanes. Thus, a fuel with high n-alkane content (15% n-heptane) gave more carbonyls (699 ppm or 59 g/48 hours) than other fuels with less n-paraffins. Toluene gave the least carbonyl compounds (59 ppm or 4.7 g/48 hours). n-Alkanes, which gave the most carbonyls, are known to cause preignition.[20]

The principal carbonyl compound from different fuels was formaldehyde, except for the olefin-rich Light FCC where acetone and propionaldehyde predominated. This suggests that the increased varnish-forming tendencies of olefin-rich fuels are related to higher carbonyl compound intermediates rather than formaldehyde. The addition of 10% by volume of methanol to Light FCC not only increased the formaldehyde content in the blow-by but also changed the distribution of other carbonyl compounds. The acetone-propionaldehyde peak was diminished relative to the acetaldehyde peak and less deposits were formed, as shown in Table II.

Fuel as a Source of Deposits

The fuel is, generally, the major source of deposits in the moderate temperature range of engine operation.[5] In this study, several fuels of widely different hydrocarbon compositions[6] were compared with respect to deposite formation, using as lubricants: (a) a mineral base oil which is relatively easily oxidized (240 Neutral), (b) an oxidatively stable synthetic base oil (6 cs polyalphaolefin or PAO "A"), and (c) the above mineral oil, inhibited with 1.25% of zinc dialkyldithiophosphate. The deposit data (Tables II, III, and IV) showed that, using the same lubricant, the fuel hydrocarbon composition causes great variations in the varnish and sludge ratings, as well as in the actual amounts of deposits formed. The olefin content of the fuel was found to have the greatest effect on varnish-forming tendencies in the above oils. Fuels with higher alkylaromatics, which are known to yield olefins by fragmentation of the alkyl groups in incomplete combustion, were also recognized as significant contributors to varnish formations.

When these data in Table II are put in the order of (a) increasing CRC varnish ratings and (b,c) decreasing varnish and pentane-insolubles weights, which are all indicators of increasing engine cleanliness, the following variations for different fuels with the 240 Neutral oil are observed:

Table II

Effect of Fuel on Deposit Formation in 240 Neutral High Viscosity Oil

	Light FCC	Light FCC, 90% Methanol, 10%	Total Plat-Formate	Toluene	Alkylate	Isooctane	Isooctane, 85% Heptane, 15%
Varnish Rating*	3.7	5.9	8.1	9.5	7.2	8.8	8.8
Sludge Rating*	7.0	7.9	8.0	8.9	7.8	9.1	9.0
Varnish, g	15.6	13.1	8.8	4.3	5.0	3.0	3.6
Pentane-Insolubles, g	16.1	10.9	13.3	17.5	17.0	21.0	13.0
Total Weight Deposits, g	31.7	24.0	22.1	21.8	22.0	24.0	16.6

Table III

Effect of Fuel on Deposit Formation in 6 cs Polyalphaolefin "A"

	Light FCC	Total Platformate	Toluene
Varnish Rating*	8.9	9.0	9.3
Sludge Rating*	8.9	9.1	8.6
Varnish, g	4.1	1.8	2.0
Pentane-Insolubles, g	1.9	3.5	2.7
Total Weight Deposits, g	6.0	5.3	4.7

*Clean = 10

(a)	(b)	(c)
Varnish Ratings	Varnish Wts.	Pentane-Insolubles Wts.
1. Light FCC	Light FCC	Isooctane
2. Light FCC (90%), Methanol (10%)	Light FCC (90%) Methanol (10%)	Toluene
3. Alkylate	Total Platformate	Alkylate
4. Total Platformate	Alkylate	Light FCC
5. Isooctane	Toluene	Total Platformate
6. Isooctane (85%) Heptane (15%)	Isooctane (85%), Heptane (15%)	Isooctane (85%), Heptane (15%)
7. Toluene	Isooctane	Light FCC (90%), Methanol (10%)

The spread in weights of pentane-insolubles with the use of different fuels was smaller than the variations in varnish weights. Consequently, varnish weights differentiate among fuels best. The heavy varnish, due to the Light FCC (22.2% olefins and 20.1% aromatics), could be slightly reduced by addition of methanol (10%). The next most varnish-generating fuel was found to be Total Platformate which is rich in aromatics (55.9%) in the range from C_6 to C_{11}. Toluene, alkylate, and isooctane gave considerably less varnish.

Light FCC gave varnish with a higher molecular weight and petane-insolubles with a higher carbon and lower oxygen content than other fuels.[6] This may be interpreted as the effect of olefins, undergoing more extensive polymerization. Toluene, on the other hand, is not a reactive polymer precursor, since it does not readily fragment to give reactants which can polymerize.[21] As a result, deposits from toluene have a lower molecular weight and carbon content, but more oxygen.[6] Methanol, added to Light FCC, modified varnish deposits by decreasing the molecular weight and carbon content and increasing oxygen and ash content.

Isooctane caused the used oil to reach a total acid number of 12 in 24 hours. The acids, apparently, attacked engine parts, which led to a high ash content in the varnish. Light FCC (57.7% alkanes), on the other hand, caused a low used lubricant acid number (2.9 in 48 hours) and low varnish ash content. The infrared spectra and acid numbers indicated that formation of acidic materials in the oil, as well as in the deposits, is greater for fuels with higher alkane content.

The varnish ratings and weights of varnish and total deposits showed that Light FCC is a dirtier fuel than Total Platformate or toluene, with 6 cs PAO "A" (Table III). Generally, the weights of deposits formed in PAO "A" were much smaller than in 240 Neutral (compare data in Tables II and III).

The same fuel effect trends were observed for molecular weights of deposits, formed in both 240 Neutral and PAO "A". Thus, Light FCC in 6 cs PAO was found to give a higher molecular weight deposit than Total Platformate or toluene.[6] Using 6 cs PAO "A", generally, the deposit carbon content was lower and oxygen and nitrogen content higher.[6] This suggests that 6 cs PAO "A" is a poor reaction medium and solvent for free radical polymerizations, regardless of the type of fuel. Used 6 cs PAO "A" had a relatively low acid number of 2.0 in 48 hours, which is another indication of its oxidative stability.

Zinc dialkyldithiophosphate oxidation inhibitor added to the mineral base oil considerably reduced the deposit formed (Table IV). This is readily seen by comparison of varnish and sludge ratings and weights of pentane-insolubles for the tests using 240 Neutral, as the lubricant with and without the antioxidant, and Light FCC, alkylate, and toluene as fuels (compare data in Tables II and IV). Addition of n-butylbenzene (2%) or alkylbenzenes and olefins (5%) to toluene increased varnish, but decreased the weights of pentane-insolubles (Table IV).

Oxidative Stability of Base Oils

For the purpose of this study, an engine was viewed as a chemical reactor where the blow-by gases and the crankcase oil undergo gas-phase and liquid-phase reactions. The crankcase oil is gradually oxidized, nitrated and burned during operation of the engine. Products from partial combustion may contribute to deposit formation, similar to the incomplete fuel combustion fragments. When an oil is oxidized and nitrated to a low degree without molecular fragmentation, it does not contribute directly to deposits. However, in the process it has become a much more reactive reaction medium and may promote deposit formation by interactions with free radical intermediates from the engine blow-by.

Oxidative stabilities of virgin, recycled and rerefined mineral oils, as well as of polyalphaolefin, alkylbenzene and ester synthetic oils, were assessed by examination of their infrared spectra, acid numbers and varnish and pentane-insoluble deposits after completion of engine tests.

Easily oxidizable 240 Neutral (95 VI) mineral oil was used in most tests of this study. Infrared spectra of used oils from engine tests with different fuels showed the presence of acid, hydroxyl, carbonyl, nitrate and nitro groups, but with some variations in their relative intensities, which depended on the fuel used. Thus, with the use of isooctane, the 240 Neutral developed very strong carbonyl but with toluene strong nitrate absorption. Light FCC (90%) with methanol (10%) caused relatively

Table IV

Effect of Fuel on 240 Neutral Base Oil With 1.25% Zinc Dialkyldithiophosphate

	Light FCC	Heavy Platformate	Alkylate	Toluene with 5% Alkylbenzenes and Olefins	Toluene with 2% n-Butyl-benzene	Toluene
Varnish Rating*	6.4	6.6	8.6	8.7	9.1	9.3
Sludge Rating*	8.4	7.7	8.4	8.1	8.2	8.8
Pentane-Insolubles, g	6.7	11.6	3.2	8.5	9.4	10.1

Table V

Engine Deposits from 240 Neutral, Rerefined and Purified Used Oils and Light FCC Fuel

	240 Neutral High Visc. Oil	Rerefined Oil	Used 240 Neutral Clay Refined	Used 240 Neutral Purified by Precipitation in Pentane	Reused 240 Neutral Purified by Precipitation in Pentane
Varnish Rating*	3.7	6.7	7.2	9.2	8.5
Sludge Rating*	7.0	7.7	9.0	9.0	8.5
Varnish, g	15.5	15.0		7.3	10.7
Pentane Insolubles, g	16.1	8.1	7.2	38.7	38.3
Total Weight Deposits, g	31.6	23.1		46.0	49.0

*Clean = 10

297

weak infrared absorption due to oxidized and nitrated groups in the oil.

Used 240 Neutral mineral oil from single cylinder engine tests was purified by dilution in pentane and filtration of the precipitated pentane-insoluble materials. The pentane-soluble nitrated and oxidized materials remained in the oil, which was used again. The reused oil, after removal of the pentane-insolubles, showed no increase in the concentration of the oxidized materials by the infrared. The same oil was used for the third time in the engine test and purified, but no further absorption increase in the carbonyl, nitrate or other infrared region of the spectrum of the oil was observed. The used and reused oxidized oils accumulated relatively large amounts of pentane-insoluble solids (Table V). These observations suggest that the oxidatively least stable molecules in the oil were oxidized and nitrated in the first engine test and that the bulk of the oil was relatively stable and not affected much by the second and third use. The oxidized and nitrated compounds largely in the oil from the first test, however, promoted free radical reactions of the blow-by gases trapped in the oil, causing a heavy formation of pentane-insolubles.

The purified used oil gave better varnish and sludge merit ratings and also less varnish by weight than the virgin oil (Table V). Apparently, the oxidized and nitrated components in the used oil protected the engine surfaces from varnish deposits. However, they also promoted reactions and dispersancy which allowed a heavy accumulation of pentane-insolubles. The used 240 Neutral contained alkali soluble acids (TAN = 6.0) and water soluble carbonyl compounds which were of a much higher molecular weight than the blow-by carbonyl compounds and could not be identified by the gas chromatographic method.[19]

Used 240 Neutral oil from the engine tests was also purified by percolation through refinery clay. This separated from the oil the insoluble and oxidized materials, but not the nitrated constituents. In the engine test, this oil gave good varnish and sludge ratings and also a relatively small amount of pentane-insolubles (Table V).

A rerefined oil, prepared by distillation and hydrofinishing of service station waste oils, gave better CRC merit ratings for deposits and also less deposit by weight than the virgin 240 Neutral oil (Table V). The total acid number of both oils was the same (2.9) after 48 hours of engine operation.

PAO "A" and "B", when tested in the engine, gave little varnish and pentane-insolubles (Tables III and VI). The infrared

298

spectrum of these used oils showed weak absorptions due to acids, ketones and aldehydes and a strong nitrate band. The oxidative stability of the highly branched oligomers may be due to the presence of methyl groups, which are known to be resistant to oxidation.[5]

A synthetic alkylbenzene base fluid exhibited almost as high oxidative stability, indicated by the low deposit accumulation, as PAO "A" (Table VI). Infrared spectrum of the used fluid had a somewhat weaker nitrate band than PAO "A".

Four synthetic esters (pentaerythritol, diisodecyl adipate, bis-tridecyl adipate and bis-tridecyl glutarate) were found to cause little engine varnish but accumulated large amounts of pentane-insolubles, while undergoing slight oxidation and nitration themselves (Table VI).

Lubricant as a Material Source and Reaction Medium in Deposit Formation

Lubricants, invariably, are consumed to some extent in engine operation. The amount of lubricant consumed was far less than the total hydrocarbons which passed into the crankcase as a part of the blow-by or were added directly to the crankcase.[6] In controlled tests where the consumption of oil was reliably measured, 240 Neutral, PAO "A" and alkylbenzene were found to be consumed at the same rate (about 370 gms in 48 hours). This showed that variations in the amounts of deposits formed are not directly related to oil consumption. Nevertheless, a lubricant in an engine may be fragmented and oxidized to yield deposit precursors which are of similar type to those resulting from the fuel. That this may be taking place can be seen indirectly by comparing deposit compositions when fuels containing widely different chemical structures have been used. Using toluene, deposits of very low hydrogen, high oxygen and aromatic ring content could be expected. Still, we found that the pentane-insolubles, with toluene as the fuel, were not much different from those with aliphatic type fuels.[6] This indicated the lubricant is an additional source of deposits.

Synthetic base oils, both hydrocarbon and ester type, gave considerably less varnish and sludge than the 240 Neutral and other mineral oils in this study. Only polyalphaolefins and alkylbenzenes of the synthetics tested gave less pentane-insolubles than the 240 Neutral and all esters yielded more (Table VI). Thus, deposit formation is a function not only of fuel and lubricant combustion and degradation, but also of the lubricant's suitability as a medium for free radical type oxidations and condensations. Effectiveness of lubricant as a

Table VI

Effect of Oil on Deposit Formation in Synthetic Base Oils and 250 Neutral High VI Oil (Light FCC Fuel)

	6 cs Poly-alpha-olefin "A"	6 cs Poly-alpha-olefin "B"	Synthetic Alkyl-benzene	Bis-(Tridecyl) glutarate	Bis-(Tridecyl) adipate	Diisodecyl adipate	Penta-erythritol Ester	250 Neutral High VI Oil
Av. Varnish Rating*	8.9	9.0	9.2	9.3	9.5	9.3	10.0	3.7
Av. Sludge Rating*	8.9	9.3	9.3	8.7	9.3	8.7	9.3	7.0
Varnish, g	4.1	2.4	1.8	1.9	1.0		3.7	15.5
Pentane Insolubles, g	1.9	0.4	1.8	83.4	55.8	18.41	26.5	16.1
Total Weight Deposits, g	6.0	2.8	3.6	85.3	56.8	–	30.2	31.6

Table VII

Engine Deposits from Light FCC Fuel and Oxidatively Inhibited 240 Neutral High Viscosity Oil

	None	A, 1.25	B, 2.00	C, 2.00	A, 2.00 C, 1.25
Varnish Rating*	3.7	6.8	3.2	7.0	8.7
Sludge Rating*	7.0	8.1	8.7	8.7	8.3
Varnish, g	15.5	7.1		5.1	7.7
Pentane-Insolubles, g	16.1	10.6	2.9	5.1	7.8
Total Weight Deposits, g	31.6	17.7		10.2	15.5

A = Zinc dialkyldithiophosphate, B = 4,4'-Methylene-bis-(2,6-di-t-Butylphenol), C = Di-t-Butylphenols

dispersant or solvent for the blow-by and other oxidized and nitrated materials will influence deposit formation. The same need to differentiate between the oxidative stability of oil and oil as a medium for crankcase reactions and dispersion is seen also for various mineral oils which were tested. All recycled oils gave better varnish and sludge ratings than the virgin 240 Neutral oil (Table V) but recycled oils, which contained oxidized and nitrated compounds, caused heavy pentane-insolubles formation.

Addition of zinc dialkyldithiophosphate, di-t-butylphenols or their combination reduced all deposits in the engine test (Table VII). 4,4'-Methylene-bis-(2,6,-di-t-butylphenol) reduced only sludge and pentane-insolubles in the oil.

The zinc additive was tested for a carryover effect on engine parts from one test to another. The different engine covers which had been left with a deposit containing some zinc after a test of 240 Neutral with 1.25% zinc dialkyldithiophosphate, did not show any improvement in cleanliness in a followup test without any additive in the mineral oil.

CONCLUSIONS

In the engine tests of this study, incomplete combustion of fuel was found to contribute more to deposits than decomposition of lubricant. Olefins caused heavier deposits than other types of fuel hydrocarbons. Deposits were formed in oxidations and condensations where nitrogen oxides acted as radical initiators, but the structure and elemental composition of the polymeric deposits showed that low molecular weight carbonyl compounds in blow-by are more readily included in their formation than nitrated intermediates.

The crankcase oil is a reaction medium for the blow-by components. The oxidatively stable synthetic oils minimized varnish, but esters accumulated much more pentane-insolubles than polyalphaolefins or alkylbenzenes. The relatively less stable mineral oils are mixtures of many different hydrocarbon types, where the oxidative stability is determined by the least stable components. They reacted first and promoted further oxidation. Consequently, a used oil may be considered to consist of a small amount of less stable hydrocarbons which have been oxidized and a bulk of relatively stable, unchanged hydrocarbons. Thus, used oils could be purified, distilled and hydrofinished to give recycled oils which were oxidatively more stable than the virgin 240 Neutral mineral base oil.

REFERENCES

1. T. K. Hanson and A. C. Egerton, Proc. Roy. Soc., A163, 90
    (1937).
2. R. S. Spindt, C. L. Wolfe, and D. R. Stevens, SAE Paper #638,
    "Nitrogen Oxides, Combustion and Engine Deposits,"
    (November 9-10, 1955).
3. ibid, SAE Transactions, 64, 797 (1956).
4. K. L. Kreuz, Lubrication, 55, 53 (1969).
5. B. D. Vineyard and A. Y. Coran, "Gasoline Engine Deposition:
    I Blow-by Collection and the Identification of Deposit
    Precursors," ACS Meeting, Petroleum Division, New York
    (September 7-12, 1969).
6. I. J. Spilners, J. F. Hedenburg, and C. R. Spohn, "Evaluation
    of Engine Deposits in a Modified Single-Cylinder Engine
    Test," ACS Meeting, Petroleum Division, Atlanta (March 29-
    April 3, 1981).
7. J. B. Edwards and D. M. Teague, "Unraveling the Chemical
    Phenomena Occurring in Spark Ignition Engines," SAE Paper
    #700489.
8. G. S. Springer and D. J. Patterson, "Engine Emissions,
    Pollutant Formation and Measurement," Plenum Press,
    New York, 1973.
9. C. E. Welling, G. C. Hall, and J. S. Stepanski, SAE
    Transactions, 69, 448 (1961).
10. W. E. Morris and K. T. Dishart, ASTM Meeting, Toronto
    (June 24, 1970).
11. H. H. Oelert, W. Mayer-Gurr, and J. Zajontz, Erdol and Kohle,
    Erdgas, Petrochemie vereinigt mit Brennstoff-Chemie, 27,
    No. 3, 146 (1974).
12. L. S. Caretto, L. J. Muzio, R. F. Sawyer, and E. S. Starkman,
    Combustion Science and Technology, 3, 53-61 (1971).
13. Z. G. Scabo and F. Marta, J. Am. Chem. Soc., 83, 768 (1961).
14. W. A. Waters, "Mechanisms of Oxidation of Organic Compounds,"
    Methuen & Co., Ltd., John Wiley and Sons, New York, 1964.
15. N. M. Emanuel, "The Oxidation of Hydrocarbons in the Liquid
    Phase," The MacMillan Company, New York, 1965.
16. A. P. Altschuller and I. Cohen, Int. J. Air and Water Pol
    lution, 4, 55-69 (1961).
17. N. Levy and Scaife, Ch. W., J. Chem. Soc., 1093 (1946).
18. J. F. Brown, Jr., J. Am. Chem. Soc. 79, 2480 (1957).
19. "Oxygenates in Automotive Exhaust Gas; Part II. Techniques
    for Determining Aldehydes by DNPH Method," CRC Project
    No. CM-59-65 (March, 1968).
20. W. A. Affens, J. E. Johnson, and H. W. Carhart, J. Chem. and
    Eng. Data 6, 613 (1961).
21. J. A. Barnard and V. J. Ibberson, Combustion and Flame, 9, 149
    (1965).

REDUCTIVE CHEMISTRY OF AROMATIC HYDROCARBON MOLECULES

Lawrence B. Ebert

Corporate Research-Science Laboratory
Exxon Research and Engineering Company
Linden, NJ 07036

I. INTRODUCTION

In approaching the chemistry of engine combustion deposits, one is faced with several difficult problems. First of all, for any given specimen, there is a limited amount of material, with typical six cylinder engine runs giving less than 20 grams of deposit. Despite their small size, these samples are quite heterogeneous. As scraped from the walls of the combustion volume, they contain refractory polymeric carbon, physisorbed aromatic hydrocarbons from the fuel, inorganic material from the lubricant, as well as bits of metal from the engine. Different areas of the combustion volume can give rise to deposits of different properties. From the chemical point of view, the deposits are somewhat of an enigma; although arising in a combustion environment, the deposits are about 65 wgt % carbon. Even though one might expect such carbon to resemble kinetically inactive species as graphite or glassy carbon, it most resembles bituminous coals in its thermogravimetric behavior.

To gain an understanding of such an unusual material, we have tried to identify a unifying theme in the chemistry. Noting that the ^{13}C nuclear magnetic resonance spectra of solid state deposits (taken with cross-polarization to ^{1}H and magic angle spinning) show carbon aromaticities of 90%, we believe that analysis of the benzenoid chemistry of the deposit could yield insight into its identity. Because the deposit is so highly oxidized (15-25 wgt % oxygen), one should focus on chemistry which is reductive in nature, so that both the chemistry of the aromatic molecules and the substituents on the molecules might be analyzed. In this spirit, we include this review chapter on the reductive chemistry of aromatic systems.

II. BACKGROUND

As noted first by Schlenk, Appenrodt, Michael, and Thal,[1] and later by Scott, Walker, and Hansley,[2] the addition of alkali metals to solutions of aromatic hydrocarbons in ethers can lead to highly colored solutions. The chemistry of this interaction is simple, with electrons initially on the alkali metal going to pi molecular orbitals associated with the aromatic hydrocarbon molecule:

$$Ar + n\ M^\circ \rightarrow (Ar^{n-})(M^+)_n \tag{1}$$

There are at least three ways by which one may gauge the proclivity of a given aromatic hydrocarbon molecule to be reduced. From a physical point of view, measurement of the gas phase electron affinity is a good predictor of the aromatic reduction chemistry, even though that chemistry generally is studied in the solution state. From a chemical point of view, the measurement of the electrochemical reduction potential, by a technique as polarography or linear sweep voltammetry, offers a quickly established scale of ease of reduction. Finally, one may revert to first principles and create a chemical scale based upon molecular orbital calculations, as in taking the coefficient m_{m+1} derived from the energy of the lowest unoccupied molecular orbital, $\alpha_0 + m_{m+1}\ \beta_0$, in the Huckel molecular orbital approximation.[2] In Table 1, we compare these approaches for several small aromatic hydrocarbon molecules, the structures of which are given in Figure 1.

Scanning the three columns in Table 1, one sees that aromatic hydrocarbons tend to be more easily reduced as molecular size increases. This feature of benzenoid reduction chemistry is helpful in the analysis of complex mixtures of molecules, as found in engine deposits, for one can be reasonably certain of reducing all aromatic cores, provided that there are no complications. Of course, even for simple aromatic molecules, both molecular size and molecular symmetry must be analyzed for a quantitative understanding.[5,6] Specifically, for a given size, one notes that aromatic molecules with growth along the graphite

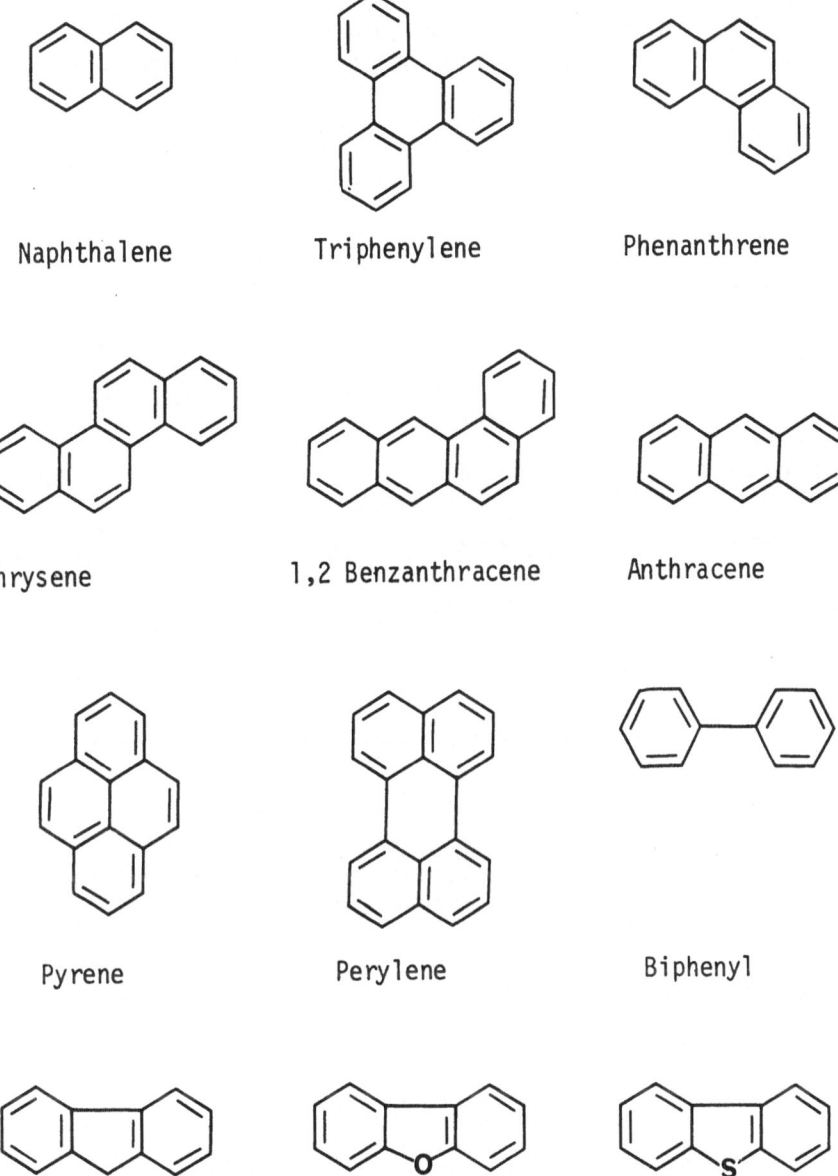

Naphthalene   Triphenylene   Phenanthrene

Chrysene   1,2 Benzanthracene   Anthracene

Pyrene   Perylene   Biphenyl

Fluorene   Dibenzofuran   Dibenzothiophene

Figure 1.   Structures of aromatic hydrocarbons. The first
eight molecules are discussed in Table 1, and
biphenyl, fluorene, dibenzofuran, and
dibenzothiophene are discussed in section III.

Table 1. Three scales of the tendency of aromatic
molecules to undergo one-electron reduction.

Molecule	EA (eV)[a]	Er (V vs. SCE)[b]	$-m_{m+1}$[c]
Naphthalene	0.074	-2.50	0.618
Triphenylene	0.251	-2.48	0.684
Phenanthrene	0.273	-2.49	0.605
Chrysene	0.516	-2.27	0.520
1,2 Benzanthracene	0.640	-2.02	0.452
Anthracene	0.653	-2.97	0.414
Pyrene	0.664	-2.04	0.445
Perylene	0.956	-1.66	0.347

[a] gas phase electron affinity, from reference 3

[b] voltammetric reduction peak potential + 30 mV in
acetonitrile solvent, from reference 3

[c] $\alpha_0 + m_{m+1} \beta_0$ is the energy of the lowest vacant orbital in
units of the standard β in the HMO approximation with all
α's and β's equal, from reference 4

[110] direction (as anthracene, tetracene, pentacene) will be more reactive than corresponding molecules with growth along the graphite [100] direction (as phenanthrene, chrysene, picene).

The numbers of Table 1 describe the single electron reduction of neutral aromatic molecules to the corresponding radical anions. Depending on size and symmetry, aromatic molecules can be further reduced to dianions, trianions, and even tetranions.[7-9] In assessing these multiple reductions, one can utilize electrochemistry and nuclear magnetic resonance. In the case of dianion formation, one studies the disproportionation equilibrium

$$2 \ Ar^{\cdot -} \ \rightleftharpoons \ Ar^{2-} + Ar \qquad K_{disp} \qquad (2)$$

As shown in Table 2, radical anions are generally favored over dianions for small aromatic molecules under typical reaction conditions. Once formed, dianions are easily detected in nuclear magnetic resonance because of the paratropic (upfield) chemical shifts caused by the antiaromatic set of electrons.[7,9,13-15]

So far, we have considered the reductive chemistry of simple aromatic hydrocarbons, in which the only atomic constituents are carbon and hydrogen. Engine deposits are more complex than this in having high levels of oxygen (atomic O/C rations of 0.16 to .27) and appreciable nitrogen (N/C ca. 0.02). In the following section, we shall illustrate the possible consequences of such complications by discussing several specific examples of reductive chemistry. More details are available in other review articles on aromatic chemistry.[16-21]

III. EXAMPLES OF THE REDUCTION OF AROMATIC
HYDROCARBONS BY ALKALI METAL

A. Naphthalene

Relative to the other hydrocarbons of Table 1, naphthalene has the smallest electron affinity and the most negative reduction potential. This implies that the radical anion of naphthalene would be capable of chemically reducing any of the other aromatic hydrocarbons in Table 1 to its corresponding radical anion:

Table 2. Disproportionation equilibrium constants
for aromatic molecules.

Molecule	Solvent	Cation	K_{disp}	Reference
anthracene	THF	Li^+	2.9×10^{-7}	10
anthracene	THF	Na^+	1.6×10^{-5}	10
anthracene	THF	K^+	6.8×10^{-7}	10
anthracene	DMF	$(CH_3)_4N^+$	1.0×10^{-13}	11
anthracene	THF	Li^+	2.0×10^{-9}	10
perylene	THF	$Na+$	1.7×10^{-6}	10
perylene	THF	K^+	2.2×10^{-6}	10
perylene	DMF	$(CH_3)_4N^+$	1.5×10^{-11}	11
1,2 bis(1-naphthalene) ethane	DMF	$(C_2H_5)_4N^+$	5.6×10^{-3}	12
1,2 bis(1-pyrene) ethane	DMF	$(C_2H_5)_4N^+$	1.1×10^{-2}	12

$$C_{10}H_8 \overset{-}{\cdot} + Ar \rightarrow C_{10}H_8 + Ar \overset{-}{\cdot} \qquad (3)$$

This concept of an aromatic hydrocarbon reduction hierarchy was recognized by Paul, Lipkin, and Weissman,[22] and was later cleverly exploited by Sternberg and co-workers to chemically reduce selected coals with naphthalene radical anion.[23-24]

Work done on the reduction of naphthalene highlights many of the general issues in aromatic reduction chemistry. The single electron reduction yields the naphthalene radical anion, which may be studied by both electron spin resonance[25] and by nuclear magnetic resonance[26] to yield insights into physical and chemical processes.

For instance, variations in alkali metal and ethereal solvent can have dramatic effects on the chemistry of the reduced aromatic. In solvents as tetrahydrofuran or dimethoxyethane naphthalene will react with alkali metals to form a strongly solvated ion pair.[18]

$$(C_{10}H_8^{-} \cdot Na^{+})_{\text{tight pair}} \rightleftharpoons C_{10}H_8^{-} \cdot + Na^{+} \qquad K_{THF} = 10^{-7} \quad (4)$$

Such pairing is of chemical significance, because the ion pair is one hundred times less reactive than the free ions.[17] Such solvation also effects the disproportionation equilibrium. Current thinking[16,17] suggests disproportionation to arise from entropy increase caused by desolvation of the ion pairs. Thus, disproportionation is favored in solvents of low dielectric constant, as diethyl ether or p-dioxane, and hindered in such solvents as tetrahydrofuran and dimethoxyethane.

The chemistry of the naphthalene radical anion in quench reactions with water, carbon dioxide, or alkyl iodides involves several steps:[22]

$$\qquad (5)$$

$$(6)$$

$$(7)$$

In this mechanism, two naphthalene radical anions will generate one molecule of 1,4 dihydronaphthalene and one molecule of naphthalene.

In general, however, one must be concerned with the possible dominance of chemistry by small amounts of dianions. Although not seen in electrochemistry, the naphthalene dianion has been reported in the literature[27-30] and could dictate the results of quench reactions. In the specific case of sodium naphthalenide in tetrahydrofuran, kinetic analysis of a water quench directly implicates the radical anion as the chemically dominant species.[31] In the case of the larger aromatic molecule, perylene, however, the dianion and not the radical anion is the species quenched.[32] With certain aromatic hydrocarbons, as 9-cyanoanthracene, the addition of water to solutions of radical anions actually promotes dimerization![33]

Quenches involving alkyl halides are more complex than those involving water. One can obtain different products depending on quench sequence (radical anion added to halide or halide added to radical anion)[34] and one can obtain products other than those predicted by equations (5)-(7). The naphthalene radical anion can reduce alkyl halides (especially iodides) to form alkyl radicals and halide ions. The alkyl radicals, once formed, can both dimerize and disproportionate.[35-37]

$$RCH_2CH_2I + C_{10}H_8^{\cdot -} \longrightarrow RCH_2CH_2^{\cdot} + C_{10}H_8 + I^- \qquad (8)$$

$$2RCH_2CH_2^{\cdot} \xrightarrow{\text{dimerize}} R(CH_2)_4R \qquad (9)$$

$$2R\,CH_2\,CH_2 \cdot \xrightarrow{\text{disproportionate}} R\,CH_2\,CH_3 + R\,CH = CH_2 \qquad (10)$$

In the reaction of benzyl chloride with sodium naphthalenide, the dimerization product is obtained in 80% yield, indicating that reactions (8)-(10) can completely overshadow reactions (5)-(7) in some cases.[38]

It is evident that the quenching of radical anions can lead to a wide variety of products. In the case of aromatic anions formed in engine deposits, one must also consider quenching by acid sites on inorganic phases, as in the reaction of naphthalene radical anion with SiO_2, ZrO_2, and zeolite Y.[39] Furthermore, the presence of substituents on aromatics, as would be expected for engine deposits, can modify the chemistry of the aromatics through both inductive[40] and steric[41] effects, as discussed in the next section.

B. Biphenyl and Fluorene

Two aromatic hydrocarbon molecules which are similar in size to naphthalene are biphenyl ($C_{12}H_{10}$) and fluorene ($C_{13}H_{10}$). Although each possesses two benzenoid rings per molecule, as does naphthalene, both molecules are harder to reduce, with biphenyl at -2.57 V vs. SCE and fluorene at -2.65 V vs. SCE in dimethylformamide.[42,43] The single electron reduction of each molecule is accompanied by the generation of an electron spin resonance spectrum, which is more persistent in the case of biphenyl (to +100°C in dimethoxyethane)[44] than in the case of fluorene (to -90°C in tetrahydrofuran).[45]

Magnetic susceptibility measurements of biphenyl/alkali metal/tetrahydrofuran solutions at room temperature show evidence only for the radical anion of biphenyl, as was the case for naphthalene,[46,47] while fluorene/alkali metal mixtures are diamagnetic at 1 to 1 molar ratios[45] and paramagnetic with the addition of extra alkali metal.[48]

The disparity in behavior between naphthalene and biphenyl on the one hand and fluorene on the other hand arises from a distinct chemical pathway which is available with the fluorene/ alkali metal mixture: metallation of an acidic carbon-hydrogen bond. Thus, fluorene, with its active methylene group (as well as other alkyl-aromatic molecules)[49-51] will react to make a diamagnetic product:

$$\text{(fluorene)} + M^o \longrightarrow \text{(fluorene carbanion)} \quad M^\oplus \qquad (11)$$

Further reduction of the fluorene carbanion leads to the dianion radical.[48] Under special conditions, one can in fact even metallate biphenyl to make o,o'-dilithio biphenyl,[52] although one will not obtain this product under "typical" reduction conditions.

Thus, in terms of room temperature chemistry, fluorene/alkali mixtures are similar to organic carbanions as benzyl lithiums,[53] with the fluorene radical anion not stable at room temperature.[54] Biphenyl/alkali mixtures will generate aromatic radical anions, the chemistry of which is similar to naphthalene systems.

As with naphthalene, the dianion of biphenyl has been claimed,[55] although the dark blue color of such solutions is commonly associated with the radical anion.[56] Consistent with radical anion intermediacy, one finds that a water quench of a lithium/biphenyl mixture yields both starting biphenyl and 3-phenyl-1,4-cyclohexadiene.[57] Furthermore, in a study of the chemistry of biphenyl/alkali metal in ammonia, the key chemical intermediate was found to be a protonated anion rather than an associated dianion.[56]

Disregarding the detailed nature of the alkali metal/biphenyl adduct, however, one notes that this species is capable of cleaving carbon-oxygen bonds (anisole or phenyl ether yielding phenol), carbon-nitrogen bonds (carbazole to yield ammonia) and carbon-sulfur bonds (dibenzothiophene to yield 3,4-benzothiocoumarin on quenching with CO_2).[56] As such, it is worthwhile to consider the interaction of alkali metals with heteroatom containing aromatic molecules.

312

C.   Dibenzofuran and Dibenzothiophene

Dibenzofuran (DBF) and dibenzothiophene (DBT) are similar to the fluorene molecule, arising by replacement of the methylene group by oxygen and sulfur respectively. In addition to the chemistry of the two benzenoid rings, one anticipates chemistry arising from the presence of the di aryl ether and di aryl thioether functionality.

As with biphenyl and fluorene, both molecules can be reduced electrochemically to the radical anion, with DBF at -2.54 V vs. SCE and DBT at -2.50 V.[58] Electron spin resonance identification has been made of the radical anion of both DBF[59] and DBT[60], as well as for radical anions of other aryl ethers[61] and aryl thioethers.[62] In comparing the hyperfine splittings between radical ions of ethers and thioethers, it found that the spin density on the heteroatom is smaller with the sulfur than with the oxygen.[62]

The resultant chemistry of DBF and DBT with alkali metal is distinct from that of naphthalene, biphenyl, or fluorene, because the ether functionality is involved. Thus, alkali metals will directly cleave the carbon-oxygen bond in dibenzofuran:[59]

$$(12)$$

Water quenching of such a species will yield o-phenyl phenol. In the case of dibenzothiophene/alkali metal chemistry, biphenyl is claimed to be the main product, suggesting the cleavage of both carbon-sulfur bonds.[63-64]

Such cleavage reactions are quite dependent on the nature of the solvent.[65] At room temperature, lithium will readily cleave dibenzofuran, but not dibenzothiophene, in diethyl ether. However, in tetrahydrofuran at room temperature, lithium will cleave dibenzothiophene, but not dibenzofuran.[65] Additionally, in refluxing dioxane, lithium will cleave dibenzothiophene to yield both biphenyl and 2-mercaptobiphenyl.[65] Further differences in chemistry between ethers and thioethers have been discussed.[66]

In the reaction of aryl ethers with alkali metals, chemical pathways other than cleavage may exist. The combination of lithium with diphenyl ether in tetrahydrofuran at room temperature leads to, on reaction with CO_2, phenol and benzoic acid (cleavage products) and 2-carboxy diphenyl ether (metallation product).[65] Metallation has been reported to occur with other aryl ethers in THF[67,68], but not in ammonia systems.[69] Additionally, the alkali metal is important in metallation yield, for sodium naphthalenide in THF will metallate dibenzofuran but potassium napthalenide will cleave dibenzofuran.[58] In alkyl-aryl ethers cleavage predominates (oxygen remaining with the aromatic system[70]), and with alkyl-benzyl ethers, one obtains the benzyl carbanion.[71]

Electrochemical reduction of dibenzothiophene has been discussed.[72,73] With proton sources, either as added ethanol[73] or as solvent,[72] one obtains hydrogenated dibenzothiophenes, as in Birch reductions, rather than cleavage products. Generally, in the reduction of aromatic systems, one must be aware of the possible role of solvent as proton donor, for an "early" protonic quench of a radical anion or dianion may alter the product slate.[74] It must be strongly emphasized that the ethereal solvent is intimately involved in the reductive chemistry,[75,76] and, in severe cases, products can arise from the decomposition of the ether.

Barring intervention of the solvent, or other proton sources, one sees that the reductive chemistry of DBF and DBT involves both aromatic chemistry (radical anions) and the chemistry of the heteroatom (cleavage). The aromatic chemistry, as measured by the reduction potential, is not strongly affected by the heteroatom. In the following, we present examples for which the aromatic chemistry is affected by the heteroatom.

D. Pyridine, Quinoline, and Bipyridine

In the case of the single ring heterocycle pyridine C_5H_5N and the two ring heterocycle quinoline C_9H_7N one has nitrogen which is part of the benzenoid system. The replacement of a CH grouping by an sp^2 hybridized nitrogen has a similar inductive effect to the placement of an electron-withdrawing substituent on an aromatic molecule.[80] This withdrawal of electron density would be expected to favor the formation of an anion of such a nitrogen heterocycle relative to the simple hydrocarbon analog, and this is in fact observed experimentally. As one example, quinoline is reduced at -1.14 V vs. SCE, while naphthalene is reduced at -2.50 V vs. SCE.[81]

This ease of reduction of nitrogen heterocycles, relative to the simple hydrocarbon analogs, causes complications in chemical analysis. For instance, the radical anion of pyridine can be formed much more easily[81] than can the radical anion of benzene,[82] and once formed tends to dimerize to yield bipyridyl derivatives.[83] This last reaction is in fact quite complex, and illustrates some of the difficulties possible when performing chemical reductions on unknown mixtures containing nitrogen heterocycles.

In the chemical reduction of pyridine by lithium/ammonia, one will obtain dialkyl tetrahydrobipyridines on quenching with alkyl iodides. If a proton source, as ethanol, is present in the ammonia solution, one obtains N-alkyl dihydropyridine on quenching, demonstrating that the radical anions will protonate more rapidly than they will dimerize. Yet, if the pyridine is reduced electrochemically in the presence of alkyl halide, one obtains merely pyridine itself and the alkane, caused by the oxidation of the radical anion by the alkyl halide.[83] The dimeric dianion of pyridine can be formed on cathodic reduction, and will revert back to pyridine on anodic oxidation. Molecular oxygen will also oxidize the dianion to pyridine, but superoxide will generate 4,4' bipyridine.[83]

Electron spin resonance can offer assistance in unravelling such complexities. In the dilute solution state, there is a scaler interaction between the unpaired electron spin and nuclear moments such that the electron spin resonance is split $2NI+1$ times, in which I is the nuclear spin and N is the number of magnetically equivalent nuclei. Thus, for the radical anion of 4,4' bipyridyl, one observes a superposition of a nitrogen quintet ($2NI+1 = 2 \times 2 \times 1+1$) and two proton quintets ($2NI+1 = 2 \times 4 \times 1/2+1$).[84] The magnitude of the splitting is the hyperfine constant, which for the nitrogen in 4,4' bipyridyl is 3.59 G, and for the protons 0.44 G and 2.37 G.[84] By use of appropriate theories, one can relate this hyperfine constant to spin densities[84,85] and even derive Hammett sigma/rho relationships.[86] Analysis of both hyperfine and g values can offer assistance in the analysis of more complex heteroatom containing molecules.[87-89]

In addition to their ease of reduction, nitrogen heterocycles can chelate a variety of metals. In some cases, one can
obtain complexes with divalent metals of the form $M(Bipy)_2$ which
are even electron triplet species.[90]

In the examples given in section III, the reader has seen
that the reductive chemistry of aromatic hydrocarbons becomes
progressively more complicated as additional functionality (such
as acidic C-H bonds, ethereal linkages or heterocyclic groupings) is considered. In fact, the reductive chemistry of unsubstituted simple aromatic hydrocarbons can involve several steps,
involving the progression from radical anion to dianion and so
on. This is considered in the next section.

E. Polyanions of aromatics hydrocarbons

The addition of more than one electron to an aromatic molecule to form a polyanion has been discussed for a long time.[91]
Electrochemically, one observes a dianion for anthracene, but
not for naphthalene, and chemical reduction with sodium produces
a dianion (slowly) with anthracene but not with naphthalene.[91]
Even with anthracene, for which the dianion is accessible, one
finds that chemical reactions commonly proceed through the radical anion.[92,93] Furthermore, there is a question of stability
in systems with excess alkali metal; although solutions of anthracene radical anion are stable for months (if not years), the
addition of more than one alkali per anthracene can produce
sample deterioration.[94]

Going beyond these early discussions, recent work has
demonstrated that dianions can be made chemically from anthracene,[9] phenanthrene,[9,95] pyrene,[96] 3,4-benzophenanthrene,[9] and
chrysene.[9] However, in agreement with early magnetic measurements,[91] the current work demonstrates that appreciable quantities of radical anions co-exist with the dianions, as can be
seen in ESR.[9,97] In kinetic studies of quench reactions with
protic acids, one finds evidence both for radical anions (rate
depends linearly on radical anion concentration) and for
dianions (rate depends on the square of radical anion
concentration).[98,99]

In other systems, a tetraanion of acepleiadylene with lithium has been reported,[100] as well as a dianion of acepleiady-
lene-5,6-dione.[101] Extended reaction times have led to tetraanions of pyrene,[7] perylene,[7] and 9,9'-bianthryl.[102] The anion

316

of 9,9'-bianthryl can decompose into anthracene and 9,10-dihy-
droanthracene.[103]

Electrochemical data is available for the 4-step reduction
of 9,9'-bianthryl-10,10'-dicarbonitrile.[104] One observes reduction peaks at -1.27, -1.538 V, -2.328 V, and -2.712 V vs.
Ag/AgCl in a solvent of proprionitrile with tetrabutyl ammonium-
PF_6 as an electrolyte. Curiously, although this molecule is
biased toward reduction by virtue of the electron-withdrawing
nitrile groups, it does show two oxidation peaks at 1.702 V and
1.878 V.[104]

In concluding this section, one notes that there is strong
evidence for dianions and even tetraanions of aromatic molecules. In judging the significance of these species in the
reductive chemistry of engine deposits, one must consider both
reaction conditions and the possible influence of substituents
on the electrochemical potential of the aromatics of the deposit. For instance, the Birch reduction of aromatic phenols is
hindered because of the difficulty in adding an electron to a
phenolate anion.[105] To more fully appreciate the influence of
substituents, we briefly consider the oxidation of aromatic
molecules in the next section.

IV. COMPARISON OF OXIDATION TO REDUCTION OF AROMATIC
     HYDROCARBONS

A.   Electrochemical Potentials

In analogy to equation (1), one may write an equation for
the removal of electrons from the pi orbitals of a benzenoid
hydrocarbon by an oxidant, Ox:

$$Ar + n\ Ox \rightarrow (Ar^{n+})(Ox-)_n \qquad (13)$$

As in the case of reduction, one may gauge the tendency of
an aromatic hydrocarbon to be oxidized in several ways.
Physically, one may measure the gas phase ionization potential
and chemically one may measure the electrochemical oxidation
potential. In Table 3, we compare the tendencies of selected
aromatic hydrocarbons to be reduced or oxidized.[3] One notes
that for unsubstituted aromatic hydrocarbons, those compounds
which are easiest to reduce are also easiest to oxidize. As
discussed in section II, these reactive molecules have their

Table 3. Reduction vs. Oxidation of Aromatic Hydrocarbons

Class	Molecule	Growth Axis	$E_{red}(V)$	$E_{ox}(eV)$	IP(eV)	EA(eV)
3 ring	Anthracene	110	-1.97	+1.37	7.97	0.653
3 ring	Phenanthrene	100	-2.49	+1.83	8.43	0.273
4 ring cata	Tetracene	110	-1.7	+1.1	7.60	1.0
4 ring cata	Benzanthracene	mixed	-2.02	+1.44	8.04	0.640
4 ring cata	Chrysene	100	-2.27	+1.64	8.19	0.516
4 ring peri	Perylene	110	-1.66	+1.06	7.72	0.956
4 ring peri	Pyrene	mixed	-2.04	+1.36	7.95	0.664
4 ring peri	Triphenylene	100	-2.48	+1.88	8.45	0.251
"Infinite"	Graphite	mixed			4.39	4.39

"growth" axis along the (110) direction of graphite, as illustrated in Figure 2.

However, the placement of substituents on the aromatics breaks down this symmetry. Consider the case of aliphatic substituents, which through inductive effects are expected to release electrons. Naively, one expects such electron release to favor the growth of a positive charge on the aromatic (oxidation) but to disfavor the growth of a negative charge (reduction). This is confirmed experimentally. With respect to the unsubstituted aromatic hydrocarbon, aliphatic-substituted aromatics are easier to oxidize[106,107] and harder to reduce.[40,108]

In the case of oxidation, there is a further subtlety. Aromatics which are more difficult to oxidize, as naphthalene, will form only dimeric radical cations (in $SbCl_5/CH_2Cl_2$), in which one electron is removed from every two aromatic molecules:[109]

$$2 \ C_{10}H_8 + Ox \rightarrow (C_{10}H_8)_2^{\overset{+}{\cdot}} \ Ox^- \tag{14}$$

As one goes to more easily oxidized aromatics, as anthracene, one encounters the monomeric radical cation. This effect is easily monitored in electron spin resonance.

There has been some interest in comparing the ESR parameters of the radical cation and the radical anion of a given aromatic molecule. With simple molecules as anthracene or pyrene, one observes larger hyperfine constants and slightly smaller g values in the cation.[109] However, the presence of substituents can alter this, as in the case of methylthio ligands.[110,111]

Nuclear magnetic resonance offers a crisper distinction between anions and cations. Whereas anions tend to have resonances at higher field than the parent hydrocarbon, cations have resonances at lower field ("deshielded"). One may observe this effect in both ^1H and ^{13}C NMR.[112,113] In the case of ^{13}C NMR, the total change in shift for all carbons is 217.3 ppm/electron,[114] although exceptions do exist.[115]

With simple aromatic hydrocarbons, ^{13}C NMR of solutions with SbF_5/SO_2ClF shows deshieldings corresponding to radical cations and, in some cases, dications.[114] Molecules with high

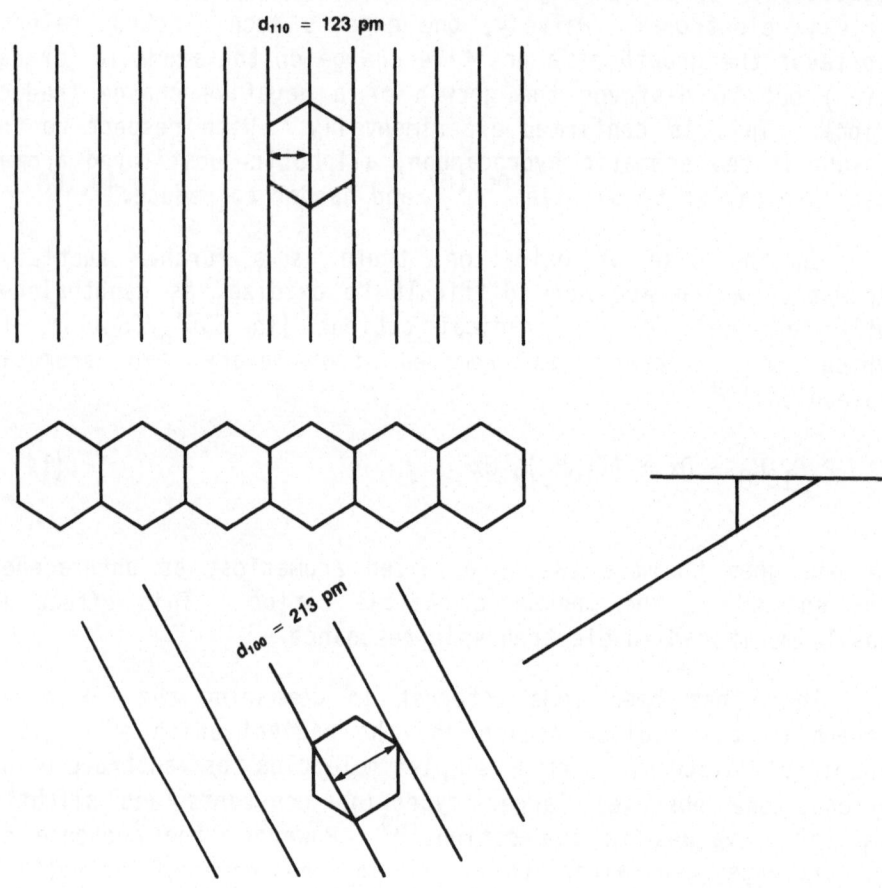

Figure 2. An illustration of the symmetry elements in the acene molecule hexacene. In terms of the "molecular" diffraction pattern, as simulated by Debye internal interference, acene molecules possess (100) diffraction peaks which get no narrower than 13.4° (full width at half maximum, Cu radiation).

electrochemical potentials (high ionization potentials), such as benzene, triphenylene, naphthalene, phenanthrene, and chrysene, show evidence only for the radical cation. Molecules with lower electrochemical potentials, such as 1,2,5,6-dibenzanthracene, pyrene, perylene, pentacene, and octamethylnaphthalene, yield the dication. The molecules picene and 1,2,3,4-dibenzanthracene are "borderline", showing mixtures of radical cations and dications.[114]

The above section illustrates the relationship between the oxidation and the reduction of aromatic hydrocarbons. With unsubstituted aromatics, those molecules which are easy to reduce are also easy to oxidize. Substituents on aromatics can break down this symmetry. Electron-releasing alkyl groups on aromatics favor oxidation and retard reduction, although the underlying reasons for this are in fact complex.[116] In any event, analysis of both oxidation and reduction does bring out the fundamentals of aromatic chemistry.

However, in the context of the chemistry of engine deposits, knowledge of aromatic oxidation may also yield insights into the mechanism of deposit formation. In the next section, we discuss some of the chemistry of aromatic cations.

B. The Chemistry of Aromatic Cations

The chemistry of cations of aromatic hydrocarbons is in large part complementary to that of the chemistry of aromatic anions, as has been discussed in review articles.[113,117-119] For instance, in the same sense that there are complications in the reaction of aromatic anions with alkyl halides (the electrophilic attack noted in equations (5)-(7) vs. electron transfer in equations (8)-(10)), one also notes competition in cation chemistry between nucleophilic attack and electron transfer.[118] There is one reaction of aromatic cations, however, which may be of direct relevance to the formation of engine deposits: their tendency to catenate.

In such a reaction, even an unreactive molecule as benzene, once oxidized, can polymerize ultimately to yield a high molecular weight polyphenylene:[120]

$$C_6H_6 + Ox \rightarrow C_6H_6^{\ddagger} \; Ox^- \tag{15}$$

$$C_6H_6^{+\bullet} + C_6H_6 \rightarrow C_{12}H_{12}^{+\bullet} \xrightarrow[+e^-]{-2H^+} C_{12}H_{10} \qquad (16)$$

More easily formed cations, as anthracenide, will react with aromatic molecules as anisole, benzene, toluene, or chlorobenzene, to give "coupled" products, which themselves may be further oxidized to give higher molecular weight products.[121] This tendency of aromatic cations to oligomerize is considered to be decisive in primary processes of chemical carcinogenesis, and thus has been well-studied.[122] Of particular interest is the reaction of the nucleophile pyridine with cations of aromatic hydrocarbons, a chemistry which thus allows the trapping and isolation of reactive cations.[122-125] As one example, the radical cation of diphenyl anthracene reacts with pyridine, is re-oxidized, and ultimately yields the 9,10 diphenyl 9,10 di-(N-pyridinium) anthracene dication.[119]

When there are no such "capping" agents present, simple aromatic hydrocarbons react with relatively weak oxidants at temperatures below 300°C to form poorly crystalline carbonaceous solids.[126-138] In the case of the combustion environment of the engine, one has the necessary components for such reactions with the aromatic hydrocarbons in the fuel and the oxygen from the air. Furthermore, oxygen-containing aromatic molecules, as we expect to find as intermediates in combustion, are especially prone to polymerization reactions of the type we have discussed.[139-146] Various other heterocyclic radical cations could also be involved in condensation chemistry.[147-151]

In concluding this section, we note that radical cations are of interest in the context of this review both because of their complementarity to radical anions and because of their possible role in engine deposit formation. We now return to our central theme of deposit characterization by reviewing the reductive chemistry of complex fossil fuels such as coal and petroleum heavy ends.

V. THE REDUCTIVE CHEMISTRY OF COALS AND PETROLEUM RESIDUA

A. Background

However satisfying one finds the study of model compounds, one must demonstrate the validity of an approach by its utility in solving real problems. In the case of reductive chemistry of aromatic molecules, an extensive literature has developed on the characterization of coal and petroleum residua. Since a recent review article has appeared,[21] our coverage will be brief, emphasizing the possible similarities between engine deposits and the fossil fuel materials.

B. The Concept of Reductive Alkylation

The underlying rationale for attempting the reduction, followed by alkylation, of coal is succinctly stated by Sternberg:

> Since coal is believed to contain clusters of condensed aromatic rings, we thought that it might be possible to introduce alkyl groups into the coal molecule by forming aromatic hydrocarbon anions in the coal molecule and then allowing these anions to react with alkyl halides.[23]

In the specific case of Pocahontas low volatile bituminous coal, the introduction of 8.8 ethyl groups per 100 carbon atoms increased the benzene solubility from 0.5 wgt.% to 95 wgt.%. This solubilization facilitated the study of the coal structure by solution-state techniques. Moreover, the mere presence of alkyl group imparted solubility led Sternberg to an interesting conclusion about structure:

> The fact that introduction of alkyl groups into lvb coal produces a benzene-soluble material points to a relation between this type of lvb coal and petroleum asphaltenes. The latter are soluble in benzene in spite of the fact that they contain a larger number of rings (8 to 9) per cluster than coal (3 to 4 rings per cluster). However, petroleum asphaltenes, in contrast to coal, contain a considerable number of alkyl groups attached to the aromatic clusters. It is these alkyl groups which import benzene solubility to the petroleum asphaltenes by preventing stacking of the aromatic clusters.[23]

This structural insight was confirmed by later work of Wachowska which showed monotonically increasing solubility as one went from methyl to ethyl to butyl to octyl side chains, even though

the number of such side chains introduced decreased monotonically as size increased.[152,153]

Of course, coals (and engine combustion deposits) contain
substantial quantities of oxygen-containing functionalities,
which will be chemically reduced under these conditions. Thus,
one must worry about the effect of this chemistry on the final
solubilization of the material. In the specific case of ether
functionality, both Sternberg[23] and Wachowska[152] determined that
the reductive cleavage of ethers, unaccompanied by alkylation,
does not greatly increase benzene or pyridine solubilities.
This conclusion was questioned by Kuhlmann, Boerwinkle, and
Orchin,[154] who reported that the reduction of Illinois bituminous coal, followed by only protonation, led to an increase of
solubility in benzene from 0% to 19%. Furthermore, a second
reduction/protonation cycle led to a solubilization of 39%.
Thus, although the chemistry of reductive alkylation is designed
primarily for aromatic hydrocarbons, one must recognize that
heteroatoms in the unknown material can influence the chemistry.

C. The Experimental Details of Reductive Alkylation

The chemical procedure for reductive alkylation involves
two distinct steps, one of reduction and one of alkylation.

In the first paper on reductive alkylation,[23] the reduction
step was performed by adding 6.17 g coal to a 24-hour old solution of 120 ml tetrahydrofuran, 112 mmol (4.4 g) potassium, and
2.4 mmol (0.304 g) naphthalene under a protective cover of
helium. In terms of ratios, this amounts to 16.7 mmol K/g, coal
and 0.358 mmol naphthalene/g, coal. After 48 hours, titration
of the liquid phase indicated a potassium consumption of 63
mmol. However, after 48 hours, a blank containing no coal consumed 11.5 mmol, so that the true consumption of potassium by
the coal was considered to be 51.5 mmol (=63-11.5). Assuming
that each potassium atom consumed leads to an anion in the coal,
one thus has 11.4 negative charges for every 100 carbon atoms in
the coal.

The alkylation step was performed by the dropwise addition
of ethyl iodide (10 ml in 30 ml tetrahydrofuran) to the coal
anion suspension at 0°C. The addition required 30 minutes,
after which the mixture was allowed to warm to room temperature,
with stirring, during the following two hours. Analysis of the

recovered product indicated an addition of 8.8 ethyl groups/100 carbon atoms, or 0.77 ethyl group per coal "anion".

In later work, one notes departures from the initial procedure. For instance, in 1974, Sternberg and Delle Donne[24] stated that larger amounts of naphthalene were required for formation of coal anions from lower rank coals, presumably because of the irreversible degradation of the naphthalene radical anion by protic acids in these coals (refer to equations (5)-(7)). Typical stoichiometries of later workers are given in Table 4; in the extreme case of the Colorado subbituminous coal, one sees that 0.67 g of naphthalene was used for every 1 g of coal!

Alemany and Stock have utilized aromatic hydrocarbons other than naphthalene, including biphenyl, 9,10-diphenyl anthracene, and anthracene.[158] They observed a correlation between the first reduction potential of the aromatic molecule and the yield of THF-soluble material following the reductive butylation of Illinois No. 6 coal. Of course, one notes[97] that the dianion of anthracene is obtained at nearly the same voltage as the radical anion of naphthalene (-2.5 V vs. SCE), so the active chemical species in the different reactions may be different (i.e. dianions of anthracene vs. radical anions of naphthalene and biphenyl). This can be significant in a reaction with an oxygen-containing substrate as coal, because[17] "the protonation of a radical anion (a relatively weak base) by a weak acid seems to require a prohibitively high activation energy, while the protonation of a dianion - a powerful base - is facile." The quenching of a solution of anthracene "radical anions" by an alcohol is second order in radical anions,[17] but the quenching of naphthalene "radical anions" is first order, again suggesting the presence of anthracene dianions but naphthalene radical anions.

With respect to the alkylation step, there has been some discussion about the length of time required for completion of reaction. Although Sternberg and coworkers[23,24] performed the alkylation step for a duration of 2.5 hours, both Ignasiak et al.[64] and Wachowska[152] allowed over ten hours at room temperature. Stock has suggested that reaction times of 48 hours are required for completion of the alkylation of polyanions prepared from bituminous and subbituminous coals.[21] This last observation[156,158] is based on the rate of precipitation of potassium iodide from solution, rather than on the disappearance of alkyl iodide, and thus could be influenced by the nature of the potassium iodide crystallization in tetrahydrofuran. After 48 hours

Table 4. Stoichiometries used in the reductive alkylation of Fossil Fuels.

Substrate	naphthalene, $\frac{mmol}{g, \, substrate}$	[naphthalene], $\frac{moles}{Titer}$	Reference
Pocahontas lvb coal	0.38	0.02	23
Bruceton hvab coal	1.7		24
Colorado subbituminous	5.2		24
Petroleum asphaltene	0.59	0.023	64
Kaiparowitz subbituminous	2.6		155
Anthracite	0.36		152
Various bituminous	1.44		152
Illinois bituminous	3.1	0.13	154
Illinois #6 bituminous	3.6	0.062	156
Illinois #6 bituminous	0.39-2.90	0.02-0.15	157

of reaction time, the extent of the alkylation reaction is the same for methylation, butylation, and octylation.[21]

Alemany and Stock have investigated the relative effectiveness in promoting coal solubility of different alkylating agents.[158] In comparing n-butyl chloride, bromide, iodide, and mesylate under equivalent conditions, they reported the mesylate to be most effective (71% solubilization), closely followed by the iodide (66% solubilization) and the bromide (50%), with the chloride noticeably poorer (10%).

In concluding this section, one notes that, although certain refinements have been made, the fundamental chemistry of the reductive alkylation of Sternberg has been preserved. There are, however, other reductive chemistries of potential utility with engine deposits, which we cover in the next section.

D. Other Reductive Chemistries Used to Characterize Fossil Fuels

In 1957, Reggel, Friedel, and Wender noted that coal could be reduced by lithium-ethylenediamine,[159] an observation which led to many further studies.[160-164] As with the reductive procedure of Sternberg, reduction by lithium-ethylenediamine greatly enhanced the solubility of the coals in common organic solvents. In the most dramatic case, the solubility of Sewell coal (West Virginia) went from 3% to 91% in pyridine following the addition of 35 hydrogen atoms for every 100 carbon atoms.[164] The mechanism for this solubility enhancement may be different from that involved in reductive alkylation. Since there is no addition of alkyl groups, there is no disruption of aromatic structure at the periphery of aromatic clusters. However, there is substantial addition of hydrogen, which will lead, in part, to rehybridization of carbon from sp^2 to sp^3, thereby disrupting aromatic structures. Unlike the reduction step of the Sternberg chemistry, lithium-ethylenediamine can reduce single ring aromatics, including phenols. In spite of this potential difference in mechanism, it is interesting to note that both methods produce the highest solubilities for coals possessing 88-90% carbon on a dry and ash-free basis (e.g. Pocahontas low volatile bituminous coal is 89.97% C (daf) and Sewell medium volatile bituminous coal is 89.4% C (daf)).

This idea of a solubility maximum, depending on the percentage carbon of the initial coal, was first suggested by Given,

Lupton, and Peover,[165] based on reductions of coal involving either lithium/ethylamine or cathodic reduction in dimethylformamide. Herein, the product coal of highest pyridine solubility (79%) contained initially 89.2% C (daf). Coals of different carbon contents, either higher or lower, consumed fewer electrons in the cathodic reduction and possessed lower solubilities. Significantly, in a later study,[166] Pocahontas coal was electrochemically reduced to yield a product that was 78% soluble in pyridine, somewhat lower than the 97% solubility found for reductive ethylation.[23] Whereas electrochemical reduction can reduce even single ring aromatics,[167] one concludes that the combination of alkylation and reduction may be more effective than exhaustive rehybridization in some cases.

Of course, there are many different ways to effect chemical reduction of aromatic hydrocarbons. One of the more active is the "solvated electron" formed by combinations of alkali metal in ammonia,[168,171] in hexamethylphosphoramide,[172] or in glymes.[173]

In 1968, Lazarov and Angelova[174] reported on the reduction of two different bituminous coals with sodium in ammonia at -77°C. Although the coal of carbon content 82.2% (daf) was relatively unchanged, the coal having 87.6% carbon showed increases in volatile matter, chloroform extract yield, and hydroxylic oxygen. The authors interpreted their results in terms of ether cleavage rather than aromatic hydrogenation. Ignasiak, Fryer, and Jadernik[175] discussed the reduction of Rasa lignite (82.0% C (daf)) with sodium in liquid ammonia followed by alkylation with octyl bromide. Mass recovery was 220%, of which 179% was insoluble in ethanol, 41% soluble. This weight increase corresponds to the addition of 10.8 octyl groups per 100 initial carbon atoms. These authors were interested in the nature of the sulfur functionality in the lignite and concluded that about one-third of the sulfur atoms occur in the form of aliphatic thioether links, with the bulk of the remainder heterocyclic. Handy and Stock reported on the reduction of Illinois No. 6 with alkali metals in ammonia followed by alkylation with alkyl halides. To produce solubility in tetrahydrofuran, they found that the combination of potassium as reductant with butyl iodide as alkylating agent was most effective.

Rabideau[177] has reported results of the reduction of model compounds by metal ammonia solutions which are of relevance to coal reductions. With respect to substituent effects, Rabideau found that single ring aromatic esters (as ethyl benzoate) could be reduced to the corresponding 1,4 dihydroaromatic compounds in nearly quantitative yield, provided THF is used as a co-solvent and a small amount of water is present before metal addition. In the case of methyl substituents on multi-aromatics, one generally finds that aromatic ring reduction occurs "away" from the ring bearing the methyl group. However, in the cases of 2-methyl naphthalene and 3-methyl biphenyl, Rabideau found that the ring bearing the methyl group was reduced. With respect to bond cleavage reactions, Rabideau[177,178] found that reduction of the isomeric 1,2 benztriptycene with lithium in ammonia led to carbon-carbon bond cleavage:

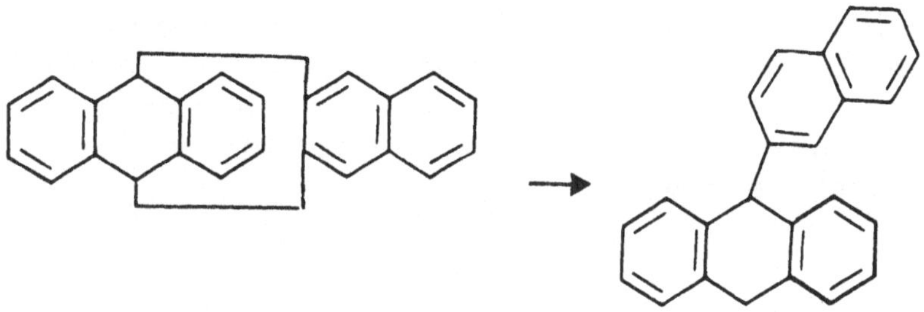

(17)

If sodium is used in place of lithium, ring cleavage does not occur, but aromatic reduction does:

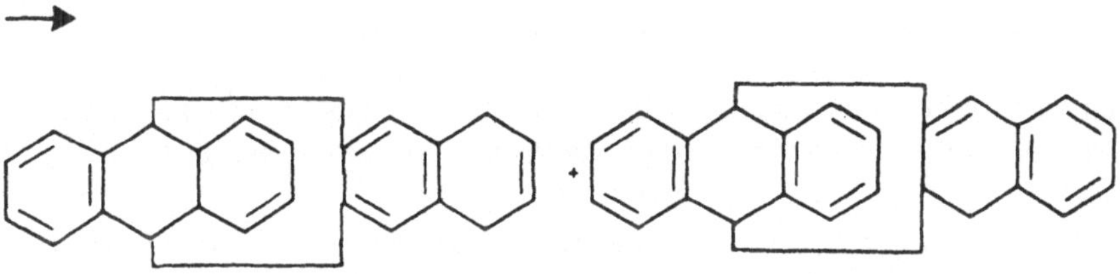

(18)

Finally, Rabideau[179] considered possible steric problems arising from the alkylation of anions of hydroaromatic molecules. Although methylation of 1-methyl-4-sodio-1,4-dihydronaphthalene is stereospecific in providing only the cis product, ethylation provides a nearly equal mixture of cis and trans. In the analogous anthracene case, just the opposite is true.[179]

Another solvated electron system capable of inducing carbon-carbon bond cleavage is that of alkali metals in hexamethylphosphoramide.[172] Ouchi, Hirano, Makabe, and Itoh[180] utilized solutions of sodium in hexamethylphosphoramide to reduce eight vitrinite-concentrated coals. The reactions were carried out at room temperature for twenty hours, with t-butanol the proton source. As with earlier reports,[164,165] those coals in the range 88-90% C (daf) had the highest pyridine extractabilities.

Solvated electrons are also produced by the dissolution of alkali metals in complex ethers as 1,2 dimethoxyethane ("glyme") and bis (2- (2-methoxy ethoxy) ethyl) ether.[180] Such solutions are quite unstable at room temperature[180] and are generally used at temperatures below 0°C.[173] Schanne and Haenel[173] investigated possible carbon-carbon bond cleavages in model compounds induced by potassium in dimethoxyethane/octaglyme at -25°C. The compounds toluene, indane, tetralin, biphenyl, 1,3 diphenyl-propane, 1,4 diphenylbutane, 1,3-bis (1-naphthyl) propane, 9,10-dihydrophenanthrene, 9,10-dihydroanthracene, and 9,9'-dimethyl-fluorene did not give products arising from carbon-carbon bond cleavage. However, benzylic bond cleavage was observed for hydrocarbons of the form

$$\text{Ar} - \overset{\displaystyle |}{\underset{\displaystyle |}{\text{C}}} - \text{R},$$

in which Ar = phenyl or naphthyl and R = phenyl, naphthyl, benzyl, naphthyl, or methyl, as discussed in Table 5. It should be stressed that carbon-carbon bond cleavages induced by alkali metals were not completely unknown at this time.[55,181,182]

With respect to coals, Niemann and Richter[183] reported reduction by potassium in poly(glymes) followed by quenching with dimethylsulfate, diethylsulfate, carbon dioxide, or para-formaldehyde. With a medium volatile bituminous coal of 89.4% C (daf), they found an increase in pyridine extractability from 12.1% to 73-75% on methylation. Niemann and Hombach[184] studied the effect of repeated reduction/proton quench cycles on the sample coal. After six such cycles, they found that the coal had gone from an H/C ratio of 0.63 (4.7% H) to an H/C ratio of 1.34 (9.2% H). They noted an increase in the number of methyl groups, which they associated with the cleavage of methylene and/or ethylene linkages in the coal. The success of this coal work generated further model compound studies.[185] Phenyl-p-tolylmethane was found to cleave to p-xylene and 4-methylbenzyl-biphenyl and 1,2-diphenylpropane yielded toluene and ethylbenzene. Returning to coal, Hagaman and Woody[186] performed ^{13}C

330

Table 5. Carbon-carbon Bond Cleavage Reactions
with Potassium in dimethoxyethane/octaglyme

Initial hydrocarbon	Quenching agent	Products
diphenylmethane	1-butanol	toluene
diphenylmethane	dimethylsulfate	toluene
2,2 diphenylpropane	1-butanol	benzene, isopropyl-benzene
1,2 diphenylethane	1-butanol	toluene
1-phenyl-2-(1-naphthyl) ethane	dimethylsulfate	toluene, methyl-naphthalene, ethyl benzene, ethyl naphthalene
1,2-bis-(1-naphthyl) ethane	dimethylsulfate	methyl-naphthalene, ethyl naphthalene
phenyl-(1-naphthyl) methane	dimethylsulfate	toluene, naphthalene

CP/MAS NMR on a sample of Illinois No. 6 coal which had been reduced with Na/K in glyme/triglyme and alkylated with CH_3I. Curiously, they noted, "The general character of the spectrum is similar to that of the chemically unaltered coal but with a broadened sp^3 envelope." Further analysis indicated the presence in the product of nine aromatic methyl ethers for every 100 aromatic carbon atoms.

In the examples given above, one performs the chemistry of reduction followed by protonation or alkylation to chemically modify the fossil fuel. The specific reductants utilized are strong, capable of reducing aromatic rings, of cleaving ethers, thioethers, and carbon-carbon bonds, and of metallating acidic hydrogen groupings. Such chemistry can modify the "substrate" beyond recognition. One possible alternative to this is to perform alkylation without reduction, as has been advocated by Ignasiak, Carson, and Gawlak.[187] Specifically, one uses sodium amide, $NaNH_2$, in solutions of ammonia, to remove protons without chemical reduction, followed by alkylation. Such chemistry will place alkyl groups on hydroaromatics[188] and does enhance pyridine extractibility of coals.[187] Similarly, the chemistry of silylation can be effective in promoting solubility when one has many acidic –OH groups.[189]

Nevertheless, even in these milder chemical schemes, one has to worry about complications involving the interaction of the solvent with the solid fossil fuel substrate. Furthermore, in reduction of aromatic-containing systems, one must contend with the problem of structural rearrangements, which might alter one's perception of the initial substrate. Most frequently involving migrations of aryl groups, such rearrangements[190-193] could well occur in fossil fuels subjected to reductive conditions. Finally, one notes the complex nature of the reductive chemistries discussed above. As one example, consider the reduction of 1,4 dihydronaphthalene by lithium compounds. Gilman and Morton noted the following reactions:

$$\text{(19)}$$

$$\text{(naphthalene dihydro)} + \text{n-BuLi} \xrightarrow{CO_2} \xrightarrow{H^+} \text{(dicarboxylic product)} \qquad (20)$$

yet, in the presence of the chelating agent N,N,N',N'-tetra-methyl-ethylenediamine (TMEDA), Brooks, Rhine, and Stucky[195] reported

$$\text{(naphthalene dihydro)} + \text{n-BuLi} \longrightarrow \text{(naphthalene)}^{-2} \qquad (21)$$

In three reactions of 1,4 dihydronaphthalene with organolithium compounds under similar conditions, we have three completely different outcomes:

1. Proton abstraction/LiH elimination/rearomatization

2. Proton abstraction only/no rearomatization

3. Double proton abstraction/antiaromatization

Furthermore, in the absence of TMEDA, the dianion of naphthalene with lithium (equation 21) is known to be unstable at room temperature.[196]

For the above reasons, we next discuss the reduction of fossil fuels directly by alkali metals.

E. The Use of Alkali Metal to Reduce Fossil Fuels

Potassium metal, in the absence of any solvent, can chemically reduce graphite,[197-200] leading to the formation of an intercalation compound. However, the direct reaction of potassium metal with aromatic hydrocarbon molecules is generally not considered to lead to anions of aromatic hydrocarbons. To quote Belser, Desbiolles, Ochsenbein, and von Zelewsky:[201]

> On the other hand, no reaction occurs when neat liquid aromatic compounds are refluxed over alkali metals. It is obviously the lack of solvation, mainly that of the metal cation, which makes reaction (1) energetically unfavorable.

Various people have claimed the isolation of solid phase aromatic radical anions,[202-204] although it should be noted that the actual step of reduction did in fact occur within the solution phase. In this context, it is interesting to note that a scheme for the direct reduction of coals by potassium metal has not only been proposed[205,206] but also has found wide acceptance in the literature.[21,186,191] Thus, it is worthwhile to consider the nature of the direct reaction of potassium metal with graphite and aromatic hydrocarbon species.

Although the customary synthesis of the graphite compounds C_8K and $C_{12n}K$ (n>2) involves vapor phase reaction at temperatures in excess of 300°C,[197-200] Podall, Foster, and Giraitis[207] discovered that C_8K could be made at 275°C under nitrogen, after which Lalancette, Rollin, and Dumas[208] found that C_8K could be made at 100°C under helium by direct combination of the reagents. Intercalation compounds formed by this last method are made quickly and lead to the correct x-ray diffraction pattern.[97] Furthermore, these compounds have a Dysonian electron spin resonance signal, in which linewidth is linearly proportional to temperature although area is relatively insensitive to temperature.[97]

The direct reaction of aromatic hydrocarbons with potassium is somewhat different. First of all, the presence of substituents on the aromatics can dominate the chemistry, as with metallation of acidic -OH or -CH groups.[207] As an example, treatment of 2,6 dimethylnaphthalene with potassium at 100°C, followed by quenching with butyl iodide, leads to ca. 90% 2,6 dimethyl naphthalene and 10% 2-pentyl 6-methyl naphthalene. With pyrene under the same reaction sequence, one again obtains ca. 90% pyrene following the quench, with the remaining 10% composed of mono-and di-butylated pyrenes and hydro-pyrenes. The ESR pattern of the pyrene/potassium adduct (2/1 on a molar basis) is completely unlike that of the graphite/potassium compound. The linewidth is 0.8 G (vs. 15 G at 25°C for C_8K[97]) and is independent of temperature. In the case of the reaction of coronene with potassium (1/3 on a molar basis), one does obtain an ESR signal with a linewidth varying linearly with temperature.

$$\Delta H(G) = 0.0161 \ T(K) + 3.0281 \qquad\qquad (22)$$

However, as illustrated in Figure 3, the absorption is symmetric, and the area of the ESR absorption obeys the Curie Law. Furthermore, quenching with CH_3I produces primarily coronene itself, although mono- and di-methylated coronene species are observed. A Debye-Scherrer x-ray diffraction photograph reveals both the lines of coronene and of a new phase, in analogy to the report of Holmes-Walker and Ubbelohde.[202] The above evidence suggests that the formation of C_8K-like products from the direct combination of aromatic hydrocarbons and potassium is unlikely. Let us consider the reaction of coals with potassium.

In 1978, Lazarov, Rashkov, and Angelov[205] reported the preparation of an ionic potassium-coal adduct by the combination of 2g (51 mmol) potassium with 2.5g of low volatile bituminous coal at a temperature of 120°C under a vacuum of 13mPa (the atomic C/K ratio of the mixture was 1.3). They stated, "During this treatment the coal sample swelled and it gradually attained a light metallic tint." Curiously, the reaction was reported not to work at normal pressure, which is in contrast to the graphite/potassium reaction.[97,208] Characterization of the product formed after a 4 hour reaction was by x-ray diffraction and by electron spin resonance. The (002) band characteristic of stacking of benzenoid planes had disappeared in diffraction and the e.s.r. spectrum consisted of a single asymmetric line of width 13G and a g factor of 2.0030.

However, with respect to the diffraction of coal, one notes that the (002) band generally arises only from dimeric structures, and that most aromatic molecules are not "stacked" in any way.[209,210] The classical intercalation of such structures is impossible, for there can be no translational periodicity.

The potassium-coal adduct was treated with methyl iodide in tetrahydrofuran, and picked up 12.2 to 13.3 methyl groups per 100 carbon atoms. This addition corresponds to a ratio of 0.17

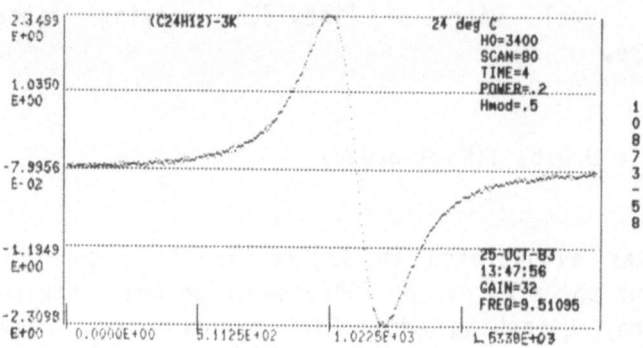

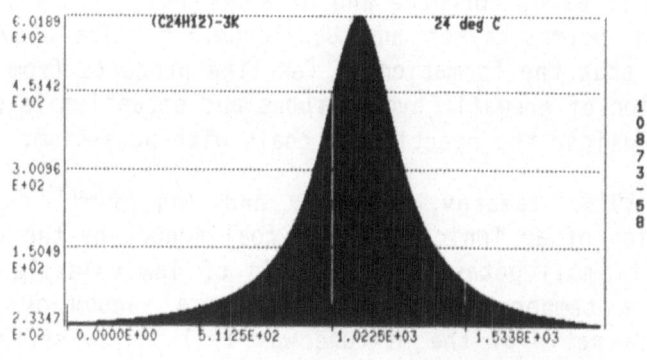

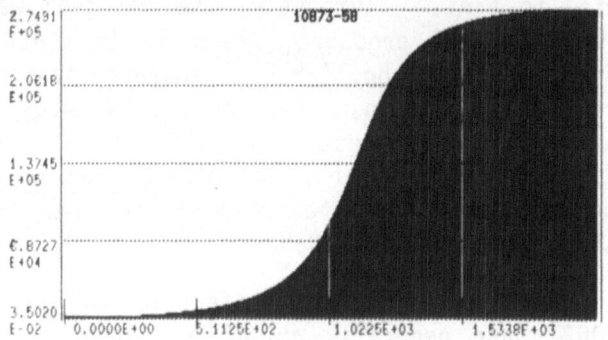

Figure 3. Electron spin resonance of the compound $C_{24}H_{12}/K_3$ recorded at 9.51 GHz at 24°C. The scan width is 80 G (= 8 mT), and the top spectrum is the derivative of the absorption mode, the middle spectrum is the absorption mode, and the bottom spectrum is the integral of the absorption mode.

methyl group per potassium, and suggests that the majority of potassium atoms did not create "alkylatable" sites. Nevertheless, the methylated coal was 61.2% extractible in benzene, an increase of 32 times over the initial value of 1.9% extractible. In a later study involving coals of varying carbon content, a "window" in the range 85-90%C (daf) was found for which extractibilities were highest.[206] This region of high reactivity/ extractibility was confirmed in a study on Yubari coal (86%C) employing a slightly modified technique.[211]

In concluding this section on reduction by potassium metal, one notes that there is noticeable, but poorly defined, chemistry which occurs. Analogies to the redox chemistry of graphite, or of thermal carbons,[212] are not well taken, and there is likely to be chemistry involving aromatic substituents during reduction and involving free radicals (equation 8) during alkylation. Nevertheless, the material _after_ potassium reduction/ alkyl halide alkylation is different from the starting material. In the next section, we discuss some of these changes.

F.   What Does Reductive Alkylation Tell Us?

One of the most striking results of the studies of the chemical reduction of coal is the discovery of a "window", defined in terms of weight percentage carbon, for which the reduction of the coal is most effective in producing extractible products. Medium and low volatile bituminous coals are affected by reduction to a greater extent than either lower rank subbituminous coals or higher rank anthracite coals. This difference in response to chemical reduction presumably arises from the difference in structures of the various coals.

Given, Lupton, and Peover[165] were the first to notice, and offer an explanation for, this effect. Low rank coals, with a relatively high abundance of cross-links between aromatic clusters will be difficult to reduce from the standpoints of steric interference (difficulty in solvating a _rigid_ network) and of inductive effects (electron release by alkyl and hydroxy substituents). They considered these effects might be sufficient to render the reduction of portions of such coals "thermodynamically impossible." With higher rank bituminous coals, there is less cross-linking, and thus reduction of aromatic clusters is more facile. For coals of carbon content greater than 89% (daf), the decreased ease of reduction is due to the rapid increase in aromatic cluster size.[165] In such an analysis, engine combustion deposits would be most like low rank coals.

Sternberg's reductive alkylation procedure is more effective than simple reduction in promoting extractibility, but it is most effective for coals in the 85-95% C (daf) range. Sternberg's explanation for this effect is slightly different from that of Given, Lupton, and Peover. Herein, the average size of aromatic clusters in the coal is directly related to the rank of coal:

Subbituminous	single ring aromatics
High volatile bituminous	two ring aromatics
Low volatile bituminous	three ring aromatics
Anthracite	twelve ring aromatics

Theoretically, the ease of reduction should be directly related to rank, with the multi-cyclic anthracite the easiest to reduce and the single-ring subbituminous the most difficult. Experimentally, one does find increased extractibility up to coals of 90% C, in accordance with this prediction. To Sternberg, the failure of anthracite coals to be solubilized by either simple reduction or reductive alkylation arises from an inability to "quench" the anions of anthracite with either protons or alkylating agents. Thus, although one can add electrons to anthracite (11 charges per 100 C atoms for anthracite, comparable to the 12 charges per 100 for Pocahontas), one is not able to alkylate these anions (0.6 ethyl group added/100C for anthracite vs. 9 ethyl groups added/100 C for Pocahontas). Without alkylation, there is no solubility, and thus anthracite coal is an "exception" to the ring size/reduction correlation for purely steric reasons.

Sternberg[23] analyzed the possible influences of the cleavage of cross-links in imparting solubility. With Pocahontas coal, reduction followed by protonation yielded solubilities in benzene and pyridine of only 3% and 13%, but reduction followed by ethylation gave a solubility of 95% in both solvents. Since bond cleavage, whether of C-O, C-S, or C-C bonds, occurs during the reduction step, it would appear that alkylation is a necessary condition for the high solubilization of Pocahontas. It is, however, curious to note that the reduction of Pocahontas with the sequence naphthalene/potassium/water is so much less effective in promoting solubility than the electrochemical reduction.[166]

In the case of low rank coals, which according to Given, Lupton, and Peover are not solubilized because of cross-links, Sternberg[24] reported a high degree of anion formation (24 charges/100 C atoms for subbituminous, higher than the 12 charges/ 100 C atoms for Pocahontas!). Only 14 ethyl groups/100 carbon atoms could be added, but none of these ethyl groups could be removed by a later reduction by naphthalene anion, which indicates these ethyl groups to be attached to carbon, rather than oxygen, atoms. In fact, much oxygen was lost from the subbituminous coal during reductive alkylation, with the initial coal 21.2% oxygen but the ethylated product (86% recovery) containing only 11.6% oxygen. Nevertheless, even the loss of this amount of oxygen (presumably from cross-links) and the addition of 14 ethyl groups/100 C atoms (only 9 ethyl groups are added to Pocahontas) are sufficient to create only a 43% solubility in pyridine for the subbituminous coal. There is something in the structure of these low rank coals which makes them resistant to solubilization.

Clearly, the chemistry of reductive alkylation does tell us something about the molecular structures of the aromatic hydrocarbons in the different coals. For the subbituminous coals of Sternberg, a consumption of 13.7 mmol K/(g, dry coal) goes in large part to the cleavage of oxygen functionality from the coal, with alkylation occurring at the corresponding carbanion fragments remaining on the coal skeleton. Although much chemistry occurs, it does little to promote the solubility of the alkylated subbituminous coal skeleton in common organic solvents. With Pocahontas coal,[23] a consumption of 8.83 mmol K/(g, dry coal) produces 4.27 mmol C-ethylated and 2.21 mmol O-ethylated groups per gram dry coal. Although this Pocahontas coal consumes less potassium and less ethyl iodide than did the subbituminous coal, the solubility of the alkylated Pocahontas coal skeleton is 95% in benzene and in pyridine. The aromatics in Pocahontas, relatively small (e.g. three-rings) but not highly substituted, are dramatically affected by the reductive alkylation treatment. In contrast, the larger aromatic molecules of anthracite consume 8.60 mmol K/(g, dry coal) but yield only 0.5 mmol alkyl group/(g, dry coal), indicating that only ca. 5% of the anthracite coal anions are "alkylatable." It is probably simply the size of the anthracite molecules, as measured by the ratio of internal carbon atoms to peripheral carbon atoms,[213] that renders them more likely to reduce alkyl halide molecules than to be alkylated by them. Furthermore, the graphite anion is not attacked by water,[200] either external or intercalated, suggesting that large benzenoid anions may have unanticipated stability to water.

In addition to this information about different coals, the reductive alkylation procedure can give valuable information about a given material, especially when that material, once reductively alkylated, has a high solubility in some solvent. For instance, in the case of Pocahontas coal, Sternberg performed vapor pressure osmometry on his alkylated materials and found weights of <u>ca.</u> 3000 daltons (alkyl-free basis), somewhat lower than that of a petroleum asphaltene from California (4200 daltons). The Bruceton coal was 2000 daltons, and the subbituminous coal was 700 daltons. In the case of petroleum asphaltenes, one can observe the effect of reductive alkylation in changing the molecular weight. Ignasiak, Kemp-Jones, and Strausz[64] found that the VPO molecular weight of Athabasca asphaltenes went from 5920 to 580 following reduction with potassium naphthalenide and alkylation with octyl iodide. If no naphthalene were used (thus relying on the aromatics of the asphaltene to form radical anions directly), the final molecular weight was 1900. In other chemistries, a diazomethane/ asphaltene product had a molecular weight of 2950 and a hexamethyl disilazane/asphaltene product had a molecular weight of 3200. From the ten-fold reduction in molecular weight caused by reductive alkylation, Ignasiak and co-workers inferred that the Athabasca asphaltene has a sulfur polymeric framework, in which average carbon moieties, consisting of alicyclic diaromatic structures with some alkyl substituents, are held together by sulfide linkages. Nevertheless, in this and other studies, one must take care in the interpretation of molecular weight data from vapor pressure osmometry, for it is greatly affected by the presence of low molecular weight contaminants.

One example of this problem can be found in the work of B. Ignasiak and co-workers on the molecular weights of alkylated coals of varying ranks. The octylation of Balmer mvb coal (89.1% C) yielded a product 87% extractible in pyridine, which possessed a VPO-determined molecular weight of 680 on an alkyl-free basis. Octylation of the same coal using a sodium ammonia reduction led to a product 75% extractible in pyridine, which possessed a VPO-determined molecular weight of 590 on an alkyl-free basis.[214] However, when these two different reductive alkylation procedures were applied to Rasa lignite (77.7% C), different number average molecular weight figures were obtained.

The treatment with sodium naphthalenide, followed by octylation, yielded a molecular weight of 1020, but the sodium ammonia reduction followed by octylation led to a molecular weight of 500.[175] The authors[215] found that the Sternberg alkylation is accompanied by "incorporation of substantial quantities of solvent (up to 25 w/v% of thw substrate) into the coal." As such, they concluded "Sternberg alkylation, GPC [gel-permeation chromatography], and vapor pressure osmometry contain inherent shortcomings which make it necessary to treat results obtained by these methods with caution."

Another technique which takes advantage of the solubilization induced by reductive alkylation is nuclear magnetic resonance. Herein, analysis of the proton and the carbon nuclear magnetic resonance spectra of alkylated products in solution yields direct information about organic functionality in these materials. Chemical logic may then be applied to make inferences about the structure of the initial substrate. As one example of this approach, Alemany, King, and Stock[216] used magnetic resonance to characterize sixteen distinct gel-permeation chromatography fractions arising from the reductive butylation of Illinois No. 6 coal.[217] The spectra from the various molecular weight fractions were different, showing variations in the degree of aromaticity, the presence of linear alkanes, the ratio of C-butylation to O-butylation, the extent of butylation on aliphatic and aromatic carbon atoms, and the amount of carbonyl and vinyl derivatives. This initial work was followed by a reductive alkylation study involving ^{13}C-labeled methyl iodide and butyl iodide, which greatly facilitated spectrum interpretation.[156,218] Between 60%[218] and 90%[156] of the alkylated coal was soluble in tetrahydrofuran, and the coal product was again separated by gel permeation chromatography. In the proton NMR, two signals at 3.7 δ and 4.2 δ were assigned to α-methylene proton resonances in aryl butyl ethers.[218] These "sterically uncrowded, non-heterocyclic aryl butyl ethers are present in abundance in the soluble alkylated coal and the related aryloxides constitute the most abundant oxygen functional group in the polyanion." In the carbon NMR, distinct resonances were observed for O-butylation in the region 60-73 δ and for C-butylation in the region 25-50 δ. The ratio of C-butylation to O-butylation increased with decreasing molecular weight, ranging from 0.76 in cut 12 to 1.77 in cut 18. Nitrogen-alkylation signals at 50 δ and 58 δ for the butylated coal were most prominent in the higher molecular weight fractions. Further analysis of peaks in the O-alkylation region suggested un-hindered butyl ethers (67-69 δ) to be more predominant than butyl carboxylates (64 δ), with the ratio of ethers to carboxyl-ates estimated to be about eight.[218] This, and related,[219-220] nuclear magnetic resonance work is of relevance to analysis of engine combustion deposits, for it demonstrates that not only carbon chemistry but also oxygen and nitrogen chemistry may be addressed by the procedure of reductive alkylation.

One specific oxygen functionality which may be abundant in engine deposits is the ether functionality, and ethers have been well-studied in reductive alkylation.[221-222] One area is in the oxidative weathering of coals.

The oxidation of coal, under relatively mild conditions which simulate weathering, is known to destroy the coking pro-perties of a coking coal.[221] There are several different mecha-nisms which can be envisioned:

1. During the period of mild oxidation, reactive oxygen func-tional groups as peroxides, ketones, aldehydes, and hydroxyls might be created within the coal. On heating the coal to the coking temperature, these functionalities would generate polymerization reactions, thereby inhibiting the ability of the vitrinite to fuse and bind the other macerals.

2. During the period of mild oxidation, ether bonds are formed directly, thereby generating a refractory, poly-meric matrix.

If the first hypotheses is correct, treating the oxidized coal with reagents that react with the active functionality might restore the coking properties of the coal. Specifically, peroxides could be destroyed by sodium in methyl alcohol, sodium sulfite, or sodium borohydride. Hydroxyl groups could be blocked by barium hydroxide or hexamethyldisilazane. Carbonyl groups could be derivatized with hydroxylamine hydrochloride.[221]

If the second hypothesis is correct, ether groups could be eliminated by treatment with alkali metal/naphthalenide.

In fact, in studies on a vitrinite concentrate of Balmer No. 10 medium-volatile bituminous coal (88% C, dry basis) and on Moss No. 3 high-volatile bituminous coal (81.3% C, dry basis), hypothesis two seems to be correct.[221] The removal of the groups and the blocking of the hydroxyl groups had little influence in restoring the swelling properties. On the other hand, cleavage of ether linkages using potassium naphthalenide in THF resulted in a lowering of the melting point and a large increase in dilatation.[221] Thus, the use of chemical reduction offers key insights into the type of oxygen linkages present in the coals.

In a later study, Wachowska and Pawlak[222] took advantage of this finding to characterize the amount of ether functionality in a series of coals of varying rank. They were especially interested in the number of ethers cleaved during treatment with potassium naphthalenide and the correlative lowering in melting point. Some of their results are grouped in Table 6. The greatest increase in hydroxyl groups after naphthalenide reduction occurred in sample K-IV, and this coal showed a dramatic change in softening point on reduction, going from 404°C to 250°C. Additionally, reduction caused increased dilatation in the coal samples.

Wachowska and Pawlak inferred structural characteristics from their data on reductive chemistry and fluidity. Lower rank coals, which tend not to fluidize and coke, have large numbers of ethers, both before and after chemical reduction. Anthracite coal, which also does not coke, also has a high percentage of unreactive ethers, although the absolute abundance is smaller than for the lower coals. Difficulties in coking for anthracite presumably arise from the large aromatic cluster size. Coking coals, on the other hand, have smaller aromatic clusters and more reactive ether functionalities.

The chemistry of oxygen in these reduction reactions was not completely straightforward. Coals K-I and K-II lost oxygen, but the higher rank K-III, K-IV, and anthracite actually gained oxygen. Carbonyl oxygen, which should be reduced by naphthalenide, was unchanged in abundance after reduction. To more fully study this chemistry, Wachowska investigated these same coals by reductive alkylation.[152]

Comparing the solubilization produced by naphthalenide reduction with that produced by reductive alkylation, Wachowska found that reduction alone was rather ineffective relative to reductive alkylation.[152] As an example, coal K-III had a benzene solubility of 0.8%; on reduction the solubility increased to 10.1%, but on reductive methylation the solubility

Table 6. Oxygen distribution in a series of Polish coals before and after reduction with potassium naphthalenide in THF

Coal	%C (daf)	As-Received		Anions Added	After Reduction	
		$O_{reactive}$	O_{ether}		$O_{reactive}$	O_{ether}
K-I	78.2	3.61	10.91	11.7	4.01	8.25
K-II	81.05	2.07	8.72	9.8	2.10	7.06
K-III	86.95	0.54	4.60	11.4	1.48	4.62
K-IV	87.85	0.63	3.22	13.5	2.66	2.81
Anthracite	92.6	0.16	3.25	7.6	1.10	3.83

The numbers of the last five columns expressed as atoms (anions) pa 100 C atoms. All numbers expressed as atoms (anions) per 100 C atoms. Reactive oxygen functionality includes hydroxyl, carboxyl, and carbonyl.

went to 42.1%. As noted in section V-B above, Sternberg had reached similar conclusions about the advantages of alkylation. Wachowska stated that the reason for the considerable increase in the yield of the benzene extracts of alkylated coals had to be sought in the degree of substitution of the coal with alkyl groups and in the chain length of the alkylating agent. She thus performed a systematic study of the effects of different alkylating agents in the reductive alkylation of her five coals. The results are given in Table 7.

One notes that the parameters "degree of substitution" and "chain length" appear to be inversely related. That is, the longer chain alkylating agents are less effective in adding to the "coal anion" than are the shorter chain alkylating agents. However, the best benzene extractibility arises from the longer chain species, in spite of their lesser abundance. Although Wachowska had no ready explanation for this behavior, Sternberg's analogy between petroleum asphaltenes and reductively alkylated coal may offer guidance. Thus, coal itself is insoluble in benzene because of its ether and carbon-carbon linkages, because of its hydrogen bonding network from phenolic and carboxylic linkages, and because of its aromatic stacking. Reduction alone is sufficient to eliminate the first problem, and reductive methylation the second. Reductive octylation generates paraffin-paraffin interactions which compete with the other interactions, thereby generating solubility. Contrary to common belief, organic moieties in the liquid state do show order on the scale of 15-30 A, and this tendency to order must be considered in assessing solubilities.[223] Thus, in reductive octylation, solubilization may allow an optimum of paraffin-paraffin and aromatic-aromatic interactions. Although aromatic-aromatic interactions in coals are already well-appreciated, it should be noted that paraffin interactions are sometimes observed in diffraction. Ironically, the "gamma band" (indicative of paraffin order[223]) was first observed, and named, in a study of coals.[224]

In concluding this section, we have seen several examples of what reductive alkylation can tell us. With respect to coals, information is available on the amount of cross-linking, the chemistry of the cross-linking, and the manner in which this structure may be perturbed. In the next section, we see how closely engine deposits resemble coals.

Table 7. Products reductive alkylation of different rank coals

Coal	Alkyl Introduced	Alkyl Per 100 C	Wgt. % Oxygen	Benzene Extractibility
K-I	Methyl	10.5	11.1	38.6
	Ethyl	9.6	10.8	44.5
	Butyl	7.2	9.8	48.2
	Octyl	6.3	9.1	49.5
K-II	Methyl	9.6	9.4	34.4
	Ethyl	8.0	8.8	44.5
	Butyl	6.2	8.2	52.4
	Octyl	5.4	7.6	58.1
K-III	Methyl	9.7	4.9	50.3
	Ethyl	8.9	4.0	69.9
	Butyl	7.1	4.8	73.6
	Octyl	6.8	3.2	78.4
K-IV	Methyl	3.2	1.7	1.2
	Ethyl	1.8	3.4	1.5
	Butyl	1.3	2.5	2.9
	Octyl	1.3	3.6	5.4

VI. THE REDUCTIVE CHEMISTRY OF ENGINE DEPOSITS

A.   The Analogy to Coal and Other Carbonaceous Materials

Throughout this chapter, we have compared engine combustion deposits to coals, most specifically to subbituminous coals. The similarity between the two with respect to elemental composition is striking. Compare Sternberg's subbituminous coal[24] to a typical base fuel/premium quality oil A deposit:

	%C	%H	%N	%S	%O
coal	68.74	4.46	1.49	0.28	21.22
deposit	66.89	4.14	0.75	0.51	23.52

As noted in an earlier chapter, there is also a similarity in thermogravimetric behavior; a base fuel/ashless oil deposit loses between 3 to 8% more weight than does a high volatile bituminous coal at all temperatures between 300°C and 800°C. Of course, not all aspects are the same. Engine deposits have low surface areas (<1 m^2/g) and no tendency to swell in the vapor of amine solvents, characteristics distinct from coals.[225] Most importantly, engine deposits are composed of about 90% aromatic carbon, whereas typical subbituminous coals contain only 40 to 50% aromatic carbon.[226] One wonders what kind of aromatic clusters exist in engine deposits -- small, weakly-ordered but cross-linked clusters as in subbituminous coals, or large, but highly-oxidized, clusters, as in a hypothetical oxidized anthracite coal.

To answer this question, we performed x-ray diffraction on a deposit derived from base fuel and ashless oil. The ashless oil was employed so that diffraction peaks from the inorganic portion of a commercial lube additive package would not complicate the pattern. As seen in Figure 4, the base fuel/ashless oil deposit shows broad (002) and (100) diffraction peaks, similar in character to those of a prototypical "poorly crystalline" carbon, Spherocarb,[97] whose pattern is illustrated in Figure 5. The linewidth of the (002) band of the deposit is 13° 2θ(Cu), corresponding to a crystallite size of 7-8 Å in the Scherrer equation. Such a linewidth is consistent with low rank coals rather than with anthracite.[227] The (100) band of the deposits, near d=2.05 Å, appears as a composite lineshape, suggesting different aromatic molecules of different size. In Figure 6a we

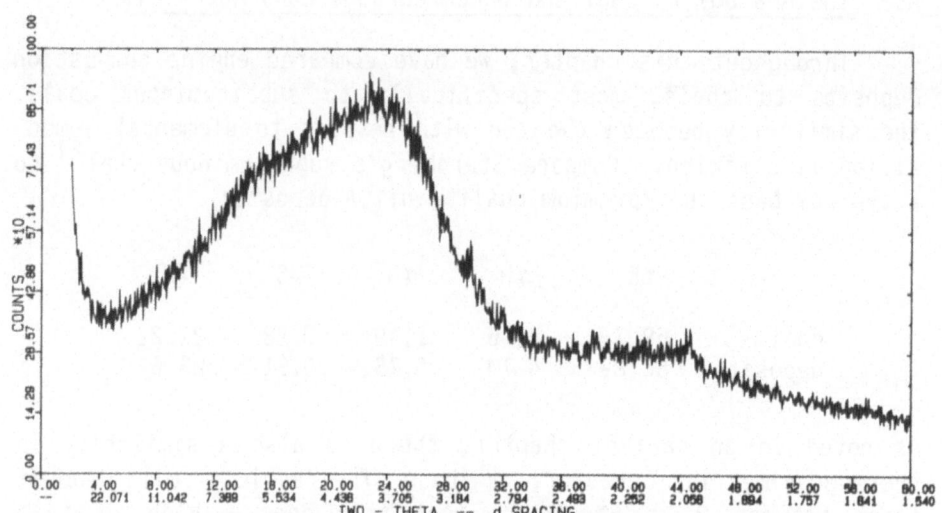

Figure 4. X-ray diffraction pattern of a base fuel/ashless oil
deposit. The x-axis of the figure runs from 0 to
60 degrees 2θ (Cu radiation). The major peak between
12 and 32 degrees is a "(002)" band, and the peak
near 44 degrees is a "(100)" band.

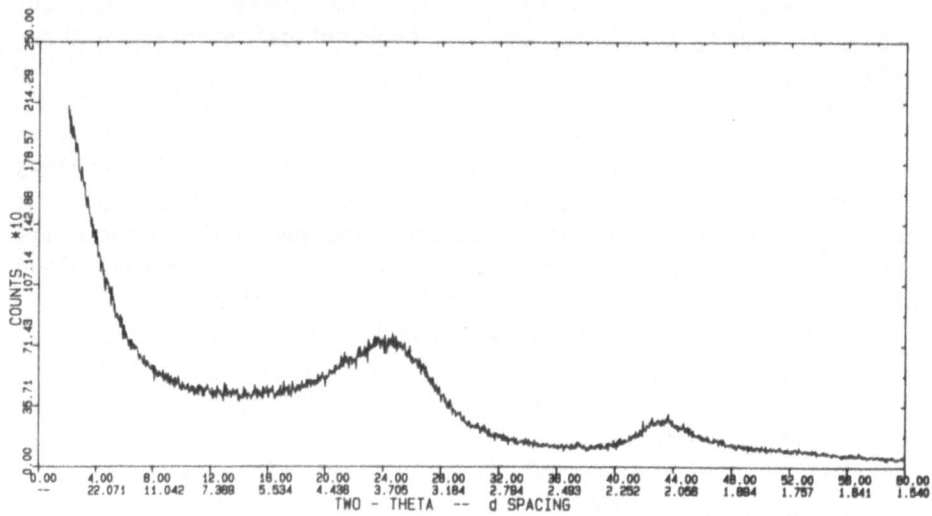

Figure 5. X-ray diffraction pattern of the "poorly crystalline"
material Spherocarb, made by Analabs. One sees the
same (002) and (100) features.

present a Debye internal interference simulation of the diffraction pattern of the deposit based on a motif containing two parallel coronene molecules at a distance of 3.51 Å (refer to reference 223 for methodology). With respect to the (002) band, the linewidth obtained (13.7°) and the d value (3.79 Å) are close to the experimental values (13°, d=3.77 Å). In Figure 6b, we present Debye internal interference patterns which address the complexity of the (100) band in engine deposits. Although the different peaks in this region (d=2.1 Å, k=3.1 Å^{-1}) may arise from aromatic molecules of different diameters, they may also arise from modulated aromatic structures, one example of which is the family of 1,2 di-aryl ethanes. Although simple aromatic structures, such as pyrene, give a (100) peak whose linewidth is roughly inversely proportional to size, molecules as 1,2 di-anthryl ethane or 1,2 di-tetracenyl ethane give "split" (100) peaks. This arises from an incommensurate modulation of the aromatic periodicity (1.42 Å) by aliphatic distances (1.47 Å, 1.54 Å). Many coals show small angle diffraction peaks arising from their macropore structure.[228] Consistent with its low surface area, the engine deposit shows no such peaks, as illustrated in Figure 7.

We thus conclude that there is some basis for comparing engine deposits to subbituminous coals. The two are similar in having high O/C ratios, large mass losses on pyrolysis, and relatively small aromatic clusters. However, the engine deposits seem to lack macropore structures. We consider next the chemistry of the deposit with potassium to assess possible chemical analogies. In all that follows, the reader should recognize that we are using alkali metals only as a means of characterizing deposits following removal from the engine.

B. The Direct Reaction of Potassium with Engine Deposits

If the "analogy" between low rank coals and deposits extends to chemistry, direct reaction with potassium should be relatively ineffective in generating aromatic anions. In reacting a lignite (68.9% C (daf)) with potassium, Lazarov, Stefanova, and Petrov[206] found little change in the ESR pattern:

	g value	linewidth (G)	radical density (s/g)
initial coal	2.0042	6.9	1.2×10^{18}
coal + 35% K	2.0039	7.1	1.1×10^{18}

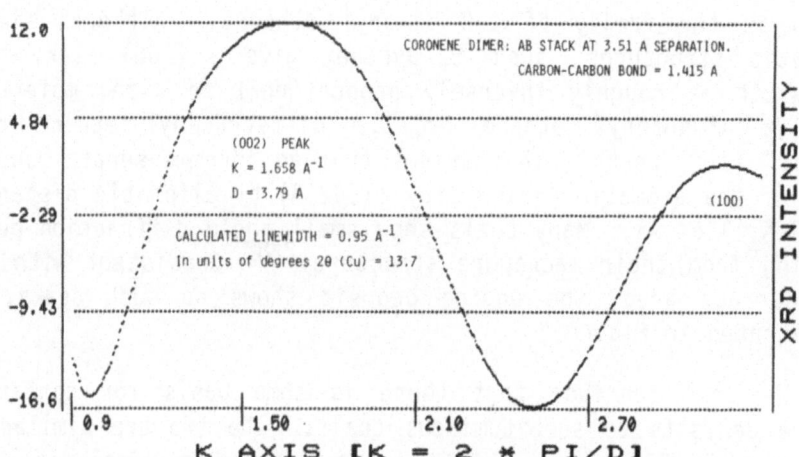

Figure 6a. A diffraction pattern of a hypothetical coronene
dimer simulated by the method of Debye internal
interference. The fundamental scattering motif
is taken to be two parallel coronene molecules,
separated by a distance of 3.51 A.

The similarity of this simulated diffraction pattern
to the pattern observed experimentally for the
deposit, with respect to the d value and linewidth
of the (002) band, indicates little tendency of the
aromatic molecules to arrange in parallel stacks.

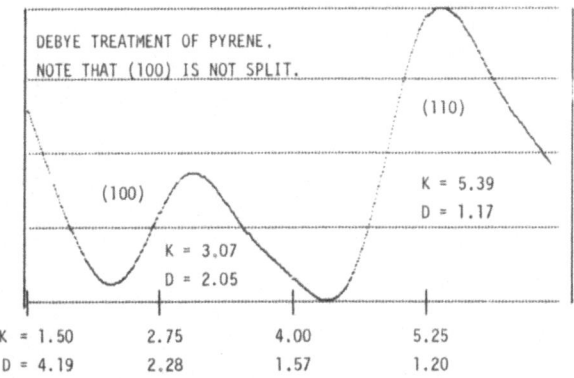

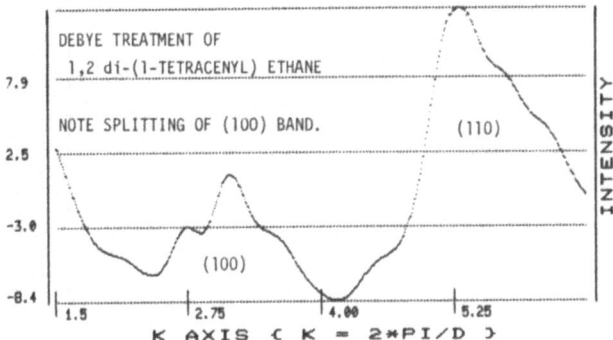

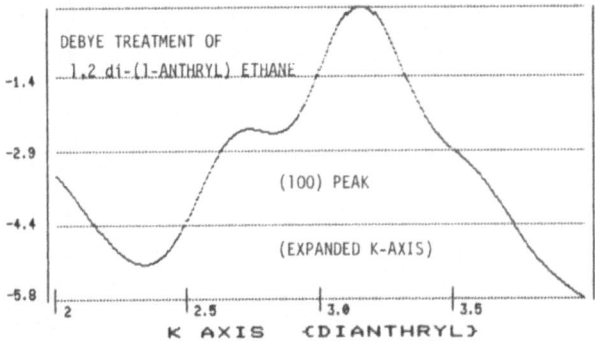

Figure 6b. Debye analysis suggests that cross-links between
aromatic molecules may create modulated (100) bands.
In the above the aromatic-aromatic C distance is 1.42 A,
aromatic-benzylic is 1.47 A, and aliphatic-aliphatic is
1.54 A. The modulation is still present if the 1.47 A
value is replaced by 1.54 A.

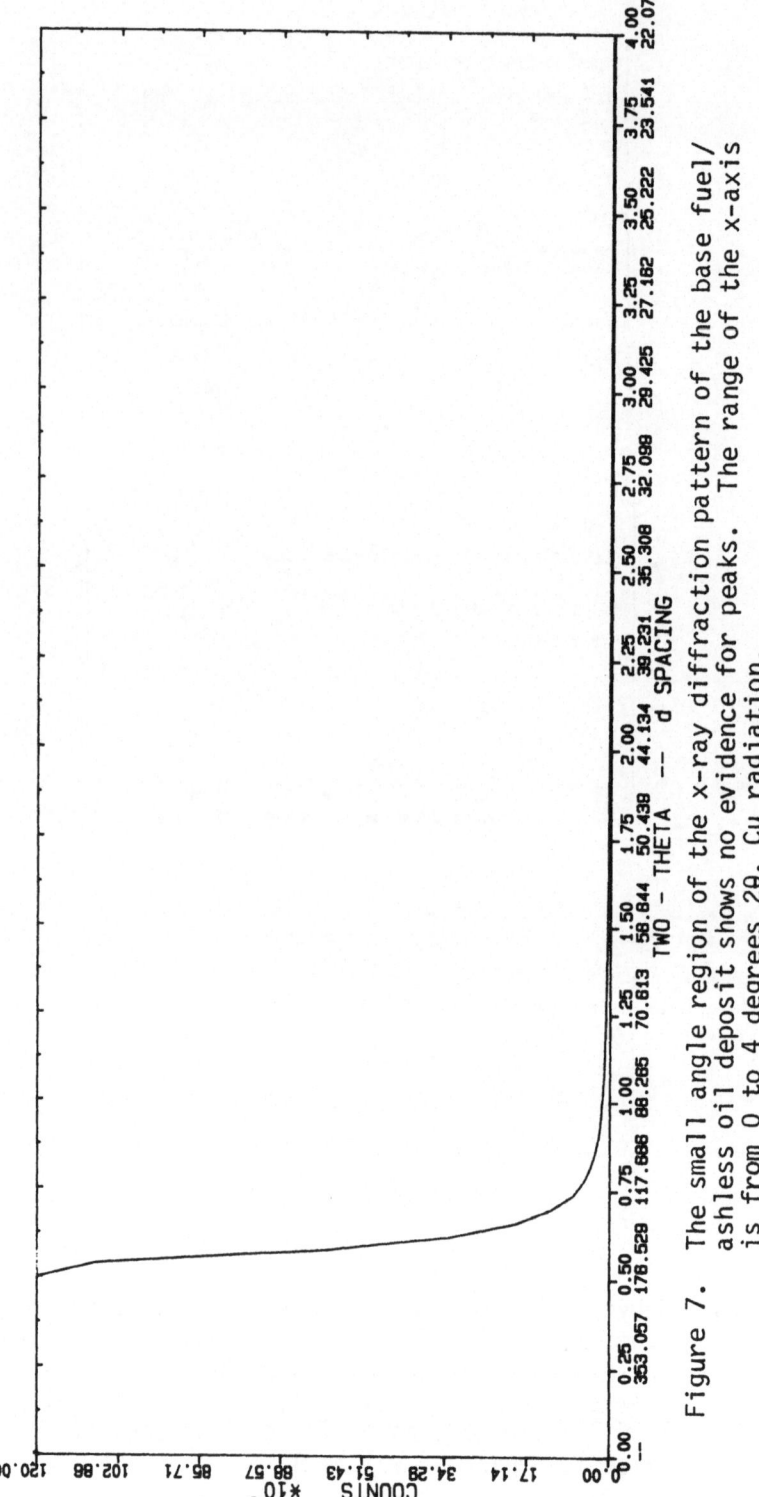

Figure 7. The small angle region of the x-ray diffraction pattern of the base fuel/ashless oil deposit shows no evidence for peaks. The range of the x-axis is from 0 to 4 degrees 2θ, Cu radiation.

Furthermore, on reacting the potassium adduct with butyl iodide, no weight increase was observed in the recovered product.

We indeed find similar results in the reaction of a base fuel/ashless oil deposit with potassium metal at 115°C (1.156 g deposit (68.60% C, 4.16% H, 24.27% O) + 0.343 g K heated for 3 hours under He. ESR tube sealed at 18 μ. The ESR pattern of the product, illustrated in Figure 8, is almost identical to that of the starting material:

	g value	linewidth (G)	radical density (s/g)
initial deposit	2.0031	4.8	3.9×10^{18}
deposit + 23% K	2.0031	4.9	6.9×10^{18}
C/K = 7.5			

Thus, the chemistry of deposits with potassium metal supports the analogy to low rank coals. The lack of reactivity is consistent with small, rather than with intermediate or large, clusters. Consider next the chemistry with potassium (+ naphthalene) in tetrahydrofuran, which whould have a dramatic effect on oxygen functionality.

C.   The Reaction of Potassium/(Naphthalene)/THF with Engine Deposits

Since insertion probe mass spectroscopy indicates deposits formed from conventional fuels to contain aromatic hydrocarbon molecules, one might expect Sternberg's procedure to work in the absence of naphthalene. To investigate this, we combined 3.031 g of deposit (whole fuel/premium quality oil A, with 59.64% C) with 2.062 g Na in 10 ml of tetrahydrofuran. After three days, the consumption of Na was 1.33 mmol/g, much lower than the n-BuLi titration equivalent of 8.01 meq/g found for this material. To investigate "classical" Sternberg behavior, we added 0.590 g (4.6 mmol) of naphthalene and an additional 5.298 g Na at this point, and found a consumption of 17.38 mmol Na/g after eight days. Such behavior is similar to that reported Ignasiak, Kemp-Jones, and Strausz,[64] in which reduction does occur without naphthalene, but is much facilitated with naphthalene.

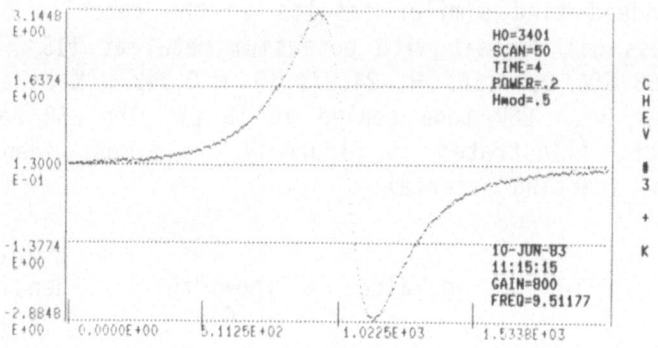

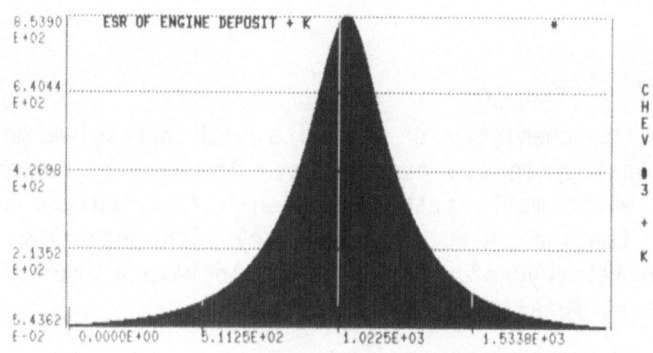

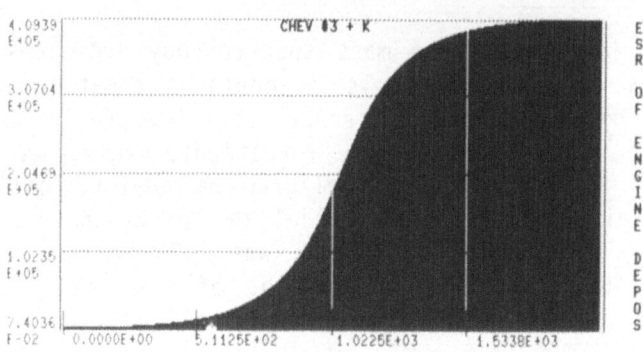

Figure 8.   Electron spin resonance of an adduct of potassium
with base fuel/ashless oil deposit.   The scan range
is 50 G (= 5 mT).

354

A second experiment was carried out with a different deposit (base fuel/ashless oil, with 69.23% C and an n-BuLi titration equivalent of 5.24 meq/g) and with a different alkali metal, potassium. Using 0.517 g deposit and 2.000 g potassium in 25 ml THF, the alkali metal consumption was 1.58 mmol/g after 24 hours. Although both this and the previous deposit react with alkali metal in the absence of naphthalene, the consumption of alkali is low relative to n-butyl lithium reactivity. For this reason, we decided to perform a normal Sternberg reduction, in which naphthalene is present at the beginning of the reaction.

Within a VAC atmospheres glove box under He, 1.223 g deposit (base fuel/ashless oil, with 68.60%C) was combined with 0.466 g (3.6 mmol) naphthalene in 50 ml tetrahydrofuran with 1.179 g potassium. The consumption of potassium is as follows:

Reaction time (hours)	Potassium consumed (mmol/g)
47	17.0
120	23.3
145 (more K added)	24.11
169	25.5
211	27.2

At the 169 hour point, a small sample (0.108 g) was placed in a stoppered ESR tube, and the ESR spectrum was quickly recorded (Figure 9). The hyperfine pattern of the naphthalene radical anion is evident, although the majority of the radical density is contained within the broad unresolved underlying envelope. On syringing methyl iodide into the ESR specimen, the radical anion spectrum is destroyed, as seen in Figure 10. After 211 hours, 3 ml of methyl iodide is added to the deposit solution at 0°C (excess potassium removed), and the mixture is allowed to stand for ten days. The solution is filtered through a fine porosity frit. The resulting fluid is rotary-evaporated at 23°C and 46 mm for one hour, and the solid is dried in a vacuum oven to constant weight. The fluid yields 1.514 g of a red-brown liquid, and the solid weighs 5.111 g. If methyl iodide reacted stoichiometrically with the consumed potassium (27.16 mmol/g), one would have 5.5 g KI and 0.5 g CH_3 added.

Characterization was performed on the liquid phase product. Elemental analysis showed 73.32% C, 7.71% H (H/C=1.25), 0.48% N, and 0.20% S, with the unaccounted balance (oxygen?) equal to

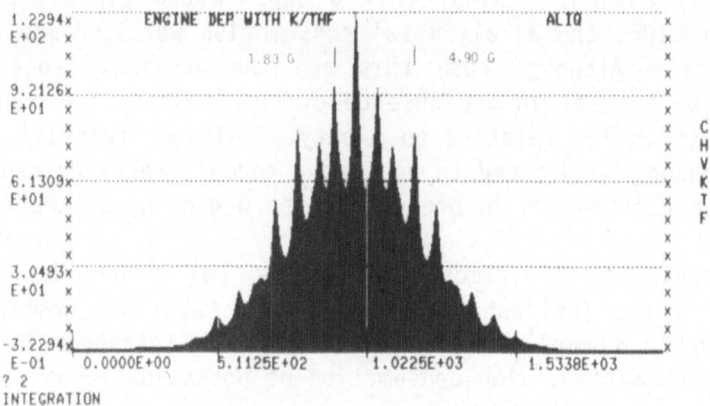

Figure 9. Electron spin resonance absorption mode of a mixture
of engine deposit, naphthalene, potassium, and
tetrahydrofuran after 169 hours at room temperature.
The observed hyperfine splittings of 1.83 and 4.90 G
arise from the radical anion of naphthalene. The
bulk of the intensity arises from the broad, underlying
resonance.

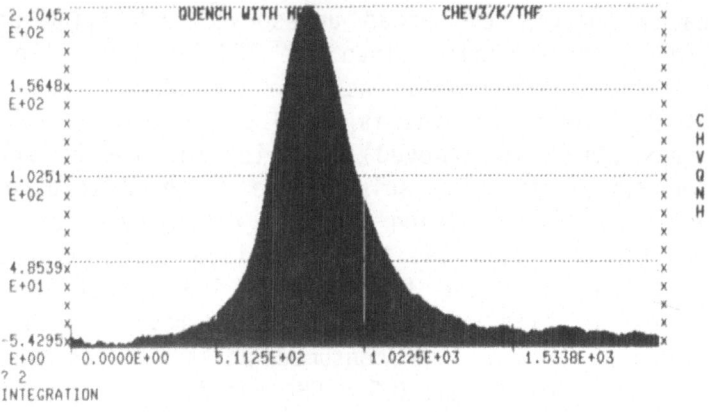

Figure 10. Electron spin resonance absorption mode of the above
mixture following the addition of methyl iodide.

356

18%. Proton nuclear magnetic resonance, recorded at 60 MHz in CS_2, reveals aromatic protons (6.8 to 8δ), vinylic protons (5.7δ), and a complex mixture of aliphatic resonances extending from 0.8 to 4.3δ, as illustrated in Figure 11. The ratio of aliphatic protons to (aromatic + olefinic) is 3.9. The most evident difference between this spectrum and that of an extracted deposit is in the 3 to 4.2δ region. Although extracted material has little intensity there, the reductively alkylated material does. In terms of anticipated chemistry, this is logical, for methyl ethers and some reductively methylated aromatics have absorptions in this region.[216,218]

In analysis of such spectra, one must worry about solubility of the material in the chosen solvent. Figure 12 illustrates the spectrum of the liquid specimen after 5 days, and one observes line broadening associated with unaveraged dipolar interactions, presumably arising from a loss of true solubility. In Figure 13, the proton spectrum of a fresh specimen is illustrated, demonstrating that reproducible results can be obtained with sufficient care.

In this section, we have discussed results of a few experiments on the reductive chemistry of engine deposits. As with coals and petroleum materials, both reduction and reductive alkylation do appear to "work" with engine deposits and thus may offer an attractive way to characterize the deposits.

Furthermore, as reductive techniques are applied to a greater variety of substrates, more information about the reductive chemistry mechanism will also be learned. Although the initial hypothesis of Sternberg concerning the reduction of aromatics is, in part, correct, the detailed manner by which electrons from the potassium naphthalenide are transferred to the substrate is not yet clear. Engine deposits, in being largely "polymeric" in nature, are particularly attractive materials for study of this issue. The intrinsic free radicals of the deposit, which involve aromatic heterocyclic molecules containing oxygen, are apparently on this "backbone" with sufficient separation to preclude exchange narrowing.[229,230] (See Figure 14). The manner in which the ESR linewidth changes, as electrons are introduced into the substrate, will reveal the distribution of the initial electron acceptors in the substrate. Furthermore, shifts in the g value will allow some assessment of the identify of the acceptors.

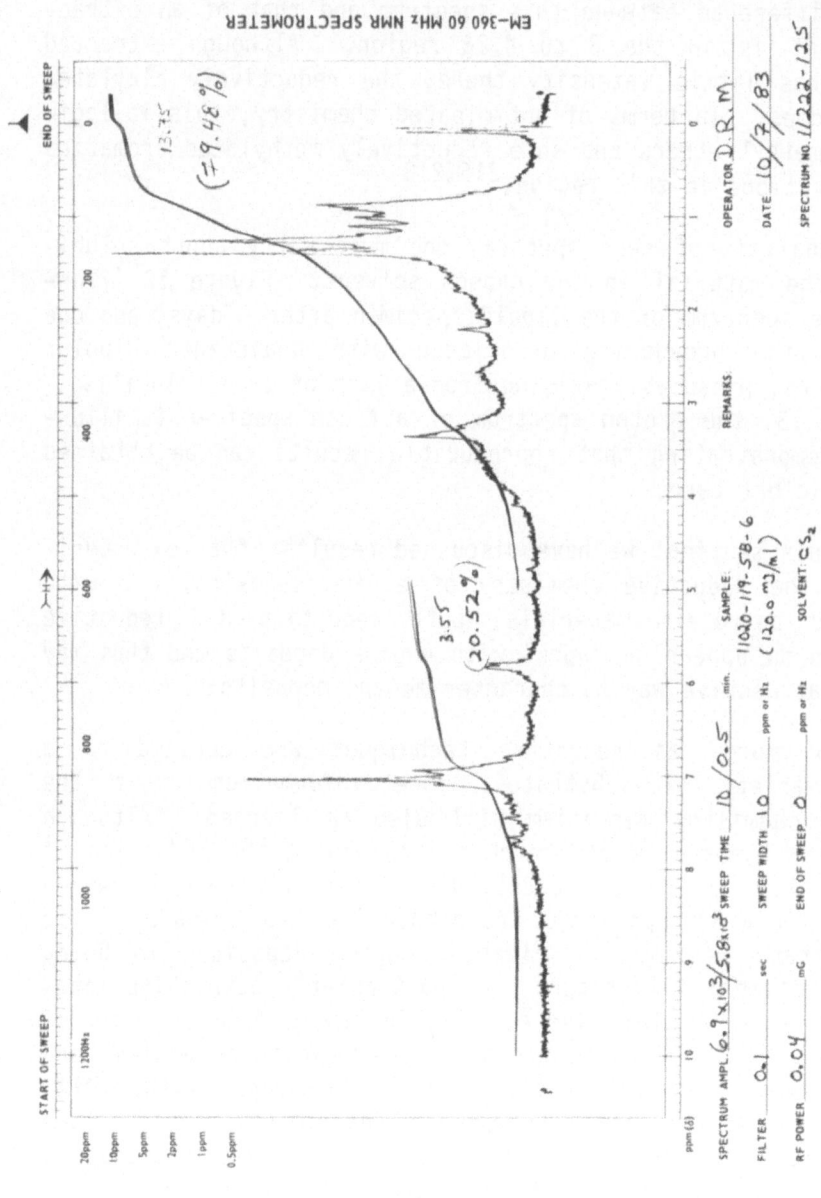

Figure 11. Proton nuclear magnetic resonance spectrum (60 MHz, 23°C) of a reductively methylated engine deposit. The concentration is 120 mg deposit per ml CS_2. Note region between 3 and 4 delta.

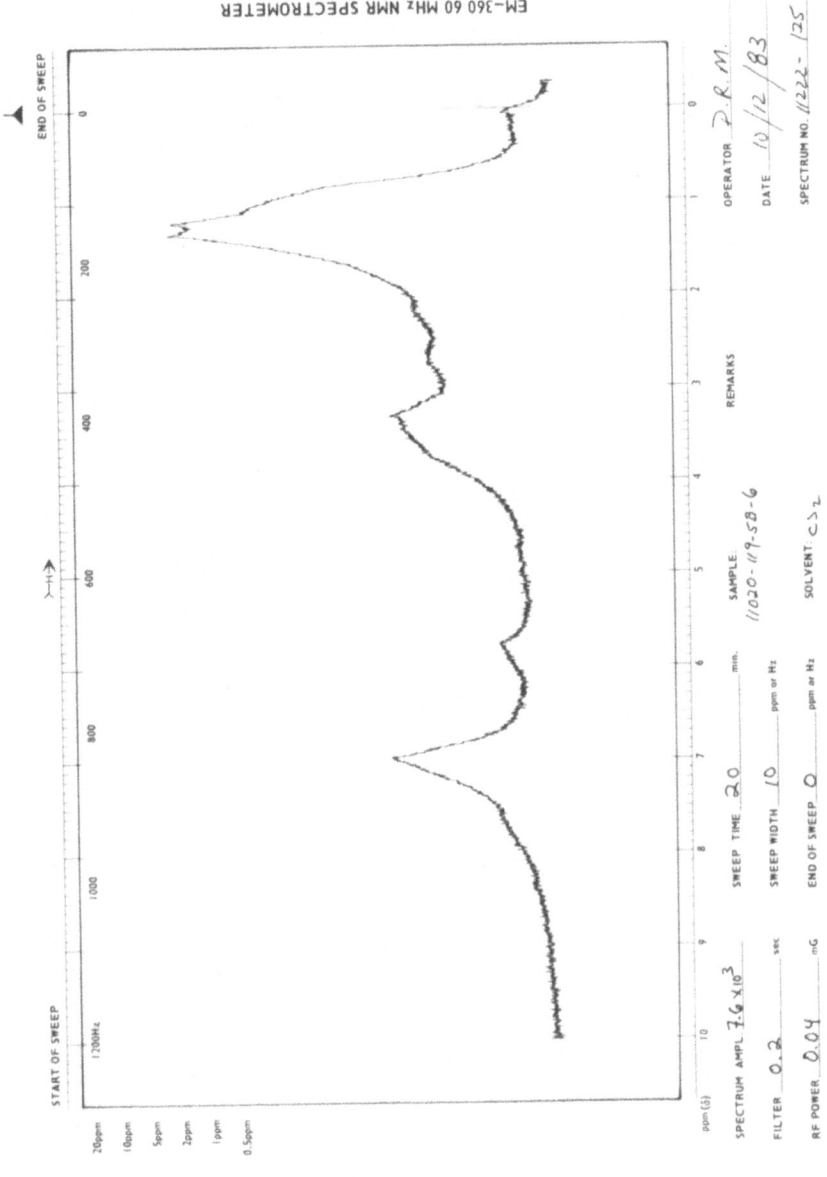

Figure 12. Proton nuclear magnetic resonance of the specimen described in Figure 11 after five days in a stoppered NMR tube.

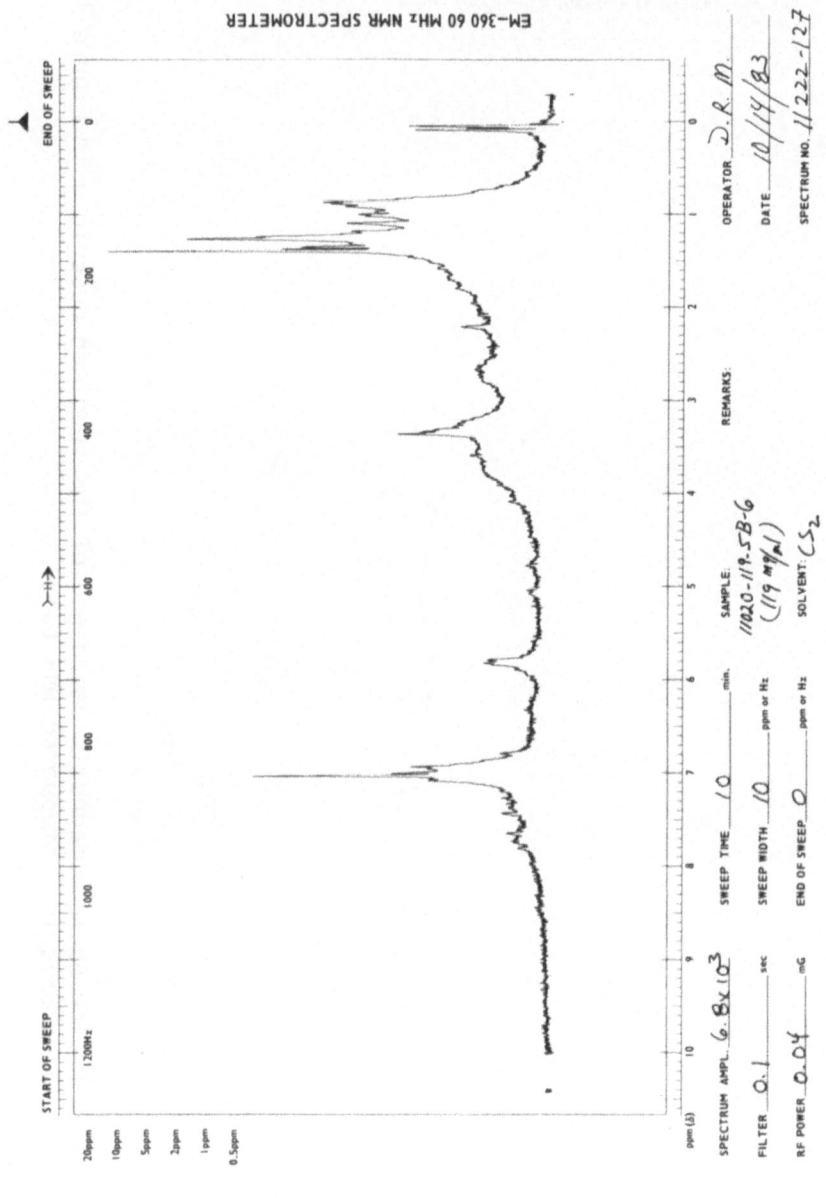

Figure 13. Proton nuclear magnetic resonance of a freshly prepared specimen of reductively methylated deposit.

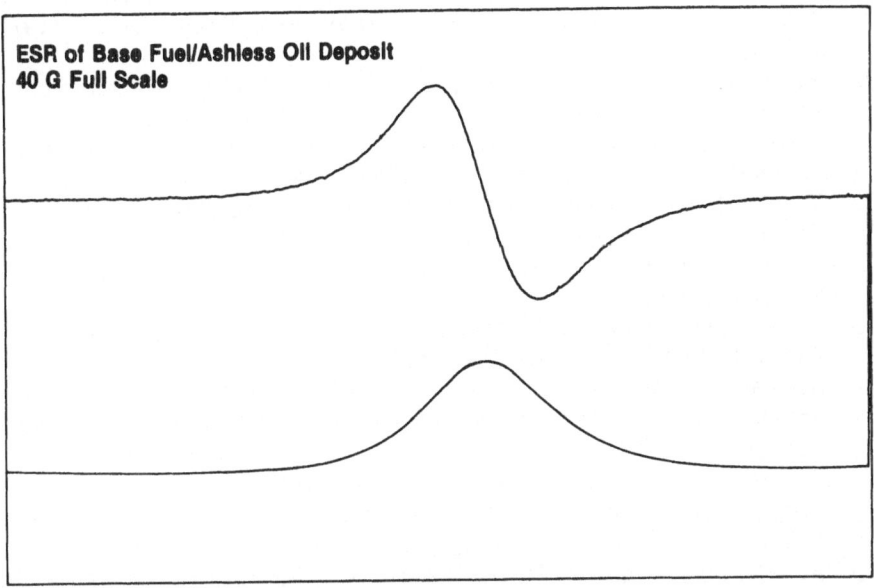

ESR of Base Fuel/Ashless Oil Deposit
40 G Full Scale

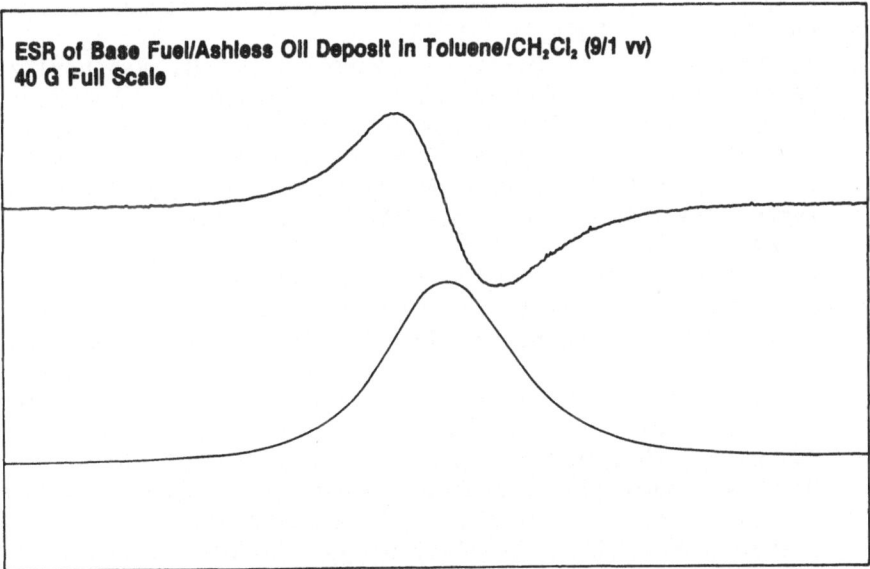

ESR of Base Fuel/Ashless Oil Deposit in Toluene/CH$_2$Cl$_2$ (9/1 vv)
40 G Full Scale

Figure 14. The electron spin resonance pattern of the solid deposit is unchanged on contact with a toluene/ methylene chloride solvent. The radicals are probably on the "backbone".

REFERENCES

1. W. Schlenk, J. Appenrodt, A. Michael, and A. Thal, Uber Metalladditionen an mehrfache Bindungen, Chem. Ber. 47:473 (1914).

2. N. D. Scott, J. F. Walker, and V. L. Hansley, Sodium Naphthalene. I. A New Method for the Preparation of Addition Compounds of Alkali Metals and Polycyclic Aromatic Hydrocarbons, J. Am. Chem. Soc. 58:2442 (1936).

3. V. D. Parker, Energetics of Electrodes Reactions. II. The Relationship Potentials, Electron Affinities, and Solvation Energies of Aromatic Hydrocarbons, J. Am. Chem. Soc. 98:98 (1976).

4. A. Streitwieser, "Molecular Orbital Theory for Organic Chemists", Wiley, New York (1961).

5. M. Rabinovitz and I. Willner, Novel Aromatic Cations and Anions: Aromaticity-Structure Relationships, Pure and Appl. Chem. 52:1575 (1980).

6. M. Randic, Relative Stability of Cations and Anions of Conjugated Polycyclic Hydrocarbons, J. Phys. Chem. 86:3970 (1982).

7. A. Minsky, J. Klein, and M. Rabinovitz, Aromatic Polycyclic Benzenoid Tetranions: Pyrene and Perylene Anions Revisited, J. Am. Chem. Soc. 103:4586 (1981)

8. A. Minsky, A. Y. Meyer, and M. Rabinovitz, Super-Charged Polycyclic π Systems: Pyrene and Perylene Tetraanions, J. Am. Chem. Soc. 104:2475 (1982).

9. A. Minsky, A. Y. Meyer, R. Poupko, and M. Rabinovitz, Paramagnetism and Antiaromaticity: Singlet-Triplet Equilibrium in Doubly Charged Benzenoid Polycyclic Systems, J. Am. Chem. Soc. 105:2164 (1983). See also B. Eliasson, T. Lejon, and U. Edlund, A ^{1}H and ^{13}C N.M.R. Study of Pyrene Dianion and Proposed Tetra-anion, Chem. Commun. 591 (1984).

10. A. Rainis and M. Szwarc, Disproportionation of the Lithium, Sodium, and Potassium Salts of Anthracenide and Perylenide Radical Anions in DME and THF, J. Am. Chem. Soc. 96:3008 (1974).

11. B. S. Jensen and V. D. Parker, Reactions of Aromatic Anion Radicals and Dianions. II. Reversible Reduction of Anion Radicals to Dianions, J. Am. Chem. Soc. 97:5211 (1975).

12. D. J. Williams, J. M. Pearson, and M. Levy, Anion Radicals of a Series of [2.2] Paracyclophanes and a,ω-Diarylalkanes. II. An Electron Spin Resonance Investigation, J. Am. Chem. Soc. 93:5483 (1971).

13. R. G. Lawler and C. V. Ristagno, Nuclear Mangetic Resonance Spectra of the Dianions of Anthracene and Other Polynuclear Aromatic Hydrocarbons, J. Am. Chem. Soc. 91:1534 (1969).

14. R. H. Cox, H. W. Terry, and L. W. Harrison, A. ^{7}Li NMR Investigation of the Structure and Ring Currents in Some Aromatic Dianion Systems, Tet. Lett. 4815 (1971).

15. L. M. Tolbert and M. Z. Ali, 1-Phenyl-3,4:5,6-dibenzocycloheptatrienyl Anion. A Stable Antiaromatic Carbanion, J. Org. Chem. 47:4793 (1982).

16. M. Szwarc, Chemistry of Radical-Ions, Prog. Phys. Org. Chem. 6:323 (1968).

17. M. Szwarc, Radical Anions and Carbonions as Donors in Electron-Transfer Processes, Acc. Chem. Res. 5:169 (1972).

18. J. F. Garst, Electron Reactions of Organic Anions, in: "Free Radicals," J. K. Kochi, ed., Wiley, New York (1973).

19. N. L. Holy, Reactions of the Radical Anions and Dianions of Aromatic Hydrocarbons, Chem. Rev. 74:243 (1974).

20. R. B. Bates, "Comprehensive Carbanian Chemistry" Elsevier, New York (1980).

21. L. M. Stock, The Reductive Alkylation Reaction, in: "Coal Science", Vol. 1, M. L. Gorbaty, J. W. Larsen, I. Wender, eds., Academic, New York (1982).

22. D. E. Paul, D. Lipkin, and S. I. Weissman, Reactions of Sodium Metal with Aromatic Hydrocarbons, J. Am. Chem. Soc. 78:116 (1956).

23. H. W. Sternberg, C. L. Delle Donne, P. Pantages, E. C. Moroni, and R. E. Markby, Solubilization of an lvb coal by reductive alkylation, Fuel 50:432 (1971).

24. H. W. Sternberg and C. L. Delle Donne, Solubilization of Coal by Reductive Alkylation, Fuel 53:172 (1974).

25. T. R. Tuttle, R. L. Ward, and S. I. Weissman, Spin Density in Naphthalene Negative Ion, J. Chem. Phys. 25:189 (1956).

26. B. M. P. Hendriks, G. W. Canters, C. Corvaja, J. W. M. de Boer, and E. de Boer, NMR Investigations on the Alkali Naphthalene Ion Pairs, Mol. Phys. 20:193 (1971).

27. G. Henrici-Olive and S. Olive, Uber Mono- and Dianiones des Naphthalins IV. Stabilitatsverhalten der Dianiones in Tetrahydrofuran, Zeit. Phys. Chem. 43:340 (1964).

28. J. Smid, A Stable Dianion of Naphthalene, J. Am. Chem. Soc. 87:655 (1965).

29. J. C. Carnahan and W. D. Closson, Reaction of Naphthalene Dianions with Tetrahydrofuran and Ethylene, J. Org. Chem. 34:4469 (1972).

30. H. B. Gia, R. Jerome, and Ph. Teyssie, New Observations on the Metalation of Naphthalene and β-ethylnaphthalene by Potassium, J. Organomet. Chem. 190:107 (1980).

31. S. Bank and B. Bockrath, Reactions of Aromatic Radical Anions. VI. Kinetic Study of the Reaction of Sodium Naphthalene with Water, J. Am. Chem. Soc. 93:430 (1971).

32. G. Levin, C. Sutphen, and M. Szwarc, Protonation of Perylene Radical Anions by Alcohols and Water in Tetrahydrofuran, J. Am. Chem. Soc., 94:2652 (1972).

33. C. Amatore, J. Pinson, and J. M. Saveant, The Role of Water in Organic Electroreductive Dimerization in Aprotic Solvents, J. Electroanal. Chem. 139:193 (1982).

34. P. W. Rabideau and E. G. Burkholder, Metal-Ammonia Reduction and Reductive Alkylation of Polycyclic Aromatic Compounds, J. org. Chem. 43:4283 (1978).

35. G. D. Sargent, J. N. Cron, and S. Bank, Reactions of Aromatic Radical Anions. I. Coupling of Alkyl Free Radicals Generated by Electron Transfer to Alkyl iodides, J. Am. Chem. Soc. 88:5363 (1966).

36. G. D. Sargent and G. A. Lux, Reactions of Aromatic Radical Anions. III. Evidence for an Alkyl Radical-Radical Anion Combination Mechanism for Alkylation of Sodium Naphthalenide with Alkyl Halides, J. Am. Chem. Soc. 90:7160 (1968).

37. J. F. Garst, R. D. Roberts, and B. N. Abels, Solvents Effects on Reactions of Sodium Naphthalene with Hexyl Fluoride, J. Am. Chem. Soc. 97:4925 (1975).

38. H. Gusten and L. Horner, Wurtz Syntheses with Naphthalene-sodium, Angew, Chemie, Int. Ed. 1:455 (1962).

39. H. A. Dirkse, P. W. Lednor, and P. C. Versloot, Alkali Metal-Naphthalene Adducts as Reagents for Neutralizing Oxide Surfaces, Chem. Commun. 814 (1982).

40. L. H. Klemm and A. J. Kohlik, Polarographic Reduction of Some Alkyl-, Alkylene-, and Polymethylnaphthalenes, J. Org. Chem. 28:2044 (1963).

41. S. Suganan, Steric Effect of Alkyl Groups, Current Science 52:124 (1983).

42. K. D. Bartle, C. Gibson, D. Mills, M. J. Mulligan, N. Taylor, T. G. Martin, and C. E. Snape, Differential-Pulse Voltammetry at the Hanging-Mercury-Drop Electrode for Identification of Aromatic Structures in Coal, Anal. Chem. 54:1730 (1982).

43. K. J. Borhani and M. D. Hawley, Electrochemical Studies of Weak Carbon and Nitrogen Acids: Fluorene and p-Cyanoaniline in Dimethylformamide, J. Electroanal. Chem. 101:407 (1979).

44. G. W. Canters and E. de Boer, Alkali N.M.R. experiments on the radical ion pairs of biphenyl and fluorene. Part I. Analysis of NMR shifts, Mol. Phys. 26:1185 (1973).

45. R. L. Kugel, W. G. Hodgson, and H. R. Allcock, The Formation of Radical Anions in Fluorene Metallation, Chem. and Ind. p. 1649 (1962).

46. T. L. Chu and S. C. Yu, The magnetic Susceptibilities of Some Aromatic Hydrocarbon Anions, J. Am. Chem. Soc. 76:3367 (1954).

47. C. Takahashi and S. Maeda, Raman Spectra of Biphenyl Negative Ion in Tetrahydrofuran Solution, Chem. Phys. Lett. 24:584 (1974).

48. E. G. Janzen and J. L. Gerlock, On the metalation of Fluorene, J. Organomet. Chem. 8:354 (1967).

49. H. Pines, J. A. Vesely, and V. N. Ipatieff, Sodium Catalyzed Reactions. II. Side-chain Ethylation of Alkyl Aromatic Hydrocarbons Catalyzed by Sodium, J. Am. Chem. Soc. 77:554 (1955).

50. H. Hart and R. E. Crocker, A. Quantitative Study of the Acidity of Certain Hydrocarbons, J. Am. Chem. Soc. 82:418 (1960).

51. G. B. Trimitsis, A. Tuncay, R. D. Beyer, and K. J. Ketterman, α,α'-Dimetalations of Dimethylarenes with Organosodium Reagents. The Catalytic Effect of Certain Tertiary Amines, J. Org. Chem. 38:1491 (1973).

52. P. V. R. Schleyer, Dimerization and Intramolecular Association in Li Synthetic Reagents, Pure and Appl. Chem. 55:355 (1983).

53. G. Fraenkel and J. M. Geckle, Influence of Substituents on NMR and Barriers to Rotation in the Tert-Benzyllithium Compounds, J. Am. Chem. Soc. 102:2869 (1980).

54. B. J. Tabner and T. Walker, Radical-anion Intermediates. Part V. Electron Spin Resonance Spectra of Radical Anions and Dianion Radicals of Some 9-Substituted Fluorenes. J. C. S., Perkin Trans. II 2:2010 (1972).

55. J. J. Eisch, Chemistry of Alkali Metal-unsaturated Hydrocarbon Adducts. III. Cleavage Reactions by Lithium-Biphenyl Solutions in Tetrahydrofuran, J. Org. Chem. 28:707 (1963).

56. D. F. Lindow, C. N. Cortez, and R. G. Harvey, Metal-Ammonia Reduction. XII. Mechanism of Reduction and Reductive Alkylation of Aromatic Hydrocarbons, J. Am. Chem. Soc. 94:5406 (1972).

57. P. J. Grisdale, T. H. Regan, J. C. Doty, J. Figueras, and J. L. R. Williams, Phenylcyclohexadienes, J. Org. Chem. 33:1116 (1968).

58. L. B. Ebert, unpublished results.

59. A. G. Evans, P. B. Roberts, and B. J. Tabnor, The Reactions of Radical Anions. Part I. The Cleavage of the Radical Anion of Dibenzofuran, J. Chem. Soc. B, p. 269 (1966).

60. D. H. Eargle and E. T. Kaiser, The Effect of Changes in the Oxidation State upon the e.p.r. Spectra of Dibenzothiophene Anion-radicals, Proc. Chem. Soc. p. 22 (1964).

61. D. H. Eargle, The Cleavage of Aryl Ethers by Alkali Metals in Aliphatic Ether Solvents. Detection by Electron Spin Resonance, J. Org. Chem. 28:1703 (1963).

62. R. Leardini and G. Placucci, Dibenz [b.f.] oxepin and Thiepin Radical Anions. Conjugative Properties of Sulfur in its Different Oxidation States, J. Heterocycle Chem. 13:277 (1976).

63. H. W. Sternberg, C. L. Delle Donne, R. E. Markby, and S. Friedman, Reaction of Sodium with Dibenzothiophene. A Method for Desulfurization of Residua, Ind. Eng. Chem., Process Res. Dev. 13:433 (1974).

64. T. Ignasiak, A. V. Kemp-Jones, and O. P. Strausz, The Molecular Structure of Athobosca Asphaltenes. Cleavage of the Carbon-Sulfur Bonds by Radical Ion Transfer Reactions, J. Org. Chem. 43:312 (1977).

65. H. Gilman and J. J. Dietrich, Lithium Cleavages of Some Heterocycles in Tetrahydrofuran, J. Org. Chem. 22:851 (1957).

66. L. Brandsma and J. F. Arens, The Chemistry of Thioethers; Differences and Analogies with Ethers, in: "The Chemistry of the Ether Linkage," S. Patai, ed., Interscience, New York (1967).

67. C. G. Screttas, Metallation of Aryl Ethers by Lithium Arenes, Chem. Commun. p. 869 (1972).

68. C. G. Screttas, On the Mechanism of Ring Metallation of Aromatic Compounds. Metallation of Thiophene by Lithium and by Lithium Dihydroarylides, Perk. Trans. II. p. 745 (1974).

69. R. A. Rossi and J. F. Bunnett, The Sense of Cleavage of Substituted Benzenes on Reaction with Solvated Electrons, as Determined by a Product Criterion, J. Am. Chem. Soc. 96:112 (1974).

70. M. Itoh, S. Yoshida, T. Ando, and N. Miyaura, Regioselective Cleavage of Aryl Decyl Ethers and Alkali Metals, Chem. Lett. p. 271 (1976).

71. H. Gilman, H. A. McNinch, and D. Wittenberg, Direct Preparation of Benzyllithium by Cleavage Reactions, J. Org. Chem. 23:2044 (1958).

72. H. W. Sternberg, C. L. Delle Donne, and I Wender, Similarity between the Electrochemical Elimination of Sulphur for Coal and from Dibenzothiophene, Fuel 47:219 (1968).

73. M. Miyake, Y. Nakayama, M. Nomura, and S. Kikkawa, Reduction of Some Sulfides and Ethers with Aromatic Rings by Electrochemically Generated Solvated Electrons, Bull. Chem. Soc. Jpn. 52:559 (1979).

74. F. M'Halla, J. Pinson, and J. M. Saveant, The Solvent as H-atom Donor in Organic Electrochemical Reactions. Reduction of Aromatic Halides, J. Am. Chem. Soc. 102:4120 (1980).

75. C. G. Screttas and M. Micha-Screttas, Carbon-13 Contact Solvent Shifts in Radical Anion Solutions. Mechanism of Spin Density Transfer to Solvent, Chem. Commun. p. 1168 (1982).

76. C. G. Screttas and M. Micha-Screttas, Paramagnetic Solvent Nuclear Magnetic Resonance Shifts in Radical Anion Solutions, J. Org. Chem. 48:252 (1983).

77. R. B. Bates, L. M. Kroposki, and D. F. Potter, Cyclorever- sions of Anions from Tetrahydrofurans, A. Convenient Syn- thesis of Lithium Enolates of Aldehydes, J. Org. Chem. 37:560 (1972).

78. T. Fujita, K. Suga, and S. Watanabe, The Reaction of Lithium Naphthalenide with Tetrahydrofuran, Synthesis, 11:630 (1972).

79. L. B. Ebert, J. C. Scanlon, D. R. Mills, and L. Matty, The Interrelationship of Graphite Intercalation Compounds, Ions of Aromatic Hydrocarbons, and Coal Conversion II, in: "New Approaches in Coal Chemistry," ed. B. D. Blaus- tein, B. D. Bockrath, and S. Friedman, ACS Symposium Series 169, American Chemical Society, Washington (1981).

80. R. T. Morrison and R. N. Boyd, "Organic Chemistry" 2nd ed., Allyn and Bacon, Boston (1966).

81. T. J. Lynch, M. Banah, H. D. Kaesz, and C. R. Porter, Iron Carbonyl Catalyzed Reductions of Model Coal Constituents Under Water Gas Shift Conditions, Fuel Division Preprints, American Chemical Society, 28:172 (1983).

82. M. T. Jones and T. C. Kuechler, An Electron Spin Resonance Study of the Benzene Anion Radical. A Model of Its Ion Pair with Alkal: Metal Ions, J. Phys. Chem. 81:360 (1977).

83. O. R. Brown, R. J. Butterfield, and J. P. Millington, Cathodic Reduction of Pyridine in Liquid Ammonia, Electro- chim. Acta 27:1655 (1982).

84. J. C. M. Henning, ^{14}N Hyperfine Structure in ESR Spectra of Heterocyclic Anions, J. Chem, Phys. 44:2139 (1966).

85. I. N. Jung and P. R. Jones, Bonding Studies in Group 4 Substitued Anilines, J. Am. Chem. Soc. 97:6102 (1975).

86. M. Branca and A. Gamba, An Advanced Laboratory Experiment Involving the Hammett Equation and Electron Spin Resonance Spectroscopy, Chim. Ind. (Milan) 65:174 (1983).

87. E. T. Strom and G. A. Russell, The Electron Spin Resonance Spectra of 2,1,3-Benzoxadiazole, -Benzothiadiazole, and -Benzoselenadiazole Radical Anions. Electron Withdrawal by Group VI Elements, J. Am. Chem. Soc. 87:3326 (1965).

88. H. C. Heller, Utilization of (n,π^*) Excitation Bands in the Formation of Radicals. II. Thiobenzophenone Anion Radical, J. Am. Chem. Soc. 89:4288 (1967).

89. L. J. Aarons and F. C. Adam, Electron Spin Resonance Studies of Thiocarbonyl Anion Radicals, Can. J. Chem. 50;1390 (1972).

90. J. Boersma, A. Mackor, and J. G. Noltes, ESR Study of Monoalkylazinc-2,2'-Bipyridine Complexes, J. Organomet. Chem. 99:337 (1975).

91. G. J. Hoijtink, E. de Boer, P. H. van der Meij, and W. P. Weijland, Reduction Potentials of Various Aromatic Hydro- carbons and Their Univalent Anions, Recuiel 75:485 (1956).

92. M. Maissard, J. P. Mazaleyrat, and Z. Welvart, On the Stereochemistry of the Reductive Alkylation of Anthracene, J. Am. Chem. Soc. 99:6933 (1977).

93. D. E. Bergbrieter and J. M. Killough, Polymer-Bound Alkali Metal Aromatic Radical Anions, Chem. Commun. p. 319 (1980).

94. A. H. Reddoch, Systematic Perturbations of the EPR Spectra of Anthracene and Azulene Anions in Solution, J. Chem. Phys. 43:225 (1965).

95. K. Mullen, The Dianions of Phenanthrene and 1,2,3,4-Dibenzocyclooctatetraene, Melv. Chim. Acta. 61:1296 (1978).

96. K. Mullen, The Dianions of Pyrene and Pyrene Isomers as $(4n)\pi$ Perimeters, Melv. Chim. Acta 61:2307 (1978).

97. L. B. Ebert, D. R. Mills, and J. C. Scanlon, The Interaction of Potassium with Graphite and Other Benzenois Systems, Mat. Res. Bull. 17:1318 (1982).

98. A. Rainis, R. Tung, and M. Szarc, Kinetics of Protonation of Li^+, Na^+, and K^+ Salts of Anthracenide Radical Ions in DME and THF by Methanol and Tert-Butanol, Proc. Roy. Soc. A 339:417 (1974).

99. B. S. Jensen and V. D. Parker, Reactions of Aromatic Anion Radicals and Dianions, Acta. Chem. Scand. B, 30:749 (1976).

100. B. C. Becker, W. Huber, and K. Mullen, Acepleiadylene Dianion and Tetraanion, J. Am. Chem. Soc. 102:7803 (1980).

101. J. Tsunetsugu, The Synthesis and Electrochemistry of Acepleiadylene-5,6-dione and Acepleidylene-5,8-dione, Chem. Commun. p. 28 (1983).

102. W. Huber and K. Mullen, Tetra-anion of 9,9'-Bianthryl, Chem. Commun. p. 698 (1980).

103. O. Hammerich and J. M. Saveant, Electrochemical Reductive Cleavage of Biaryls. The Formation of Anthracene and 9,10-Dihydroanthracene from 9,9'-Bianthryl, Chem. Commun., p. 938 (1979).

104. J. Heinz, 9,9'-Bianthryl-10,10'dicarbonitrile, An Aromatic π-System with Six One-electron Redox Steps, Anew. Chem. Int. Ed. 20:202 (1981).

105. J. Fried, N. A. Abraham, and T. S. Santhanakrishnan, Birch Reduction of Phenols, J. Am. Chem. Soc 89:1044 (1967).

106. K. Yoshida and S. Nagase, Anodic Cyanation. Aromatic Nucleophilic Substitution of Monomethyl- and Dimethylnaphthalenes, J. Am. Chem. Soc. 101:4268 (1979).

107. A. Ledwith and P.J. Russell, Factors Governing the Direct Reaction between Aromatic Cation Radicals and Chloride Ion, Chem. Commun. p. 959 (1974).

108. I. H. Klemm, A. J. Kohlik, and K. B. Desai, Polarographic Reduction of Some Alkyl- and Polymethylanthracenes, J. Org. Chem. 28:625 (1963).

109. O. W. Howarth and G. K. Fraenkel, Electron Spin Resonance Spectra of Monomeric and Dimeric Cation Radicals, J. Chem. Phys. 52:6258 (1970).

110. H. Bock and G. Brahler, Oxidation and Reduction of Methyl-thio-Substituted π-Systems Radical Ions, Angew. Chem. Int. Ed., 16:855 (1977).

111. H. Bock, G. Brahler, D. Dauplaise, and J. Meinwald, One Electron Oxidation of 1,8-Chalcogen-Bridged Naphthalenes, Chem. Ber. 114:2622 (1981).

112. D. G. Farnum, Charge Densiy-NMR Shift Correlations in Organic Ions, Adv. Phys. Org. Chem. 11:123 (1975).

113. A. J. Bard, A. Ledwith, and H. J. Shine, Formation, Properties, and Reactions of Cation Radicals in Solution, Adv. Phys. Org. Chem. 13:155 (1976).

114. D. A. Forsyth and G. A. Olah, Oxidation of Polycyclic Arenes in SbF_5/SO_3ClF, J. Am. Chem. Soc. 98:4086 (1976).

115. K. Lammertsma, G. A. Olah, C. M. Berke, and A. Streit-wieser, 1,4,5,8-Tetramethyl-Naphthalene Dicction and Related Radical Cations, J.Am. Chem. Soc. 101:6658 (1979).

116. A. Pross and L. Random, Does a Methyl Substituent Stabilize or Destabilize Anions? J. Am. Chem. Soc. 100:6572 (1978).

117. K. A. Bilerich and O. Yu. Okhlobystin, Electron Transfer as an Elementary Act of Organic Reactions, Russ. Chem. Rev. (Eng. trans.) 37:954 (1968).

118. L. Eberson, Z. Blum, B. Helgee, and K. Hyberg, Radical Ion Reactivity-I. Application of the Dewar-Zimmerman Rules to Certain Reactions of Radical Anions and Cations, Tetrahedron 34:731 (1978).

119. V. D. Parker, Properties of Aromatic Ions Generated at Electrodes, Pure and Appl. Chem. 51:1021 (1979).

120. P. Kovacic and W. B. England, Novel Pathway for Homopoly-merization by Nuclear Coupling via Aromatic Radical Cation Initiation, J. Polymer Sci., Poly. Lett. 10:359 (1981).

121. V. Svanholm and V. D. Parker, Kinetics and Mechanisms of the Reactions of Organic Cation Radicals and Dications. III. Arylation of Aromatic Hydrocarbon Cation Radicals, J. Am. Chem. Soc. 98:2942 (1976).

122. J. Rochlitz, Neue Reaktionen der Carcinogenen Kohlenwas-serstoffe-II, Tetrahedron 23:3043 (1967).

123. M. Farcasiu and D. Farcasiu, ESR and UV Evidence of a Donor-Acceptor Complex Present in the Pyridine Solution of Triphenylpyrylium Perchlorate, Tet. Lett., p. 4833 (1967).

124. M. Farcasiu and D. Farcasiu, Untersuchung des Elektron-transfers von Pyridin auf arylsubstituierte Pyrylium salze durch ESR Spektroscopie, Chem. Ber. 102:2294 (1969).

125. V. D. Parker, Qualitative Mechanism Analysis by Linear Sweep Voltammetry, Acta Chem. Scan. B 34:359 (1980).

126. E. Ota and S. Otani, Carbonization of Aromatic Compounds in Molten Salt, Chem. Lett., p. 241 (1975).

127. M. Morita, K. Hirosawa, and T. Sato, Interaction between Aromatics and Zinc Chloride in the Molten State. The Formation of 6-Complexes and Radicals, Bull. Chem. Soc. Jpn. 50:1256 (1977).

128. M. Morita, K. Hirosawa, T. Sato, and K. Ouchi, Interaction between Aromatics and Zinc Chloride. II. The Formation of 6-Complexes and Cation Radicals on Supported Zinc Chloride, Bull. Chem. Soc. Jpn. 53:3013 (1980).

129. J. F. Rey Boero and J. A. Wargon, Study of the $AlCl_3$ Catalytic Activity on Aromatic Hydrocarbons-I. Low Temperature Range, Carbon 19:333 (1980).

130. J. F. Rey Boero and J. A. Wargon, Study of the $AlCl_3$ Catalytic Activity on Aromatic Hydrocarbons-II. Mesophase Formation, Carbon 19:341 (1980).

131. M. Miyake, H. Sakashita, M. Nomura, and S. Kikkawa, Catalytic activities of binary molten salts composed of $ZnCl_2$ and metal chlorides for hydrocracking of phenanthrene, Fuel 61:124 (1982).

132. B. D. Flockhart, I. M. Sesay, and R. C. Pink, Perylene Cation-radicals on the Surface of Catalytic Aluminas, Chem. Commun., p. 439 (1980).

133. J. L. Garnett and A. Rainis, EPR and Chemical Studies of the Heterogeneous Reaction Between Polycyclic Aromatic Hydrocarbons and Platinum Chlorides, J. Catal. 26:141 (1972).

134. N. M. D. Brown and D. J. Cowley, Interactions of Aromatic Hydrocarbons with Heavy-metal Halides in the Solid State Studied by Electron Spin Resonance, Chem. Commun., p. 74 (1974).

135. H. H. Perkampus and E. Schonberger, Investigations about the Interaction of Aromatic Compounds with Antimony Trichloride, Zeit. Naturforschung B 31:73 (1976).

136. G. M. Muha, On the Electron Donor and Electron Acceptor Properties of the γ-Alumina Surface, J. Catal. 58:470 (1979).

137. G. M. Muha, On the Redox Properties of Certain Oxide Surfaces, J. Catal. 58:478 (1979).

138. L. B. Ebert and L. Matty, Intercalation Compounds of Graphite: Chemical Identity and Reactivity, Synth. Metals 4:345 (1982).

139. G. F. Endres, A. S. Hay, J. W. Eustance, Polymerization by Oxidative Coupling. V. Catalytic Specificity in the Copper Amine-catalyzed Oxidation of 2,6 Dimethylphenol, J. Org. Chem. 28:1300 (1963).

140. W. F. Taylor, Mechanism of Deposit Formation in Wing Tanks, SAE paper no. 680733 (1968).

141. A. S. Hay, Oxidation of Phenols, U.S. Patent 3,306,874 (1967).

142. A. S. Hay, Oxidation of Phenols and Resulting Products, U.S. Patent 3,306,875 (1967).

143. A. S. Hay, Process for Preparing Polyphenylene Ethers, U.S. Patent 3,382,212 (1968).
144. M. D. Ryan, A. Yueh, and W.-Y.Chen, The Electrochemical Oxidation of Substituted Catechols, J. Electrochem. Soc. 127:1489 (1980).
145. M. K. Eberhardt, Reaction of Benzene Radical Cation with Water Evidence for the Reversibility of OH Radical Addition to Benzene, J. Am. Chem. Soc. 103:3876 (1981).
146. D. G. H. Ballard, A. Courtis, I. M. Shirley, and S. G. Taylor, A Biotech Route to Polyphenylene, Chem. Common., p. 954 (1983).
147. L. Roullier and E. Laviron, Etude Electrochimique de Radicaux Libres-III. Etude des Radicaux Derives des Naphthyridines -1.5, -1.6, -1.7, -1.8, -2.6 et -2.7 et du Bipyridyl -4, 4', Electrochimica Acta 23:773 (1978).
148. V. S. F. Chew and J. R. Bolton, The Analysis of the EPR Spectrum of the 10-Hydro-5-methyl-phenazinium Cation Radical, J. Magn. Res. 37:231 (1980).
149. M. Shlotani, Y. Nagata, M. Tasaki, J. Sohma, and T. Shida, Electron Spin Resonance Studies on Radical Cations of Five-Membered Heteroaromatics. Furan, Thiophene, Pyrrole, and Related Compounds, J. Phys. Chem. 87:1170 (1983).
150. D. N. Ramakrishna Rao and M. C. R. Symons, Unstable Intermediates. Part 205. Radical Cations of Pyrrole, Furan, and Thiophene Derivatives: an Electron Spin Resonance Study. J. Chem. Soc., Perkin Trans. II, p. 135 (1983).
151. H. Chandra and M. C. R. Symons, The Radical-cation of p-Benzoquinone, Chem. Commun., p. 29 (1983).
152. H. Wachowska, Chemical structure of coals as indicated by reductive alkylation, Fuel 58:99 (1979).
153. N. Berkowitz, On some inconsistencies in current concepts of coal chemistry, Technol. Use Lignite 1:414 (1981) (CA 97:112278f).
154. E. Kuhlmann, E. Boerwinkle, and M. Orchin, Solubilization of Illinois bituminous coal: the critical importance of methylene group cleavage, Fuel 60:1002 (1981).
155. J. A. Franz and W. E. Skiens, Side Reactions in the Reductive Alkylation of Low-rank Coal, Fuel 57:502 (1978).
156. L. B. Alemany, C. I. Handy, and L. M. Stock, The Alkylation of Coal, in: "Coal Structure," M. L. Gorbaty and K. Ouchi, eds., Adv. Chem. 192, Amer. Chem. Soc., Washington (1981).
157. L. B. Ebert, D. R. Mills, L. Matty, R. J. Pancirov, and T. R. Ashe, Complications in the Reductive Alkylation of Coal, in: "Coal Structure," M. L. Gorbaty and K. Ouchi, eds., Adv. Chem. 192, Amer. Chem. Soc., Washington (1981).
158. L. B. Alemany and L. M. Stock, The Reductive Alkylation of Illinois No. 6 Coal. Factors Governing the Reductive Alkylation Reaction in Ethereal Solvents, Fuel 61:250 (1982).

159. L. Reggel, R. A. Friedel, and I. Wender, Lithium in Ethylenediamine: A New Reducing System for Organic Compounds,
J. Org. Chem. 22:891 (1957).

160. L. Reggel, R. Raymond, S. Friedman, R. A. Friedel, and
I. Wender, Reduction of Coal by Lithium-Ethylenediamine,
Fuel 37:126 (1958).

161. S. Ergun and I. Wender, X-ray Scattering Intensities of
Coals Treated with lithium in Ethylenediamine, J. Appl.
Chem. 10:189 (1960).

162. L. Reggel, R. Raymond, W. A. Steiner, R. A. Friedel, and
I. Wender, Reduction of Coal by Lithium-Ethylenediamine.
Studies on a Series of Vitrains, Fuel 40:339 (1961).

163. H. W. Sternberg, C. L. Delle Donne, L. Reggel, and
I. Wender, Reduction of Coal by Lithium-Ethylenediamine at
Room Temperature, Fuel 43:143 (1964).

164. L. Reggel, I. Wender, and R. Raymond, Reduction of Coal by
Lithium-Ethylenediamine. A Reevaluation of Previous Data,
Fuel 43:75 (1974).

165. P. H. Given, V. Lupton, and M. E. Peover, Ease of Reduction of a Series of Coals and its Relation to their Structure, Nature 181:1059 (1958).

166. H. W. Sternberg, C. L. Delle Donne, R. E. Markby, and I.
Wender, The Electrochemical Reduction of a Low Volatile
Bituminous Coal--Nature of the Reduced Material, Fuel
45:469 (1966).

167. R. A. Benkeser and E. M. Kaiser, An Electrochemical Method
of Reducing Aromatic Compounds Selectively to Dihydro or
Tetrahydro Products, J. Am. Chem. Soc. 85:2858 (1963).

168. J. C. Thompson, "Electrons in Liquid Ammonia," Clarendon,
Oxford (1976).

169. R. L. Harris and J. J. Lagowski, Metal Ammonia Solutions.
10. Electron Spin Resonance. A Blue Solid Containing a
Crown Ether Complexing Agent, J. Phys. Chem. 82:729
(1978).

170. T. A. Beckman and K. S. Pitzer, The Infrared spectra of
Marginally Metallic Systems: Sodium-Ammonia Solutions, J.
Phys. Chem. 65:1527 (1961).

171. J. Van Schooten, J. Knotnerus, H. Boer, and Ph. M.
Duinker, Selective Reduction by Calcium Hexammine. II.,
Rec. Trav. Chim. 77:346 (1958).

172. W. Kotlarek and R. Pacut, Novel C-C Reductive Cleavage of
Terphenyls with Alkali Mtal-Hexamethylphosphoric Triamide,
Chem. Commun., p. 153 (1978).

173. L. Schanne and M. W. Haenel, Cleavage of Carbon-Carbon
Bonds by Solvated Electrons, Tet. Lett., p. 4245 (1979).

174. L. Lazarov and G. Angelova, Treatment of Coals with Sodium
in Liquid Ammonia Solution, Fuel 47:333 (1968).

175. B. S. Ignasiak, J. F. Fryer, and P. Jadernik, Polymeric
Structure of Coal. 2. Structure and Thermoplasticity of
Sulphur-rich Rasa Lignite, Fuel 57:378 (1978).

176. C. I. Handy and L. M. Stock, Reductive Alkylation of Illinois No. 6 Coal in Liquid Ammonia, Fuel 61:700 (1982).

177. P. W. Rabideau, The Metal-Ammonia Reduction and Reductive Alkylation of Coal Tar Hydrocarbons and the ^{13}C NMR. Characterization and Conformational Analysis of the Reduced Products, Department of Energy Report ER 10339-1 (1979). (See also Energy Res. Abstracts 5:6038 (1980)).

178. P. W. Rabideau, D. W. Jessup, J. W. Ponder, and G. F. Beekman, Metal-Ammonia Reduction of Triptycene and Related Benzobarrelene Derivatives, J. Org. Chem. 44:4594 (1979).

179. P. W. Rabideau and E. G. Burkholder, Concerning the Stereochemistry of Reductive Alkylation of Anthracene and Naphthalene, J. Org. Chem. 44:2354 (1979).

180. F. Cafasso and B. R. Sundheim, Solutions of Alkali Metals in Polyethers. I., J. Chem. Phys. 31:809 (1959).

181. J. M. Pearson, D. J. Williams, and M. Levy, Anion Radicals of a Series of [2.2] Paracyclophanes and α,ω - Diarylalkanes. I. Formation and Chemistry, J. Am. Chem. Soc. 93:5478 (1971).

182. A. Lagendijk and M. Szwarc, Mechanism of Carbon-Carbon Bond Fission by Electron Transfer Leading to Dianions, J. Am. Chem. Soc. 93:5359 (1971).

183. K. Niemann and U. B. Richter, Studies in the Chemical Characterization of Coal: Reduction, Fuel 58:838 (1979).

184. K. Niemann and H. P. Hombach, Studies in the Chemical Characterization of Coal: Reduction via Solvated Electrons, Fuel 58:853 (1979).

185. C. J. Collins, H. P. Hombach, B. Maxwell, M. C. Woody, and B. M. Benjamin, Carbon-Carbon Cleavage during Birch-Huckel-Type Reductions, J. Am. Chem. Soc. 102:852 (1980).

186. E. W. Hagaman and M. C. Woody, Structure Analysis of Coals by Resolution Enhanced Solid State ^{13}C n.m.r. Spectroscopy, Fuel 61:53 (1982).

187. B. Ignasiak, D. Carson, and M. Gawlak, Non-destructive Solubilization of Coal, Fuel 58:833 (1979).

188. N. Cyr, M. Gawlak, D. W. Carson, and B. S. Ignasiak, Structural Characterization of Non-reductively Ethylated Coal by 13C and ^{1}H n.m.r., Fuel 62:412 (1983).

189. D. Seyferth, D. P. Duncan, and H. W. Sternberg, Silylation: a Method for Benzene Solubilization of Benzene-insoluble, Pyridine-soluble Coal-derived Products, Fuel 58:74 (1979).

190. E. Grovenstein and A. B. Cottingham, Carbanions. 17. Rearrangements of 2,2-Diphenyl-4-pentenyl Alkali Metal Compounds, J. Am. Chem. Soc. 99:1881 (1977).

191. D. E. Bergbreiter and J. M. Killough, Reactions of Potassium-Graphite, J. Am. Chem. Soc. 100:2126 (1978).

192. W. F. Bailey and E. A. Cioffi, Reductive Rearrangements of 4-Phenyl-1,3-dioxans to 2-Phenylbutane-1,3-diols upon

Treatment with Sodium-Potassium Alloy, Chem. Commun. p. 155 (1981).

193. A. Oku, K. Harada, T. Uagi, and Y. Shirahase, Cyclopropylidene Rearrangement in the Reduction of 1,2:3,4-Bis-(dihalomethano)-1,2,3,4-tetra hydropolymethylenophthalenes by Naphthalenides, J. Am. Chem. Soc. 105:4400 (1983).

194. H. Gilman and J. W. Morton, The Metalation Reaction with Organolithium Compounds, in "Organic Reactions", Vol. VIII, Wiley, New york (1954).

195. J. J. Brooks, W. Rhine, and G. D. Stucky, π Groups in Ion Pair Bonding. Stabilization of the Dianion of Naphthalene by Lithium Tetramethylethylenediamine, J. Am. Chem. Soc. 94:7346 (1972).

196. A. Essel, B. Graveron, G. Merle, and C. Pillot, Stabilite du dianion du naphtalene dans le dioxane et le 2.5-dimethyl-tetrahydrofuranne, C.R. Acad. Sci. Paris 275:925 (1972).

197. Yu. N. Novikov and M. E. Vol'pin, Lamellar Compounds of Graphite with Alkali Metals, Russ. Chem. Rev. (Eng. Trans.) 40:733 (1971).

198. L. B. Ebert, Intercalation Compounds of Graphite, Ann. Rev. Mater. Sci. 6:181 (1976).

199. H. Selig and L. B. Ebert, Intercalation Compounds of Graphite, Adv. Inorg. Chem. Radiochem. 23:281 (1980).

200. L. B. Ebert, Catalysis by Graphite Intercalation Compunds, J. Molec. Catal. 15:275 (1982).

201. P. Belser, G. Desbiolles, U. Ochsenbein, and A. Zelewsky, Aromatic Radical Anions in Neat Aromatic Hydrocarbons as Solvents. Direct Evidence of Through Space Spin-Density Transfer to the Ligand of the Counter Ion, Helv. Chim. Acta 63:523 (1980).

202. W. A. Holmes-Walker and A. R. Ubbelohde, Electron Transfer in Alkali Metal-Hydrocarbon Complexes, J. Chem. Soc., p. 720 (1954).

203. G. R. Stevenson and E. Williams, Solvation Enthalpies of Organic Anion Radicals, J. Am. Chem. Soc. 101:5910 (1979).

204. G. R. Stevenson, C. R. Siedrich, and G. Clark, Crystal Lattice Energies of Solid Anion Radical Salts, J. Phys. Chem. 85:374 (1981).

205. L. Lazarov, I. Rashkov, and S. Angelov, Direct Preparation of Ionic Potassium-Coal Adducts, Fuel 57:637 (1978).

206. L. Lazarov, M. Stafanova, and K. Petrov, Structural Study of Coals by Means of Directly-Prepared Potassium-Coal Adducts, Fuel 61:58 (1982).

207. H. Podall, W. E. Foster, and A. P. Giraitis, Catalytic Graphite Inclusion Compounds. I. Potassium Graphite as a Polymerization Catalyst, J. Org. Chem. 23:82 (1958).

208. J. M. Lalancette, G. Rollin, and P. Dumas, Metals Intercalated in Graphite. I. Reduction and Oxidation, Can. J. Chem. 50:3058 (1972).

374

209. P. B. Hirsch, X-ray Scattering from Coals, Proc. Roy. Soc. A, 226:143 (1954).

210. S. Ergun and I. Wender, X-ray Scattering Intensities of Anthraxylons Reduced with Lithium Ethylenediamine, Fuel 37:503 (1958).

211. M. Miyake, M. Sukigara, M. Nomura, and S. Kikkawa, Improved Method to Alkylate Yubari Coal of Japan Using Molten Potassium Under Refluxing THF, Fuel 59:637 (1980).

212. J. M. Austin, T. Groenewald, and M. Spiro, Heterogeneous Catalysis in Solution. Part 18, The Catalysis by Carbons of Oxidation-Reduction Reactions, J.C.S., Dalton Trans., p. 854 (1980).

213. L. B. Ebert, L. Matty, D. R. Mills, and J. C. Scanlon, The Interrelationship of Graphite Intercalation Compounds, Ions of Aromatic Hydrocarbons, and Coal Conversion, Mater. Res. Bull. 15:251 (1980).

214. B. S. Ignasiak and M. Gawlak, Polymeric Structure of Coal. I. Role of Ether Bonds in Constitution of High-Rank Vitrinite, Fuel 56:216 (1977).

215. B. S. Ignasiak, S. K. Chakrabartty, and N. Berkowitz, Molecular Weights of Solubilized Coal Products, Fuel 57:507 (1978).

216. L. B. Alemany, S. R. King, and L. M. Stock, Proton and Carbon N.M.R. Spectra of Butylated Coal, Fuel 57:738 (1978).

217. E. H. Burk and J. Y. Sun, Coal Molecular Weight Distributions by GPC, in The Fundamental Organic Chemistry of Coal: Proceedings of a Workshop Sponsored by the National Science Foundation, Knoxville, TN (1975).

218. L. B. Alemany and L. M. Stock, Reductive Alkylation of Illinois No. 6 Coal. ^{1}H and ^{13}C N.M.R. Spectra of the ^{13}C-enriched Alkylation Products, Fuel 61:1088 (1982).

219. R. Dogru, G. Erbatur, A. F. Gaines, Y. Yuram, S. Icli, and T. Wirthlin, Nuclear Magnetic Resonance Spectra of Two Reductively Ethylated Fuels, Fuel 57:399 (1978).

220. R. Dogru, A. Gaines, A. Olcay, and T. Tugrul, Mild Oxidation of Reductively Ethylated Solid Fuels, Fuel 58:823 (1979).

221. H. M. Wachowska, B. N. Nandi, and D. S. Montgomery, Oxidation Studies on Coking Coal Related to Weathering. 4. Oxygen Linkages Influencing the Dilatometric Properties and the Effect of Cleavage of Ether Linkages, Fuel 53:212 (1974).

222. H. Wachowska and W. Pawlak, Effect of Cleavage of Ether Linkages on Physicochemical Properties of Coals, Fuel 56:522 (1977).

223. L. B. Ebert, J. C. Scanlon, and D. R. Mills, X-Ray Diffraction of n-Paraffins and Stacked Aromatic Molecules: Problems in Using Diffraction to Determine the Average

Structure of Asphaltenes, Preprints, Petroleum Div., Amer. Chem. Soc. 28:1353 (1983). Also Liq. Fuels Tech., 2:257 (1984).

224. H. E. Blayden, J. Gibson, and H. L. Riley, An X-Ray Study of the Structure of Coals, Cokes and Chars, Proceedings Conf. Ultra-fine Structure of Coals and Carbons, p. 176 (1944).

225. B. G. Silbernagel, L. B. Ebert, R. H. Schlosberg, and R. B. Long, Magnetic Resonance Study of Labeled Guest Molecules in Coal, in: "Coal Structure," M. L. Gorbaty and K. Ouchi, eds., Adv. Chem. 192, Amer. Chem. Soc., Washington (1981).

226. N. J. Russell, M. A. Wilson, R. J. Pugmire, and D. M. Grant, Preliminary Studies on the Aromaticity of Australian Coals: Solid State N.M.R. Techniques, Fuel 62:601 (1983).

227. J. K. Brown and P. B. Hirsch, Recent Infrared and X-Ray Studies of Coal, Nature 175:229 (1955).

228. L. Cartz and P. B. Hirsch, A Contribution to the Structure of Coals From X-Ray Diffraction Studies, Phil. Trans. Roy. Soc. A 252:557 (1960).

229. A. Carrington and A. D. McLachlan, "Introduction to Magnetic Resonance," Chapman and Hall, London (1979).

230. L. T. Calcaterra, G. L. Closs, and J. R. Miller, Fast Intramolecular Electron Transfer in Radical Ions Over Long Distances Across Rigid Saturated Hydrocarbon Spacers, J. Am. Chem. Soc. 105:1505 (1983).

INDEX

Acetylene flames, 238
Additives
  alcohol, 265
  high molecular weight, 53
Alkanes, in deposit, 107
Alkylbenzenes, 109, 113, 293, 299
Air/fuel
  distribution, 39
  ratio (AFR), 40
Anthracene, 305
Aromatic hydrocarbons
  influence of boiling point, 9
  oxidation of, 318, 321
  reduction of, 318
Autoignition, 3, 10
Autothermal cracking, 264

Benzanthracene, 305
Biphenyl, 305, 311
Bright stock, 13, 15, 201, 285
BSTFA titration, 74
Butyllithium titration, 73

Calcium acetate titration, 74
Carbon-14 radiotracer, 9
Carbon deposits, (see also
    engine deposits)
  and inorganics, 59
  and knocking, 6
Carbonaceous deposits, 11
Carbon dioxide from deposits,
    73, 87, 89
Catalytic wall reaction, 264
Cathodic reduction, 328
Cetane, 264
CFR engine, 9, 10, 44, 264

Chrysene, 305
CLR engine, 146, 148, 277, 289
Coals, compared to deposits,
    93, 120, 132, 323, 324,
    338, 342, 347
Component durability, 20
Compression ratio, 44, 50
Conradson carbon, 6
Coronene, 108
Coulometry for CO_2, 73, 88
Crank angle, 218, 219, 220
Cross-linking, 337

Deposits, (see also engine
    deposits)
  combustion chamber, 3, 12, 19,
    40, 71, 101, 199
  EGR valve, 56, 68
  intake port, 40, 54, 60
  jet fuel, 245
  model combustion, 136, 230
  polymeric, 69
  valve seat, 40
Detergent/dispersant, 13
Dialysis, 66, 121
Dibenzofuran, 313
Dibenzothiophene, 313
Diesel engine, 263
Differential scanning
    calorimetry, 202
Direct injection stratified
    charge engine, 16

Electron microscopy, 11, 13
Electron spin resonance, 125
    136, 137, 145, 361

378